The Next Giant Leap

How we can solve resource scarcity and climate change to build a better world

Cameron MacPherson

Copyright 2020. Cameron MacPherson. All rights reserved.

Print Edition 1.1 (greyscale ink)

ISBN: 9781698530802

Important Notes:

The Next Giant Leap is a work published over several mediums: print, electronic document and online at https://nextgiantleap.org. A free companion PDF in full color is available at https://nextgiantleap.org/ngl.pdf

As this work contains nearly two hundred images, graphs and charts, copies printed in *greyscale ink* may contain graphics that are <u>not as clear</u> as full-color versions. In such cases, please download the free companion PDF at https://nextgiantleap.org/ngl.pdf, consult the free online copy at https://nextgiantleap.org/universal-energy or purchase a print copy in color at https://nextgiantleap.org/print-copy

As this version is intended for American audiences, all units of measure included for general explanation are **imperial**. Units of measure for specific figures in the Appendix and citations are in *both* imperial and metric notations.

For information on how to further support this project, please visit https://nextgiantleap.org/support

For a changelog of major revision changes to this work, please visit https://nextgiantleap.org/changelog

For more information about the author, please visit https://cameronmacpherson.com

The Next Giant Leap proposes solutions to resource scarcity and climate change. It outlines blueprints for a system – Universal Energy – that can accomplish this task by deploying the best energy technologies we have available into a framework that's designed to work cooperatively from the ground-up.

Universal Energy's purpose is to revolutionize how we power our civilization and acquire resources, while at the same time heal the widespread environmental damage humanity has inflicted on our planet.

Developed to remove the limitations of scarcity, resource conflict and need, Universal Energy is software for the great challenges of our time and a platform to build our civilization to greater heights.

Every gun that is made, every warship launched, every rocket fired signifies, in the final sense, a theft from those who hunger and are not fed, those who are cold and are not clothed. This world in arms is not spending money alone. It is spending the sweat of its laborers, the genius of its scientists, the hopes of its children.

The cost of one modern heavy bomber is this: a modern brick school in more than 30 cities. It is two electric power plants, each serving a town of 60,000 population. It is two fine, fully equipped hospitals. It is some fifty miles of concrete pavement. We pay for a single fighter with a half-million bushels of wheat. We pay for a single destroyer with new homes that could have housed more than 8,000 people.

This is not a way of life at all, in any true sense. Under the cloud of threatening war, it is humanity hanging from a cross of iron.

– President Dwight D. Eisenhower. April 16, 1953.

Table of Contents

A Future Worth Having	p. 13
Chapter One: Mindset	p. 35
Chapter Two: The Renewable Revolution	p. 47
Chapter Three: A Tale of New Cities	p. 71
Chapter Four: The Thorium Backbone	p. 81
Chapter Five: Water and Hydrogen	p. 133
Chapter Six: Cogeneration	p. 145
Chapter Seven: The National Aqueduct	p. 159
Chapter Eight: The World's Largest Battery	p. 173
Chapter Nine: Everybody Eats	p. 183
Chapter Ten: Materials and Recycling	p. 201
Chapter Eleven: The End of Resource Scarcity	p. 227
Chapter Twelve: Advanced Infrastructure	p. 241
The Next Giant Leap	p. 271

Appendix: p. 285

A1: Universal Energy Implementation Strategy	p. 287
A2: Collective Capitalism	p. 295
A3: Revenue Allocations	p. 307
A4: Universal Energy Cost Estimate	p. 315
A5: Citation Policy	p. 335
A6: Source Policy	p. 337
A7: Retraction Policy	p. 341
A8: Copyright Policy	p. 343
Special Thanks	p. 345
Cited Facts and Sources	p. 347

Once we realize we deserve a bright future, letting go of our dark past is the best choice we ever make.

- Roy T. Bennett

A Future Worth Having

I'd like to ask you a question: when was the last time you felt great about the world? Not the last time you were in a good mood or something happened to brighten your day, but the last time you were truly hopeful for our future. The last time you believed, legitimately, that tomorrow will be better than days past. When was the last time you could say that? For most of us, I bet it's been a while.

It seems that most everywhere we look today, our future seems destined for darker times. Even if the polar ice caps weren't melting before our eyes,[1] our planet's biodiversity wasn't rapidly going extinct by our hand[2] and wildfire wasn't laying waste to rainforests,[3] vast geographical regions[4] and entire nations,[5] our time is still deeply imperiled. Representative democracies are waning worldwide against the rise of populist, authoritarian nationalism.[6] Multicultural societies are facing crises of conscience in the face of mass migration from strife and conflict.[7] Partisan division and rampant misinformation have sown seeds of enmity throughout our social fabric.[8] The long peace[9] threatens to unravel as the drumbeats of brinkmanship[10] and war begin to sound progressively louder.[11] And that's just what's on the news.

Yet the most serious reasons our horizon grants cause for alarm are rooted deeper, into something more structural, and ultimately, something more fateful. To explain why, we'll take a brief step back and afford ourselves a high-level view of where we stand at present.

Radiological dating tells us definitively that the first page of the human story began at least 200,000 years ago.[12] From the moment we made our first steps, it took us nearly *that entire length of time* to reach a population of 1.5 billion people – a milestone achieved around 1900.[13] Yet barely a century later in 2010, we reached seven billion people, and we're well on track to hit ten billion by 2050.[14]

As it's difficult to put those numbers in perspective, the following figures help establish a more succinct context:

Our approximate time to reach 1.5 billion people: **200,000 years.**
Our approximate time to grow from 1.5 billion to 7 billion people: **110 years.**
Our estimated time to grow from 7 billion to 10 billion people: **38 years.**

This chart shows humanity's population growth since height of Rome:

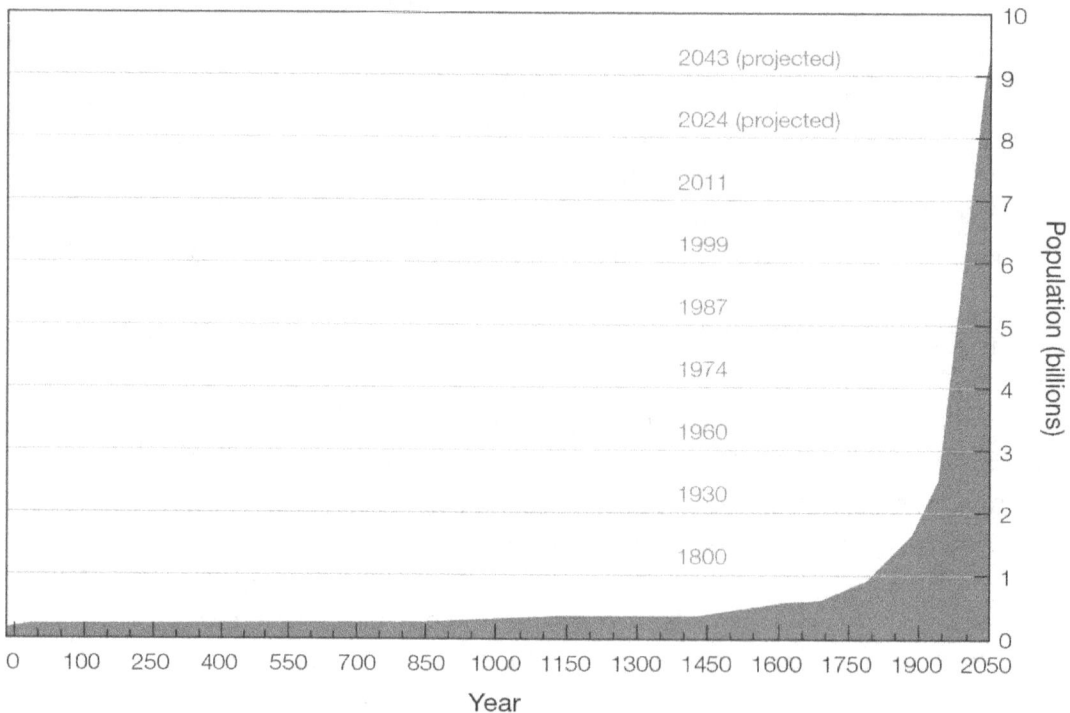

In the image below, each of the 2,000 rectangles equals **one century**. The blue rectangles, in aggregate, represent *the entire length of time it took humanity to reach 1.5 billion people*. The lone black rectangle (last) represents the time it took humanity to grow from 1.5 billion to 7 billion people.

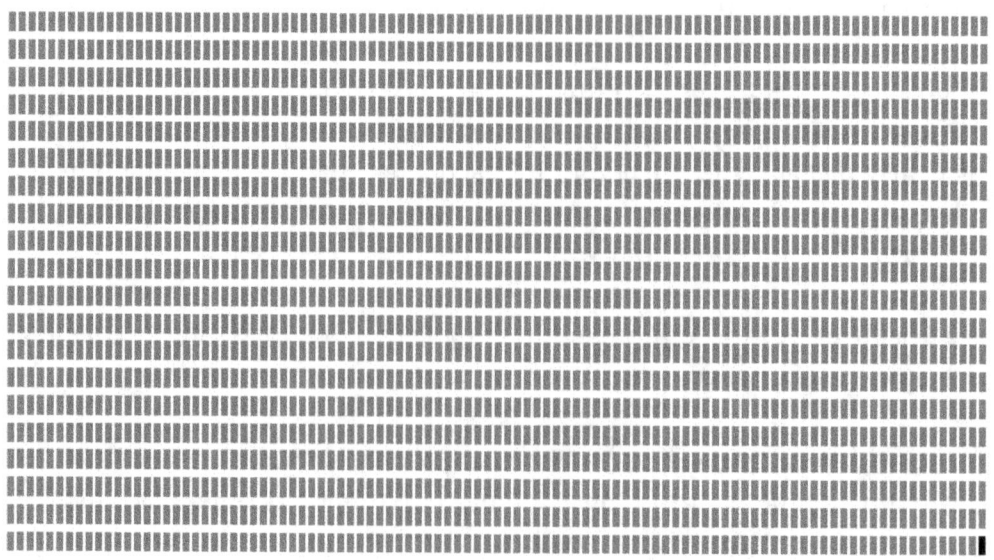

This growth – and the rise of the civilizations that encompassed within – has been powered exclusively by natural resources. As our population has rapidly expanded, so too has our rate of resource consumption – far past the thresholds that spawned Thomas Malthus' *Principle of Population* and the 1970's-era *Limits to Growth* – famous publications forecasting the dangers of rapid population expansion and its accompanying over-extraction of resources.[15-16] Our population as of 2020 has increased sevenfold from Malthus' time, and is today twice the population of the 1970's. Consequently, we are now consuming resources so rapidly and on such an immense scale that it's destroying our planet's ability to support our existence. In a world of sensationalist media, such a claim might seem exaggerative. In the world of facts, it's an understatement.

A short list frames our circumstances in sobering perspective:

1. Since the start of the Industrial Revolution, we've cut down more than half of the world's forests.[17]

2. Plant and animal species are dying off so rapidly that global scientific bodies conclude that humanity has caused Earth's sixth great extinction event,[18] which has the potential to wipe out millions of years of evolutionary history.[19]

3. Global fish stocks are overexploited by 80%,[20] and ecologists predict they may collapse entirely by as early as 2048.[21]

4. Ocean acidification is occurring at a rate not seen in the last 300 million years and Earth is estimated to have lost half of its shallow corals in the past three decades.[22] If current trends continue, this figure is estimated to rise to 90% by 2050.[23]

5. Between 1970 and 2014, Earth saw a 60% decline of its mammal, bird, fish, reptile and amphibian species – almost exclusively due to human activity.[24] Since the dawn of our civilization, humanity has destroyed 83% of all mammalian life on Earth.[25] Of the mammalian life that remains, 96% are either human or livestock.[26] A scant 4% is all that's left in the wild.[27]

The United Nations' Intergovernmental Science-Policy Platform on Biodiversity and Ecosystem Services (IPBES) released a report in March, 2019 concluding that human activity is destroying nature at an "unprecedented rate," which is "eroding the very foundations of our economies, livelihoods, food security and quality of life worldwide."[28] Its noteworthy highlights include:[29]

- Approximately 60 billion tons of renewable and non-renewable resources are globally extracted annually – twice the rate of extraction from 1980.

- Plastic pollution has increased tenfold since 1980, with 300-400 million tons of heavy metals, solvents, toxic sludge and industrial waste dumped annually into global water systems.

- Worldwide, pollinator loss risks upwards of $577 billion in agriculture every year.

- 75% of Earth's land environments and 66% of Earth's marine environments have been significantly altered by human actions, and since 1992 the world's urban areas have doubled in size.

- More than a third of Earth's land surface and nearly 75% of freshwater resources are now consumed by agriculture and livestock.

- Up to 1 million of the estimated 8 million plant and animal species on Earth are at risk of extinction, many of them within decades.

As humanity is inexorably tied to Earth's ecology, these crises alone present major risks to our long-term survival. But this problem is more complex than simple extraction, pollution and waste. Take fresh water, for instance.

As 85% of humanity lives on the driest half of the planet,[30] 2.4 billion people today lack access to clean water – and half of the world's population lacks access to the quality of water available to ancient Rome.[31] The United Nations estimates that by 2025 more than two billion people will live in conditions reflecting absolute water scarcity and five billion people will live in conditions reflecting extreme water stress,[32] thresholds indicating life-risking lack of access to our most vital resource.[33] That's respectively between 30% and 70% of the planet. The international body further estimates that global fresh water demand between now and 2050 will increase by 55%[34] with demand exceeding supply by more than 30% by 2040.[35] Private investment forecasts are similar, as Goldman Sachs estimates that global fresh water consumption is *doubling* every 20 years.[36]

To see what this looks like in practical terms, consider the Aral Sea, which was once the 4th largest lake in the world. In 1960, the body of water had a surface area of 26,300 square miles (68,000 km²) and a volume of 254 cubic miles (1,080 km³).[37] For comparison, that is 4,000 square miles larger than Lake Michigan by surface area and nearly two and a half times larger than Lake Erie by total volume.[38] Yet in a timespan of just *35 years*, the Aral Sea was depleted to become what is now known as the Aralkum Desert that comprises the borders of its eastern basin.[39]

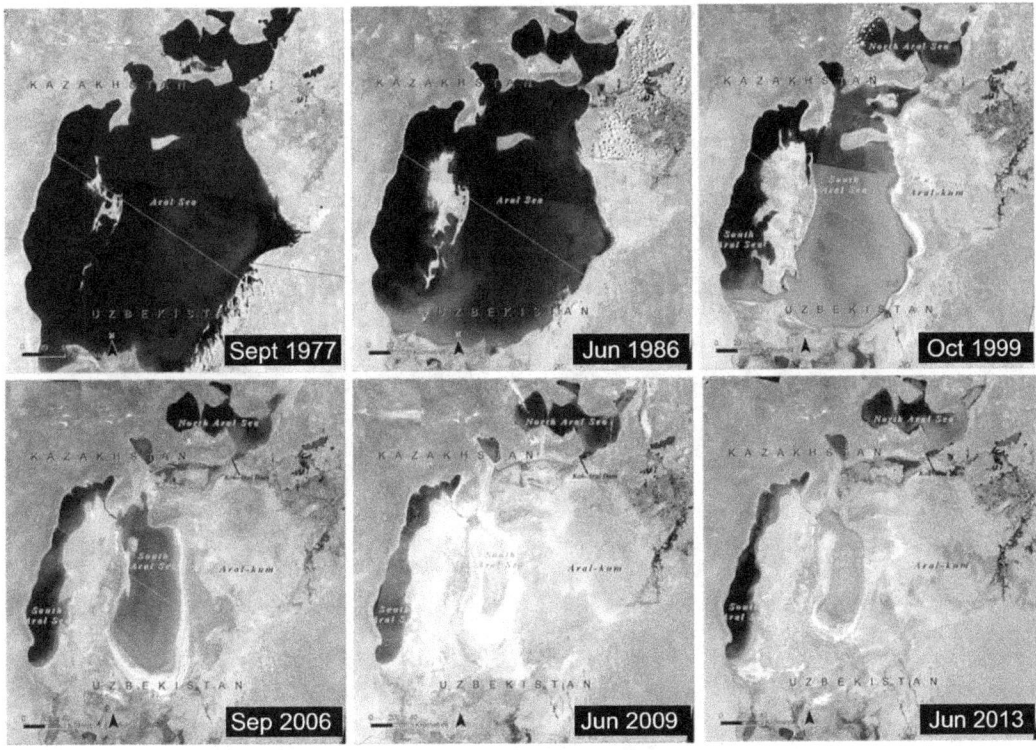

The Aral Sea falling victim to unsustainable resource extraction and overconsumption[40] is not an outlier. It's part of a consistent trend. As our water needs have increased alongside worsening global drought conditions, we've needed to tap water from aquifers – large sources of groundwater that slowly replenish from soil absorption over time. Consequently, most of them today are being depleted faster than their ability to recover.[41]

Data from NASA satellites show that 21 of the world's 37 aquifers have passed their sustainability tipping points, meaning they will eventually run dry if current circumstances continue.[42] Aquifers supply 35% of global water use, and they are among the last reliable sources of fresh water we have left.[43] California is already tapping aquifers for up to 60% of its water supply, and climate scientists expect aquifers will be relied upon to even greater extents in the future.[44]

Colorado's NCAR (National Center for Atmospheric Research) recently released a series of climate maps that model the future degree of global drought and desertification based on current trends. They aren't meant to be exact predictions as far too many factors influence climate and drought, nor do they incorporate the possibility of human activities *worsening* in terms of unsustainable extraction and ecological destruction. They only model drought and desertification if our direction remains unchanged, visualized from year 2000 to 2099.[45]

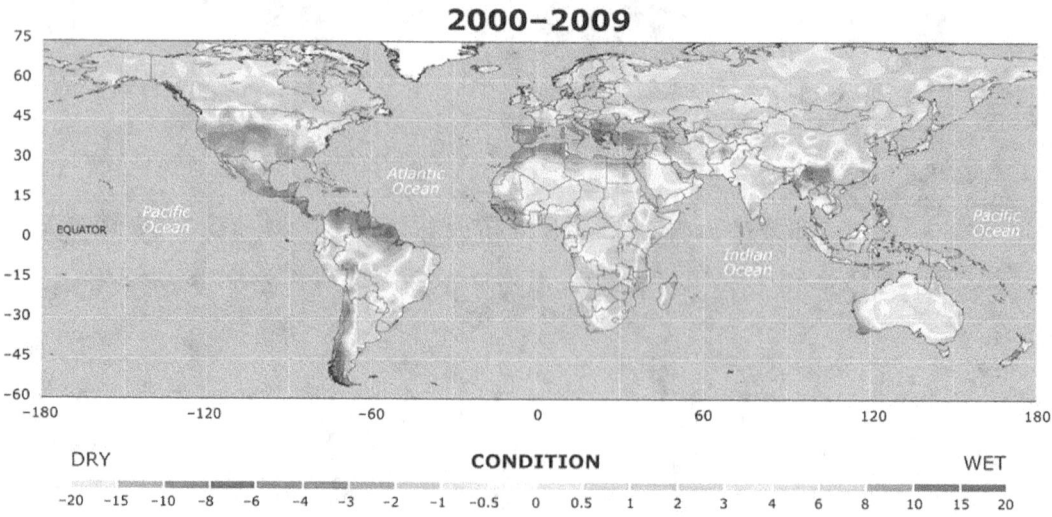

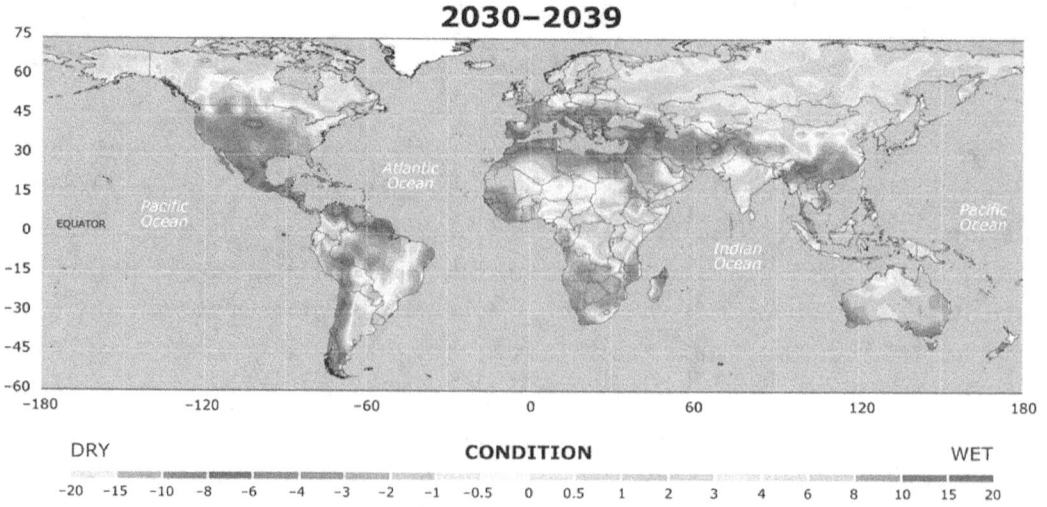

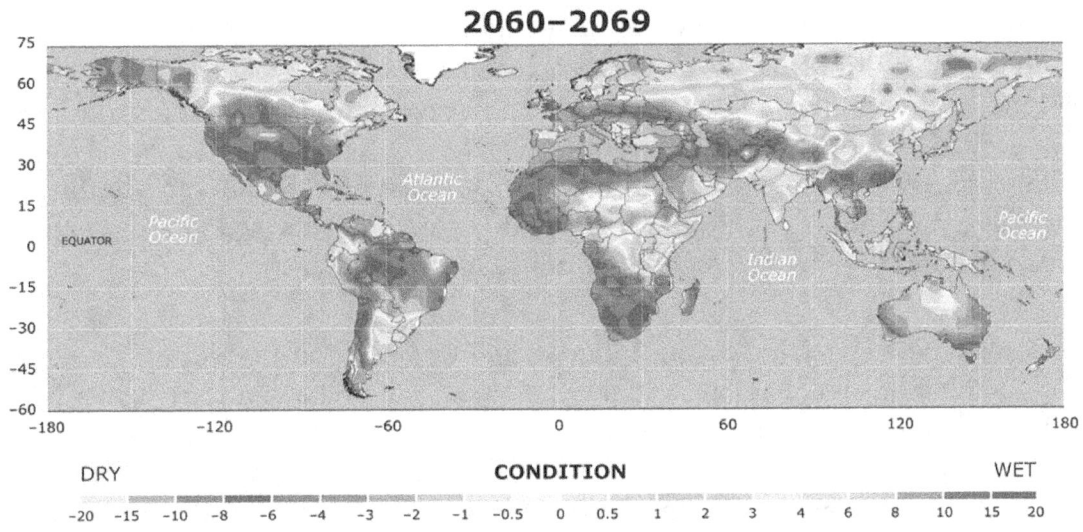

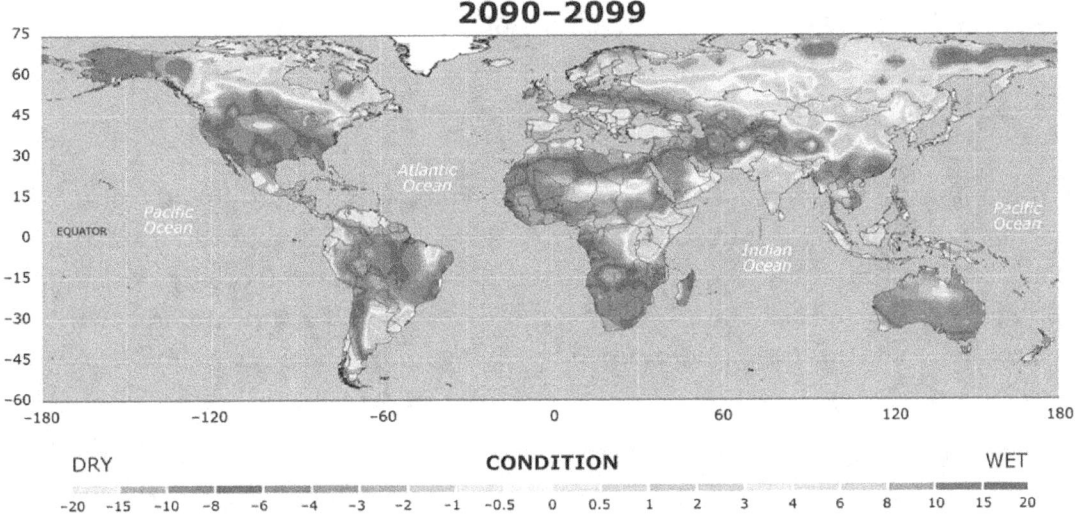

Should these models prove accurate, even fractionally, it risks the survival of billions of people – saying nothing of supporting an ecosystem, nor the conflicts spawned by such circumstances – and even wealthy nations will face major complications to life as they know it.[46]

If that wasn't bad enough, a substantial portion of our remaining fresh water supplies are too toxic for human consumption. Industrial contamination, for example, has rendered 60% of Chinese rivers unusable for drinking, bathing or agriculture according to data published by the Chinese Ministry of Water Resources.[47] That figure rises to 70% when it comes to Chinese lakes and 80% when it comes to wells that source groundwater – 90% when that groundwater is sourced near cities.[48]

Another example is India. Upwards of 50% of the country faces extremely high water stress according to a World Resources Institute study.[49] The study found that Indian drought conditions have become so severe that 330 million people – greater than the population of the United States – are living in a dust bowl. The situation has become so dire that the nation's coal-fired power plants are shutting down as there's not enough water to generate steam – with armed guards being posted at dams to prevent water theft from desperate farmers.[50]

A recent report by WaterAid, an international organization promoting greater water sanitation and hygiene, estimates that 80% of India's surface water is severely polluted and contaminated with water-borne disease.[51] A separate report from India's Centre for Science and Environment estimates that roughly 80% of Indian sewage flows untreated into its rivers, and that out of 8,000 towns surveyed by a pollution control board, only 160 had both functioning sewage systems and a sewage treatment plant.[52]

India and China's population, combined, represent 35% of humanity.

This is all before the impact of climate change, a factor that by itself adds massive gravity to our current state of affairs. As extensive lobbying and campaign "contributions" from the fossil fuel industry have turned the looming crisis into a political issue,[53] the basic principle that more carbon in the atmosphere leads to a warmer climate has been undermined by disingenuous partisan dismissals.[54]

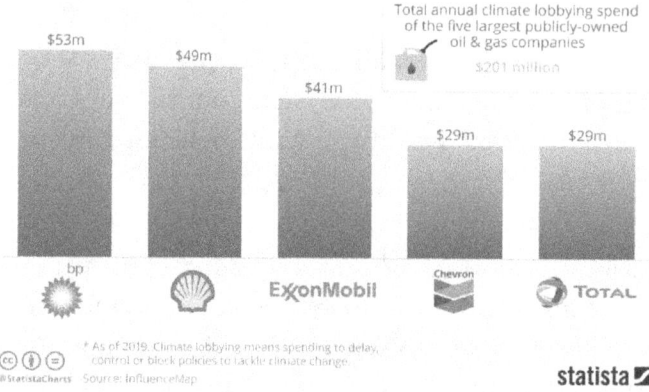

While this has hindered effective policy measures to reverse course, the rapid increase in global temperatures, accelerating loss of polar ice, increased frequency and severity of droughts, wildfires, floods, and other natural disasters have evaporated any doubts held in good faith as to the validly of the warnings echoed by the nigh-unanimous majority of the scientific community.[55] (Chart source.[56])

Carbon Dioxide Emissions Since 1850-Present (gigatonnes):

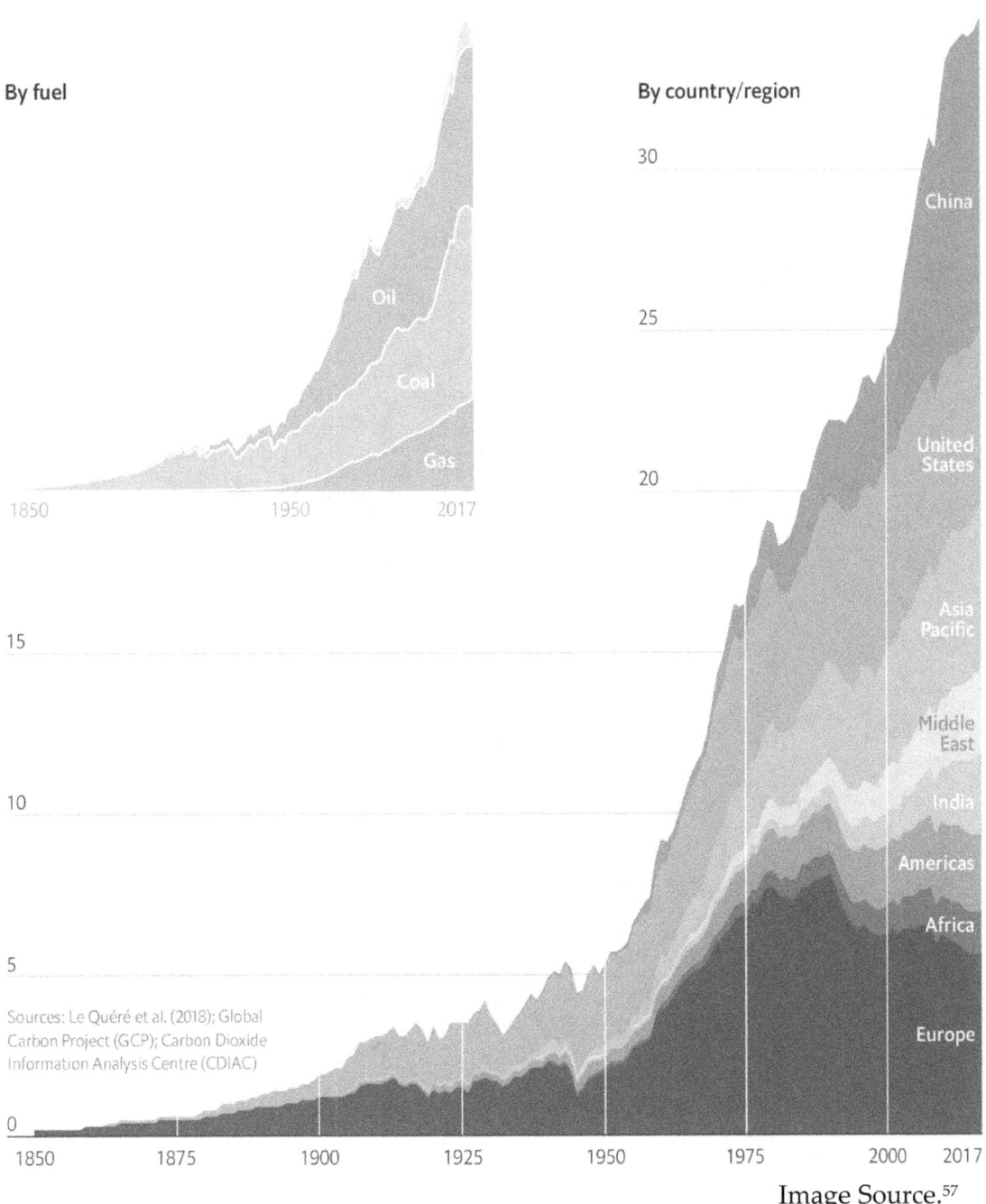

Image Source.[57]

Global Average Temperature Difference from 1850, projected until 2025 (°C)

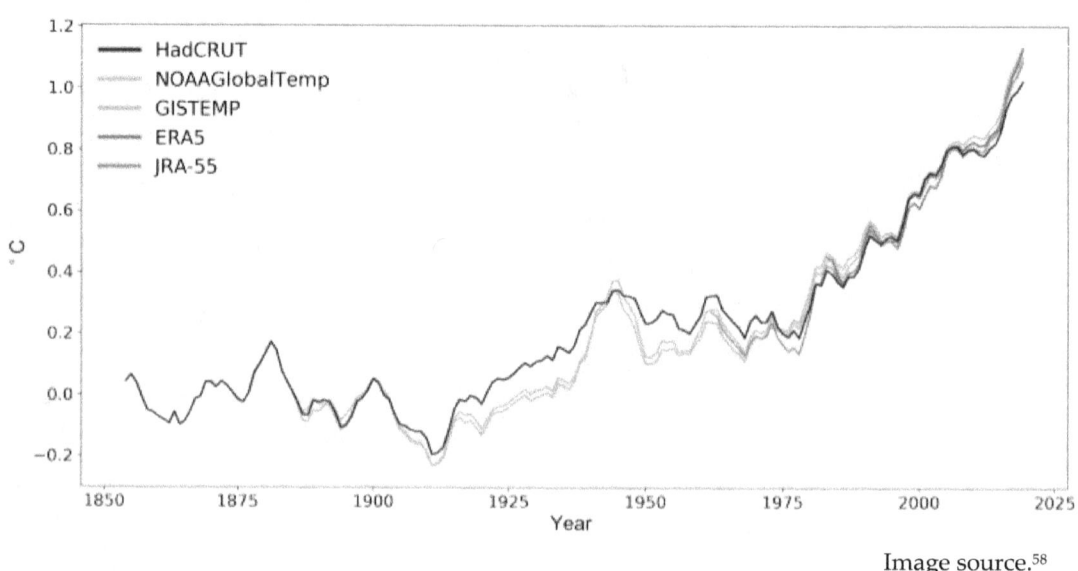

Image source.[58]

Global Average Temperatures, 1900–2015 (°F)

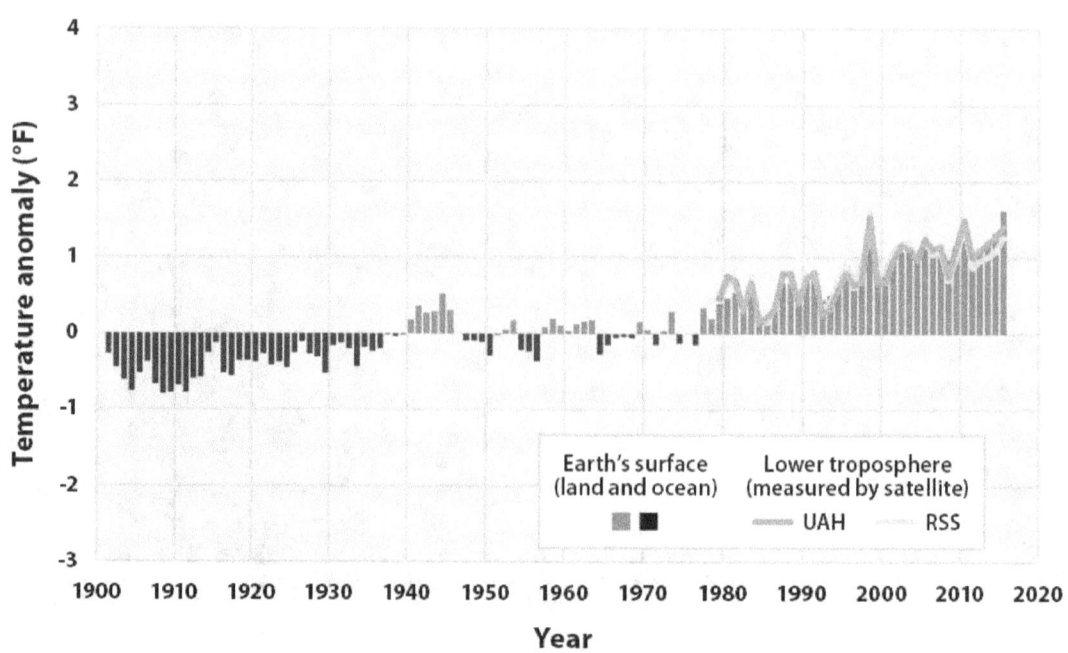

Image source.[59]

Rate of Temperature Change in the United States, 1900-2015

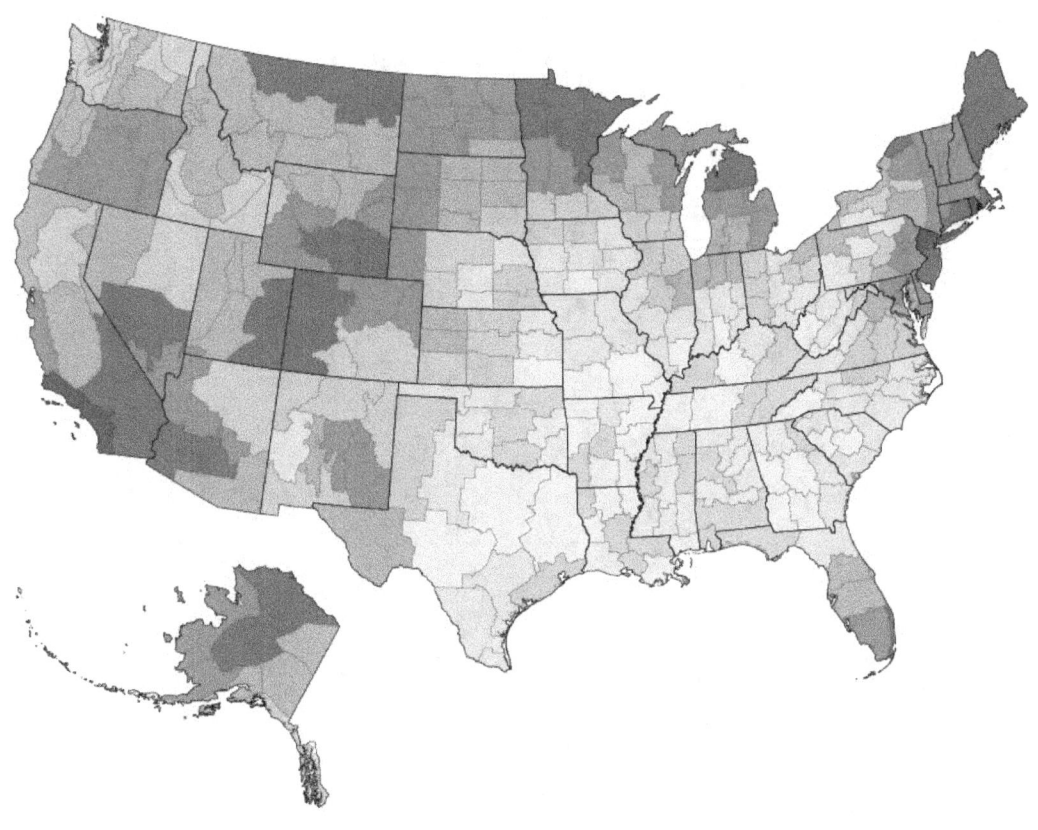

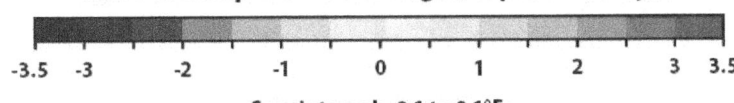

Image source.[60]

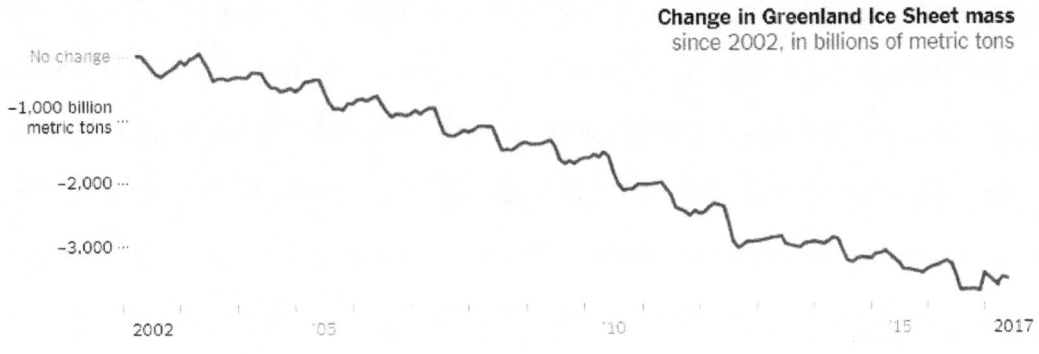

Image source.[61]

Change in Arctic Ice Size and Age from 1984-2018

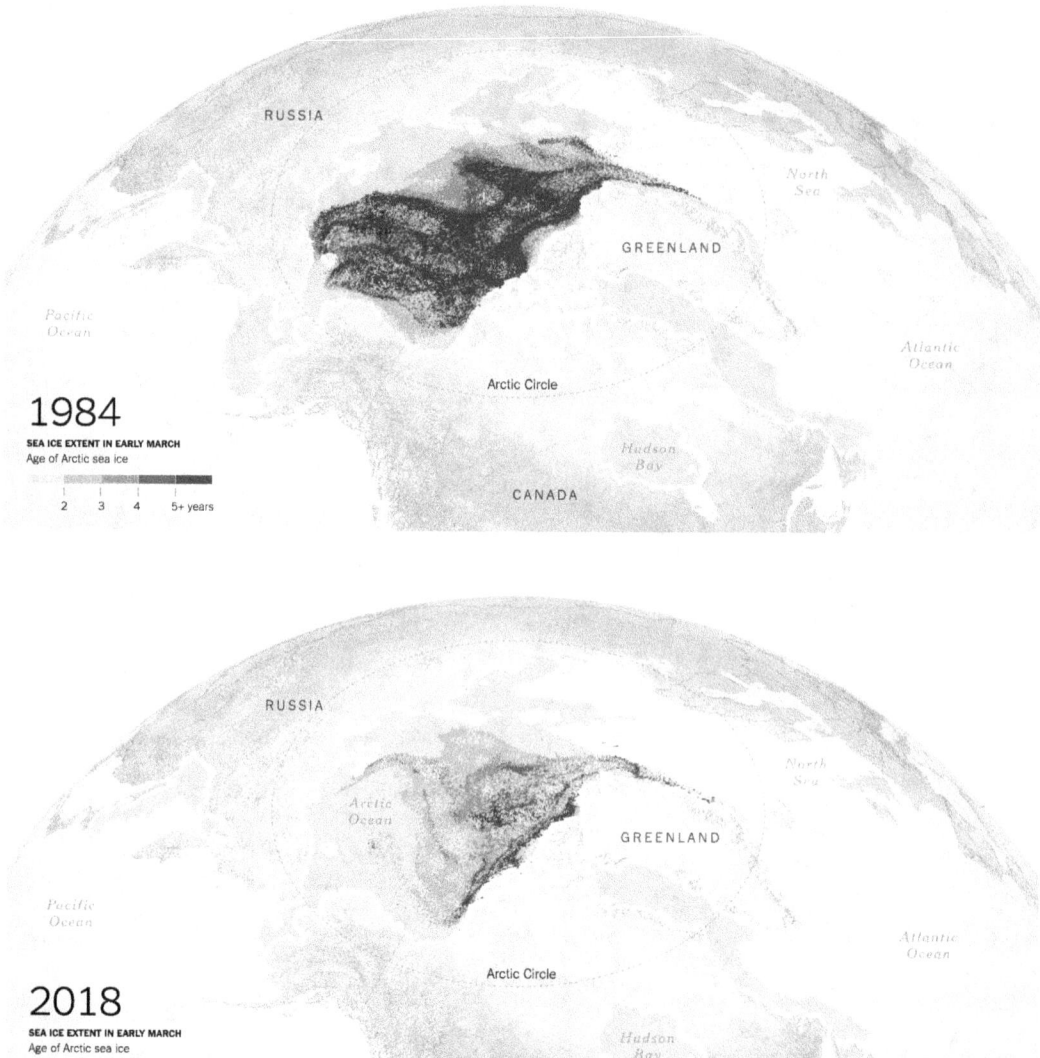

Image source.[62]

Yet the ecological consequences of a warming climate have greater fallout to our society. Their impact on trade, security, economy, public health and mass-migration all have worldwide implications.[63] The global food supply, however, remains perhaps the most imperiled. As water (and oil)[64] are vital to both producing and transporting food, their scarcity alone would substantially disrupt our ability to feed our growing world. Climate change makes that task inexorably harder, and substantially more expensive.

Between 2006 and 2008, global average food prices rose 107% for soybeans, 136% for wheat and 217% for rice.[65] A 2010 study by the International Food Policy Research Institute estimated that the price of corn, wheat and rice will rise by at least 60% by 2050.[66] A 2012 report by Oxfam – one of the largest humanitarian charities in the world – estimates that by year 2030, food prices will rise 107% for rice, 120% for wheat and 177% for corn compared to their baseline price in 2010.[67] Oxfam's report stated further that:

> *"While prices could double by 2030, the modeling suggests that one or more extreme events in a single year could bring about price spikes of comparable magnitude to <u>two decades</u> of projected long-run price increases."*[68] (Emphasis mine)

We must also contend with the fact that precious little of our situation is aided by the *realpolitik* of our time. The United States has been at war for the past eighteen years as conflicts continue and self-perpetuate over much of the globe. Faith in the promise of globalization and global institutions has retreated in many of the western cultures that gave them life. With more than 65 million people displaced around the world currently, millions of refugees are arriving in other nations as unwelcome aliens.[69] The storms brewing on our near horizon don't present mitigating effects to this dynamic, they're potent accelerants. Humanity has never before seen what happens when several hundred million people migrate in desperation – and these circumstances stand to impact billions of lives. If current trends continue, few doubts remain that our future may bear such witness.

By themselves, any one of these problems are calamitous – be it ecological collapse, unsustainable resource extraction, climate change, or the accompanying risks to the global food supply and mass migration that comes with all of the above. Yet none of them exist in vacuums. Their arrival is in unison, and they are manifesting today – and worsening – with combined effect.[70] Even a casual observation could see how their results risk spawning humanitarian crises and breakdowns in social order that could fundamentally compromise the global economy and the state of global security.

As we saw in early 2020, COVID-19, a novel respiratory virus, brought the world to its knees. Foreign and domestic supply chains were stretched so far past their breaking points that it became effectively impossible for developed nations to source even basic household items like bleach, hand sanitizer and toilet paper – let alone medical necessities such as masks, ventilators and hospital intubation equipment. Our social functions were pummeled to a standstill, and even first-rate healthcare systems were overwhelmed to the point of paralysis.

This was the result of a single infection with a relatively low fatality rate. When we consider what might happen if a billion people were to run out of food or water? Or if climate change were to displace hundreds of millions of people? Or if Earth's ecology collapsed alongside its ability to support humanity's existential foundations? It's hard to overstate the danger any one of these events would present to our social order and our logistical capabilities to deftly respond to crises. Yet the simple math of our situation shows that the likelihood of each these results coming to pass – in unison – increases as long as current trends persist.

This reality exposes a fundamental truth of human nature and the actions of nations: of the motives that cause us to take up arms against one another, few, if any, are greater than resource scarcity and the economic damage caused as a result.[71] As individuals, people may fight over any number of reasons, be it religion, nationalism, identity or pride – but nations aren't driven by causes so fickle.[72] Nations are driven by the resources that sustain their existential basis – the resources that power the vast unseen functions which enable a modern, interconnected society. And should they not manifest in either perception or reality, the uncompromising nature of need beats the drums of war to unleash the horror that follows. Most every conflict, occupation or atrocity on a large scale can be attributed to this fact.[73] The entirety of human history, even if varnished through a rose-colored sheen, provides a bitter testimony.

What might ultimately result from the combination of these circumstances is of course not yet known. Yet what remains known is that the darkest examples of human nature manifest in times of ecological, geopolitical or economic strife.[74] Further known is the possession of at least 15,000 nuclear weapons in the hands of nine countries – thousands of which can be launched in minutes.[75]

Some macabre trivia: a single *Ohio*-class United States Navy submarine is capable of raining thermonuclear warheads on as many as 288 targets within a range of 7,000 miles – *each* with 2,500% greater destructive power than the atomic bomb dropped on Nagasaki.[76] The U.S. Navy has *fourteen* of such submarines. The Russians, Chinese, French, British, Indians and Israelis[77] each have their own, and that's on top of the land-based missiles and aircraft that can be used to deliver thousands of nuclear munitions to any corner of the globe within an *hour or less*.[78]

There are fewer than 4,500 cities on the planet with more than 150,000 people.[79] Humanity possesses an equivalent of four nukes for each.[80]

It's a hardened truism that society is intimately acquainted with prophecies of doom. It's also true that people have been harking about things "going to hell in handbaskets" since we had words for either. Fear sells. That's why the boy cries wolf – a lesson well-known to any politician, theologian or journalist worth their title. And when dark nights eventually turn to dawn and there remains no wolf to be seen, we become numb to future bells tolling its arrival.

Yet the quintessential point of the fable is that the wolf does one day arrive, to either be defeated or devoured by. That day is now.

The multitude of problems facing our future place the foundations of our civilization in existential peril. The continued survival of not only our way of life, but our very species and the planet we call home is incumbent on their solution. Our hands, in our time, are tasked with either developing that solution, or failing to. Failure in this context is not quantified by a loss of money, power, prestige or reputation. It's quantified by the immolation of our civilization as we know it. The extinction of Earth's natural beauty. And, ultimately, the ashes of what we love and hold dear. That is the stone-cold reality of our present state of affairs.

But *The Next Giant Leap* wasn't written to bow to that reality. It was written to change it. And it starts with a story of technology.

Ever since we invented machines that could perform at greater levels of speed and precision than human hands, we've rapidly increased our capabilities to build advanced systems: jet aircraft, smartphones, supercomputers, satellites and spaceships. And while these capabilities have caused problems, they can derive even greater solutions.

Because of technology, billions have been brought out of poverty, devastating diseases have been eradicated, vast distances have been spanned, veritable wonders have been built, and transformational new ways to process and communicate information have been developed.

Indeed, as you read this, a single smartphone-wielding bartender in Mombasa, Kenya today has instantaneous access to a wealth of information the world's governments, universities and corporations **combined** didn't have when *The Simpsons* first aired.

Technology made all of this possible. Yet we've now reached the point where sophisticated systems – systems which, thirty years ago, took millions of dollars and months to build – can be inexpensively manufactured on automated assembly lines in hours. This capability has been extended to cutting-edge aircraft, advanced electronics, even large-scale infrastructure. But it can now also be extended to energy-generating systems of all forms – small modular power plants, systems for energy transfer, large-scale storage and energy management, mass-produced wind turbines and solar panels – enabling us to generate energy on far higher scales than was ever before possible.

But what if we took things a step further? What if instead of just mass-producing energy technologies, we also built them to work together by design? In the era of the smartphone or smart car, why not have a smart power plant or smart power grid? And, having done that, what happens if we apply all of the technological advancements we've recently made within and outside of energy generation, and combine them into one intelligent system for energy and resource production that can be easily scaled and deployed worldwide?

We don't yet have a system today that can answer these questions. So, we're going to take the opportunity to build one that can – right here, in this book.

The Next Giant Leap outlines blueprints for a system to solve resource scarcity and climate change. This system is called "Universal Energy" and it is an open-source framework of the best energy technologies we have available, designed to work together in a way that makes them greater than the sum of their parts in terms of output, efficiency and flexibility. As a framework, Universal Energy is designed to be deployed rapidly on a large scale to turn the tide against the ecological, climate and resource problems of our time. Its goal is to shift the foundations of how we generate energy to in turn shift the foundations for how we acquire and distribute resources – and, in the process, permanently erase the concepts of scarcity and need.

As the underlying conceptual structure for a system, frameworks apply in varied contexts, but their greatest applications are in software. In software, frameworks are virtual tools that enable integrated, cooperative deployments of different technologies in a way can be both flexible, modular and standardized. They're what made 0's and 1's into the interconnected world we live in today. Software can be developed and applied for the ever-better management of data, goods, currency, ecosystems and even people. It can also do the same for energy and resources in a new paradigm of global abundance.

Universal Energy's DNA is software for energy generation and resource production through an indefinitely scalable method. By heavily leveraging sophisticated mass-production and modular, standardized deployment strategies, this approach allows us to lower the price of energy to the extent where it becomes feasible, for the first time in our history, to synthetically produce critical resources on an effectively unlimited scale. No matter how much energy or resources are consumed, the system can always produce more at a rate higher than that of consumption – a feature by design.

This framework is environmentally friendly.
This framework is affordable.
This framework is sustainably powered.
This framework is built with technologies that can exist today.

And, this framework can be deployed anywhere in the world to functionally end resource scarcity – and do so with finality.

If we can unchain ourselves from the eternal problem of resource scarcity and the untold human potential it consumes, it frees up the immense resources devoted to extinguishing its fires within a scarcity-dominated world. We can then invest those resources in social advancement and ecological healing, powering a perpetually ascendant course. This wouldn't cost us more, and would ultimately cost us less – not just in terms of money, but also in terms of focus.

If we are no longer forced to surf a never-ending tsunami of social maladies, we can devote greater attention to improving our lives and human civilization as a whole, solving long-vexing problems and addressing humanitarian crises that have plagued us for millennia.

With the technologies behind Universal Energy and the framework they power, we can change the world.

And we can build it better, stronger and brighter than before.

Universal Energy

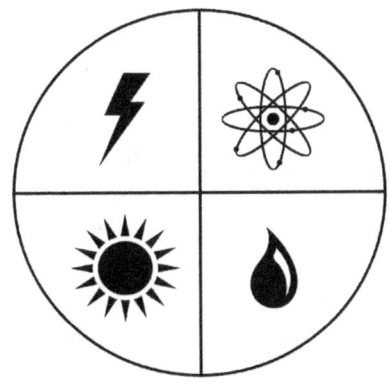

FRAMEWORK

1. An essential supporting structure of a system, building or idea.

2. A foundation underlying a concept, philosophy or mindset.

You never change things by fighting the existing reality. To change something, build a new model that makes the existing model obsolete.

- R. Buckminster Fuller

Chapter One: Mindset

Since the dawn of human time, civilization and resources have been inexorably linked, powering and making possible every part of our existence. As our existence has evolved and expanded, so too have our needs, making resources ever-more critical for the advanced, global societies we seek to continue building.

Resources have been the key to nearly every social and technological advancement we have ever achieved, and resource scarcity has conversely been the cause of major problems both in the past and in our time today. Throughout history, the societies and nations of mankind have all attempted to mitigate scarcity through varied constructs: laws and social policies; ideologies and political movements; technological innovations; the rewriting of borders; and, ultimately, war. Yet these approaches have almost universally sought to avoid resource scarcity by addressing its varied symptoms – rarely, if ever, have they dealt with the core problem itself.

It is for that reason why I believe they have failed.

A true solution doesn't cure the symptoms. It cures the disease. In the case of resource scarcity, our cure comes from technology – and more importantly, how we can use it.

Technology provides the solution to resource scarcity because it allows us to extract resources more efficiently and with less expense. It also allows us to advance the means in which we acquire resources in terms of scale, sophistication and potential. Over time, we have developed and depended on technology to solve resource scarcities – which has led to breakthroughs that have changed the world, even if we didn't realize it at the moment.

For example:

- The years following WWII gave rise to the threat of the first global resource crisis: food scarcity. Humanity was rapidly expanding in population and feeding the planet was becoming progressively more difficult. This crisis was detailed in *The Limits to Growth*,[81] a 1971 paper that predicted catastrophic consequences for humanity should it fail to curb population expansion. These predictions were well-reasoned, yet they never came to fruition. Why not? Technology came to the rescue through industrialized farming techniques, high-performance fertilizers and genetically modified crops, all of which increased food production to the extent that Earth now supports 7.5 billion people and counting – twice the population of when *The Limits to Growth* was published.

- In the 1800s, aluminum was extremely rare, considered to be one of the most valuable metals in the world.[82] Today we throw it in waste receptacles. What made the difference? A method called electrolysis, which allowed us to inexpensively extract aluminum from its naturally occurring form, bauxite.[83] This method made aluminum extraction easy and inexpensive, dropping its cost to almost nothing. (Next time you throw away that soda can, though, realize that not 150 years ago it was worth its weight in silver.)

- The need to obtain water by traveling to a location and carrying it back used to be a massive time expenditure for everyone within society, a problem that still exists within much of the developing world. Yet for the developed world, the invention of modern plumbing brought water to us on-demand. This collectively saved people trillions of hours in free time and removed a major impediment to cascading economic growth.

- Sugarcane was introduced to Mediterranean regions around the 7th century and thereafter remained a major luxury commodity. As a valuable cash crop, sugar was heavily taxed and was a revenue source for government, making it a driver of the slave trade. Yet when technology introduced the steam engine and methods of vaporization in the late 1800s, the cost to refine sugar plummeted to less than 5% of its former price.[84]

In each of these examples, a once-scarce resource was made both abundant and inexpensive as a function of technology, for technology has the unique ability to expand the scale of resource production while also lowering costs. But in the past, technology only really improved our ability to *extract* resources that were naturally present – and, over time, extraction has proven to be unsustainable as our natural resource supplies eventually dwindle. But what if we shifted gears to develop systems that could instead *synthesize* resources at scale?

For the first time in history, that's a capability we now possess today.

The past three decades have seen transformational breakthroughs in several critical industries. Information technology has been transformed by the advent of high-performance computing at low cost, which alongside similar advances in networking, has ushered in an unprecedented capability to collaborate on state-of-the-art initiatives with sophisticated virtual modeling. It's further enabled to-the-second global logistics and a degree of operational reliability that would have been unthinkable even twenty years ago. Polymer and material sciences have been transformed through the creation of synthetic substances that rival hardened steel in strength at a fraction of its mass, yet also present revolutionary benefits in terms of conductivity and flexibility of form.[85] Large-scale manufacturing can now rapidly build complex machinery on assembly lines at a level of precision that would have been nigh-impossible until the latest decades of our modern era. Combined, these advances allow us to engineer and manufacture solutions to problems on much larger scales than ever before.

To put this in perspective, most nuclear power plants in the United States were built between 1970 and 1990.[86] That means a good deal of them were designed and built *without the aid of a calculator*.[87] The same is true with other types of power plants and larger-scale social infrastructure. Yet today, we have the capability to design a power plant on a computer and build it on an assembly line – much as we build a toaster.

To be sure, we can build many things with these increased capabilities. But the starting point is to build a system that can sustainably produce resources. And not just any resources, but the five most critical:

WATER, FOOD, ELECTRICITY, FUEL AND BUILDING MATERIALS.

Above all else, these are the most important resources for our civilization to operate. Without water, nothing grows and nothing lives. Without food, we starve. Electricity is the currency of capability and information, and is the glue that holds our modern social framework intact. Fuel provides high-density energy in standalone contexts where electricity is not present, and building materials enable us to advance, repair and extend the infrastructure that enables our civilization to flourish.

These are the resources that are most essential to powering our advanced economies, and these are the resources most likely to spark conflict when they become scarce.[88]

The purpose of Universal Energy is to act as this resource-producing system, and it works by leveraging three critical concepts: *standardization*, *modularity* and, with these two in place, *cogeneration*.

Standardization, in this case, is a way of building something to a universal standard that's adopted society-wide. (For example, all of your electronic devices are charged by connecting a standardized type of plug into a standardized type of wall outlet.) *Modularity* is simply a way of deploying something that features the ability to rapidly change configuration or scale using a *standardized* means (think Legos™ that enable you to build whatever you like, a model city or model ship, with pieces that connect to each other in the exact same way).

Standardization and modularity allow us to take a technology and deploy it in a way that can be mass-produced, providing easy replacement of parts and driving down costs. Recent advances in technology enable us to apply these concepts on larger scales, especially within energy generation.

Our energy infrastructure today (and, by extension, our civilization as a whole) is powered by a hodgepodge of sources: oil, coal, solar, wind, uranium, natural gas, geological heat, hydroelectric, corn ethanol, and biomass.[89] Few of these energy production systems work with each other,[90] they barely even talk to one another.[91] And they are all implemented ad-hoc, meaning they were *designed and built to order as unique systems* with only minimal standardization and even less modularity in design. Each may reflect compliance with relevant building codes, but unlike most every other sophisticated product in our society – no one power plant is identical to another.

These factors make power-generating systems highly expensive to build and operate. It further prevents us from rapidly scaling them in size or extent of deployment, which limits the volume of energy available and thus raises its price.

There is a better way.

By designing the technologies within Universal Energy to incorporate modularity and standardization, we can leverage the concept of *cogeneration*, which is to use the waste energy of one technology to power something else. For example: diverting the waste heat energy from a coastal power plant to power a nearby facility that desalinates seawater into fresh water. Desalination is presently a costly and energy-intensive process,[92] yet when it's powered primarily by waste energy, energy requirements and costs drop drastically.

Each technology within the Universal Energy framework is designed to easily connect and work cooperatively with others from the ground up, while maintaining the ability to rapidly scale in size on-demand. This empowers us to push the bounds of cogeneration, allowing our energy infrastructure to efficiently produce *both* energy *and* resources in the same footprint. This will do to energy and resource production what technology has done to most other consumer products: increase availability, lower prices, and advance quality over time.

The essential part to making this approach successful is identifying technologies that can generate energy in the way we need them to, which we'll define as meeting the following criteria:

1. **The technology must be able to generate immense energy at low cost.** In order to synthesize enough resources to satisfy all of our requirements, we'll need to generate an effectively unlimited supply of energy. This means that energy will always be regenerated at a rate faster than that of consumption, regardless of how much energy is consumed. This requirement will set an initial target of 300% of annual electricity consumption in the U.S. (3,860 terawatt-hours as of 2018),[93] coming to a total of **12.5 trillion kilowatt-hours** generated annually. (If you're a little unfamiliar as to what a "kilowatt-hour is," there is a helpful guide next chapter).

   The cost of this energy is intended to be no more than **two cents per kilowatt-hour**, down from today's 10.53-cent average.[94] Reaching this target would provide enough energy at a low enough cost to allow large-scale synthetic production of civilization's five most critical resources. (For those inclined, a detailed pricing model is included in this writing's Appendix on page 315).

2. **The source of this energy must be abundant and sustainable.** If an energy source and its corresponding extraction methods aren't sustainable available after widespread adoption, we'll eventually find ourselves in the same position we're in now. As such, any technology we

employ within the Universal Energy framework will need to be long-term viable, quantified for our purposes to be 1,000 years or longer.

3. **The technology must be safe and environmentally friendly.** The energy production system, its fuel, and any waste must have negligible environmental impact and must further be carbon-neutral, meaning it does not emit carbon dioxide or methane which adversely impacts climate change. Additionally, it must not leave toxic waste that cannot be rendered inert in 300 years or less.

4. **The technology must be affordable to develop and use.** Whatever benefits are brought by advances in energy technology will be irrelevant if they are not affordable, presenting the requirement for all energy-generating systems to have a realistic price tag.

5. **The technology must be flexible in where it can be located.** There are already energy technologies that can fit the previous four requirements, but many can only function in limited areas and thus cannot be deployed in areas that are geographically remote and/or have rugged terrain. Universal deployability is vital for a modular and standardized energy framework.

6. **The technology must be deployable rapidly.** Considering the state of the world today, the solution to resource scarcity and climate change needs to get here soon. If we don't have these problems solved in the next 20-30 years, other solutions may not matter in the end.

Universal Energy meets these requirements through a strategic deployment of five technologies: solar, wind, thorium and hydrogen, all interconnected through a revolutionary use of water. But before we see how they all function and work together in-depth, here's an overview of Universal Energy's intended deployment:

Integrated Renewables – The energy potential presented by renewable power is revolutionary, yet common problems with its use today are expense, land requirements, piecemeal deployment and material/carbon throughput in manufacturing. Renewables may be adopted by individual businesses, landowners or cities as they wish, but there's not really a standardized method to deploy them nationwide on a large scale. Yet integrating renewables within urban municipal infrastructure – buildings, bridges, highway medians and road canopies – gives us a unique location for installation that *critically eliminates the expense of buying additional land*. Not only does this allow us to use the high costs of constructing public infrastructure to offset the expenses of renewable energy – it helps build a smarter and more resilient electric grid.

Liquid Fluoride Thorium Reactors (LFTR) – Thorium is a unique type of nuclear fuel that avoids nearly all complications inherent to our current approach to atomic power. Instead of uranium, which is traditionally enriched within highly pressurized reactors, thorium generates energy as a high-temperature liquid within advanced reactor designs that are not pressurized.

These reactors are far cleaner, far safer and more resistant to weaponization than uranium reactors used today. Thorium is also abundant – about as common as lead – which makes it thousands of times more sustainable than fuel-grade uranium. The waste footprint of LFTRs is minimal and short-term, rendered safe in decades as opposed to millennia. And because they can be built to small size and don't operate under pressure, they can cost a fraction of what other approaches to atomic energy run. Just as importantly, they present a carbon-neutral energy source to manufacture and recycle renewables (and supplemental resources) on a large scale.

Water and Hydrogen – Our ability to extract fresh water and hydrogen fuel from seawater is well-known to science and industry.[95] The problem is that the process has thus-far been energy-intensive and thus expensive. Using the near-unlimited supply of inexpensive heat and electricity from thorium changes that, enabling us to desalinate billions of gallons of seawater at minimal cost. Of this water, Universal Energy devotes a portion to producing hydrogen fuel, with the remainder being transported nationwide through another central component of Universal Energy: The National Aqueduct.

The National Aqueduct – As a proposed nationwide delivery system for desalinated seawater, the National Aqueduct also doubles as a power plant and battery for renewable energy. Deployed alongside the pre-cleared and publicly owned land at the sides of our highways and high-tension power lines, it would be built via prefabricated, modular pipelines with embedded solar panels and hydroelectric turbines – allowing these water pipelines to passively generate immense levels of energy. Most critically, any excess energy generated by the system can be used to keep billions of gallons of water *at high temperature*, permitting us to use our fresh water supply as a giant "battery" by way of converting heat energy into electricity (thermoelectric charge).

How It All Fits Together

Universal Energy harnesses the limitless potential of **renewable energy** and significantly increases its utility by integrating it within public infrastructure to both increase scale and reduce costs. **Liquid Fluoride Thorium Reactors (LFTR)** are deployed second to safely provide an immensely powerful parallel source of energy to complement renewables within a self-reinforcing framework. In doing so, these reactors act as a base load energy backbone. They also work as integrated power sources meant to **desalinate seawater** on a massive scale – a portion of which is then used to produce **hydrogen fuel**. This desalinated water is then pumped nationwide throughout the **National Aqueduct**, which further functions as both a power source and an eco-friendly storage battery for renewable energy.

Combined, these technologies are designed to work together in a decentralized, co-generating system that's both modular and standardized – maximizing efficiency, flexibility and reliability. With Universal Energy, our power network isn't built from piecemeal technologies that neither work together nor communicate with one another; it's instead designed to work in tandem from the ground-up, dramatically increasing our capability to generate energy on a nationwide scale at substantially lower cost. An open-source and indefinitely extendable operating system for energy generation, one suited for the 21st century.

With this abundant supply of energy and water at our fingertips, we solve the problems of resource scarcity. We can grow food indefinitely in indoor farming networks near urban centers.[96] These indoor farms can be extended further to grow the organic substances needed to create sophisticated synthetic materials, materials that can revolutionize how things are built and recycled.[97] In aggregate, Universal Energy gives us nigh unlimited supplies of energy, water, food, electricity, fuel and building materials – enabling us thereafter to advance our manufacturing capabilities and build our civilization to ever-greater heights in a world spared of scarcity and need.

What About New Technologies?

As current technologies evolve and new ones emerge, the Universal Energy framework will adapt to include them. The goal is to generate enough clean energy at a low enough cost that we can inexpensively synthesize resources to effectively unlimited scales at minimal environmental impact. Just as players change roles in a football game, technologies will do the same within Universal Energy. As an evolving framework, Universal Energy will update to reflect the greatest available potential for clean energy generation.

The mindset behind Universal Energy and all it seeks to achieve is to rewrite the rules of our existence by systematically dismantling the challenges that have held us back since the dawn of time. If successful, we can position ourselves for a future where humanity – us, our children, and generations hence – not only survives on this planet, but permanently thrives. Where we can reach goals that were never before achievable and bypass the limitations of resources as we once knew them. The technical capability is here today. Universal Energy is how we can make it real.

The word "energy" incidentally equates with the Greek word for "challenge." I think there is much to learn in thinking of our federal energy problem in that light. Further, it is important for us to think of energy in terms of a gift of life.

- Thomas Carr

Chapter Two: The Renewable Revolution

The Universal Energy framework starts with renewable energy. Not just because it's versatile and easily accessible, nor because it boasts a historically unprecedented capability to generate energy from a truly unlimited source. Rather, Universal Energy leverages renewables – specifically solar, wind and artificial hydroelectric – because of their ability to both integrate within large-scale infrastructure and power resource production as a *municipal function*.

This works across two areas of focus.

The first is the integration of solar and wind power directly within urban infrastructure to remove the obstacles and cost limitations they face today. This is a critically important distinction to our current approach towards these technologies, which only comprise 10% of energy generation nationwide.[98] While the usual suspects[99] in fossil energy lobbies certainly play a role in this,[100] the justified condemnation of their undue influence often glosses over the significant logistical and economic challenges to implementing renewables over a large scale. As we'll see shortly, an integrational approach inherently avoids those challenges.

The second area of focus is **The National Aqueduct**, a vital function of Universal Energy intended to transport desalinated water anywhere in the country by piggybacking on the pre-cleared and largely publicly owned land of our interstate highway system and high-tension power line networks. Further, the National Aqueduct acts as *both* a power plant and a battery by deploying solar, wind, hydroelectric and thermoelectric functions as a single system. The National Aqueduct is detailed within Chapters Seven and Eight, but it's useful to keep it in mind as it's referenced several times before then.

As Universal Energy seeks, by design, to bypass the challenges inherent to renewable energy, we'll take a minute to review these challenges and why they stand in the way of a clean energy future.

Location and transmission. Electricity, like sound, weakens over distance due to resistance in transmission mediums – in our case, power lines.[101] As a general rule, the farther electricity must travel, the harder it becomes to transmit. For example: there is enough open space in the American southwest for solar panels to power the entire planet.[102] Yet transmitting electricity from the southwest to locations thousands of miles away is both difficult and expensive. It's possible to generate the energy, but we can't efficiently get it from point A to point B once we start spanning large distances. Even if more efficient power lines emerged, they would still cost millions of dollars to build per-mile.[103] *This means renewables are best deployed in locations close to where their generated energy is consumed.*

Deployment expense and physical space. Solar and wind power require relatively large areas of land to be useful. Given that they are best deployed in locations close to energy consumption, this presents a secondary problem: land costs generally increase with population density. If land needs to be purchased for installation, the cost effectiveness of renewables proportionally reduces the closer they get to population centers.

While eco-conscious households and businesses can install renewables by choice, that option becomes harder for governments and power companies when they need to buy land at top dollar. So even as manufacturing costs of renewables continue to fall,[104] expenses of their start-to-finish implementation can remain high, both fiscally and politically.

Lack of standardization and prefabrication. As with many fledgling industries, our current approaches to solar and wind power remain unstandardized. Today, solar panels have extensive sizing options and are designed to be installed on varied locations: roofs, soil, rock, motorized platforms, etc. Wind turbines, likewise, come in all shapes and sizes for both residential and commercial applications. While flexibility in deployment is generally a good thing, there is not yet a modular and standardized method to implement renewables on a nationwide scale. This complicates the manufacturing of prefabricated, turnkey systems, increasing end unit cost and limiting overall usability.

Material throughput. Renewables require large volumes of materials to manufacture (10,000-17,000 metric tons per terawatt-hour),[105] materials that further need to be extracted, transported and processed in a carbon-emitting supply chain. This manufacturing process also carries its own drawbacks in terms of toxic waste.[106] Absent a massive carbon-neutral baseload power source such as thorium (outlined in Chapter Four), the large-scale manufacture and recycling of renewable energy and the costs therein would present tremendous carbon emissions and ecological toxicity – even if they themselves generate carbon-neutral energy.

These obstacles have slowed down the adoption of renewable power nationwide, and while advances in research and development have mitigated their impact to various degrees, they still remain substantial. Universal Energy seeks to solve this problem by targeting large-scale infrastructure for renewable deployment – especially public infrastructure.

Public infrastructure is the ideal location to install renewables, as it solves four of the five largest obstacles to their implementation: land, location, standardization and scale, with thorium solving the fifth (carbon-neutral manufacturing).

Here's why:

No need to buy land. Municipalities can install solar and wind on publicly owned property by their own volition. They don't need to submit bids or seek permission to buy expensive private land – cities can simply use city funds to deploy renewables on city-owned property to generate municipally provided electricity as they deem fit.

Close to population centers. In most cases, city-owned property tends to be close to cities. Public infrastructure, then, zeroes out the distance between generation and consumption. No expensive power lines need to be constructed, there is no need to transport electricity over long distances. And as advances in material science now allow us to build solar panels that are 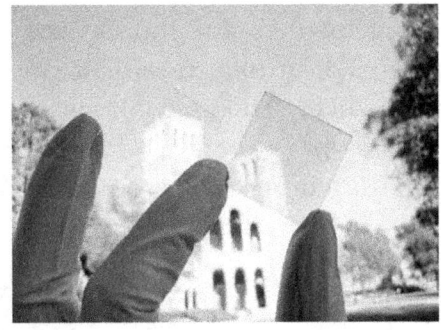 completely transparent (image to right), we don't need to sacrifice aesthetics to take this approach.

Think of the streets in your nearest city. The sides, windows and rooftops of every public or government building, every highway overpass and bridge, and every stadium built in part or full with public money; even privately-owned buildings. Imagine all of this infrastructure integrated with solar panels from the moment they were constructed – and all without spending so much as a nickel to buy extra land for their deployment. This image shows only *one section* of *one area* of *one city*.

Now imagine every city in America taking this approach. Applying this mindset across every municipality in the country gives us billions of square feet of surface space to deploy renewables that's both close to consumption and doesn't require us to buy expensive land.

Standardized manufacturing. Because renewables are so well-suited for large-scale infrastructure, this affords opportunities to build systems that can be mass-produced to a modular standard. Once you know exactly how something's going to be implemented, it removes variables that add to unit cost and complexity – the end goal of any effective deployment strategy. That's in part why several solar companies likely choose rooftops as locations for residential solar deployment: most homes have roofs. The following images show Tesla corporation's solar shingles, although several companies (RGS Energy, CertainTeed and SunTegra) have released similar products to market.[107]

As these surfaces both retain the aesthetics of traditional roofing tiles while boasting the functionality of solar panels, these companies identified a *modular standard* to deploy solar panels and manufactured their products to that standard. Yet they still require the personal investment of individual homeowners in order to adopt them on a large scale. By integrating renewables within municipal infrastructure, large-scale adoption can be accomplished with the stroke of a pen. This incentivizes manufacturers to invest in renewable technologies that can be built to a standard for wide and consistent deployment nationwide.

Aside from the National Aqueduct, Universal Energy looks to three general types of large-scale infrastructure to deploy renewables. They include standing structures, road medians, and road canopies, which we'll review in depth after taking a quick aside to make mention of some necessary assumptions.

A Quick Note:

The Next Giant Leap uses some terms surrounding energy that you might be a little unfamiliar with. We'll take a minute to explain these terms (and their use in some basic arithmetic) more clearly to make it easier to grasp how Universal Energy works.

Energy vs. Power: Power quantifies the potential amount of energy that could be exerted *in a given moment*. **Energy** is the aggregate total of power exerted *over time*. Both power and energy are denoted in "watts," with power denoted *in-moment* and energy *over time*. Watts is a metric unit of measure (1,000 watts = 1 kilowatt = 0.001 megawatt).

For example: if you turned on a 60-watt light bulb for ten seconds, that light bulb would have had 60 watts of **power** flowing through it. Yet it consumed 600 watt-seconds of **energy** (60 watts x 10 seconds). If it was on for an hour, it would have consumed 60 watt-hours (3,600 watt-seconds). If it was on for one year (8,760 hours) it would have consumed a total of 525,600 watt-hours of **energy**. That's equivalent to 525.6 kilowatt-hours. At any given moment, though, it only had 60 watts of **power** flowing through.

Solar Power: In determining the effectiveness of solar power, we're going to need to establish some baselines so we're measuring consistently. Solar panels vary in size, sophistication and effectiveness, and there are several factors at play that determine both.

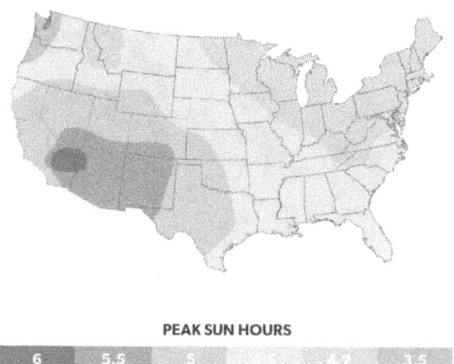

PEAK SUN HOURS

Peak sun hours is a unit of measure for solar panel generating capacity. It quantifies the aggregate total of solar energy that reaches a region in a day, denoted in hours of maximum solar intensity. Let's say the sun shines for 13 hours a day over a region, depending on the season. That solar intensity may be equivalent to a certain number of "peak sun hours" of the sun at its maximum output (say at noon). Most regions of the country are exposed to an annual average of between 4.5-5.5 peak sun hours of solar energy per day. For our purposes here, we'll split the middle and assume a total of *five peak sun hours per day* when calculating for solar.

Solar panel output varies by sophistication, size, region and efficiency. So, we'll have to establish some assumptions based on the myriad solar options available. The SunPower E20 solar panel is a solid quality solar panel available on the market today. With a nominal power output of 327 watts, it is 61.4 inches wide and 41.2 inches tall, coming to a surface area of about 17.5 square feet.[108] That translates to roughly 18.7 watts of power per square foot. Now let's convert that to energy. If we circle back to the energy vs. power section on the prior page, 18.7 watts of power exerted over an hour is measured as 18.7 watt-hours. Assuming five peak sun hours in a day, that totals 93.5-watt hours generated per day. Extrapolated over a calendar year, that totals 34,127.5 watt-hours.

Denoted in kilowatt-hours (the standard most energy is measured in), that comes to 34.12 kilowatt-hours of energy generated per square foot, per year. To play it safe, however, we'll cut that figure by 10%, and estimate that one square foot of solar panel surface can reliably generate 30 kilowatt-hours per year. As there are 27.88 million square feet in one square mile, we will conclude that one square mile of solar panels would reliably generate 836.5 million kilowatt-hours in a year.

Assumption Totals:

One square foot of solar generates: **30 kilowatt-hours per year.**

One square mile of solar generates: **836.5 million kilowatt-hours per year.**

According to the Energy Information Administration as of 2017, the average U.S. household consumes **10,400 kilowatt-hours per year.**[109]

With these assumptions explained and these figures established, we'll circle back to the revolutionary benefits of integrated renewables and how they can help build a clean energy future.

Standing Structures

Universal Energy's first target for large-scale renewable deployment are standing structures with a primary emphasis on solar power. As we've invented solar panels that are both clear and visually unobtrusive, virtually any type of construction can be harnessed to generate clean energy from the sun.

This approach is ideal when it comes to location for solar deployment: standing structures are directly integrated within (and thus close to) population centers, and, as such, don't require the purchase of additional land that might otherwise cost billions of dollars in aggregate. Further, in many cases, integrating solar can be accomplished as a seamless upgrade of the building's existing features.

As a base concept, this idea isn't new. Rooftops have been outfitted with solar panels since they were invented, and their use has increased as the costs of renewable energy have dropped over time. But recent advances in manufacturing and material science empower us to more deeply integrate solar into architectural designs. Solar glass now allows buildings to replace exterior windows with transparent solar panels. Solar-thermal HVAC uses excess solar energy to heat water and the building itself, generating electricity while at the same time slashing energy costs. Rooftop solar panels have never been cheaper, and for residential applications, solar shingles give homeowners the ability to use solar power with zero aesthetic sacrifice. Next-generation energy storage technologies tie all of the above together in a reliable, closed-loop system – an important feature that we'll review later in this writing.

When applied to existing buildings on the scale of skyscrapers or large office complexes, these advances can easily fractionalize energy use several times over  – if not turn commercial buildings into mini power plants. To explain how, let's circle back to windows. Consider the skyscraper to the left for a moment. Now imagine if every single one of those windows were replaced with solar glass – which again can be made aesthetically identical to normal glass. Just for one skyscraper, the energy potential would exceed hundreds of thousands of kilowatt-hours, even at moderate efficiencies – which would account for shadows or cloudy days as solar still works under such conditions.[110]

According to SolarWindow, a leading manufacturer of solar glass for commercial applications, a 50-story skyscraper in downtown Manhattan would generate 1.12 million kilowatt-hours annually if completely outfitted with solar windows (equivalent to the annual electricity usage of 108 single-family homes).[111] That number would rise to 1.38 million annually in Denver, and rise further to 1.57 million in Phoenix.[112] With present energy costs, the time it would take for each of those installations to pay for themselves is estimated to be roughly one year.[113] As solar glass can auto-tint based ambient brightness, the potential dividends for insulation and indoor climate management can be extended further.

Functional prototypes of solar window glass:

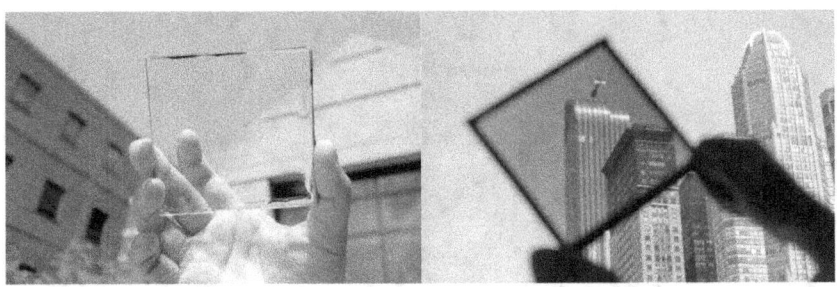

A large building such as the skyscraper on the previous page could likely generate more energy than it could reliably consume in a given day. What to do with this extra energy? It could be held locally in municipal storage (including the National Aqueduct), sold back to a local or municipal electric utility, or it could be diverted directly into its heating and cooling system.

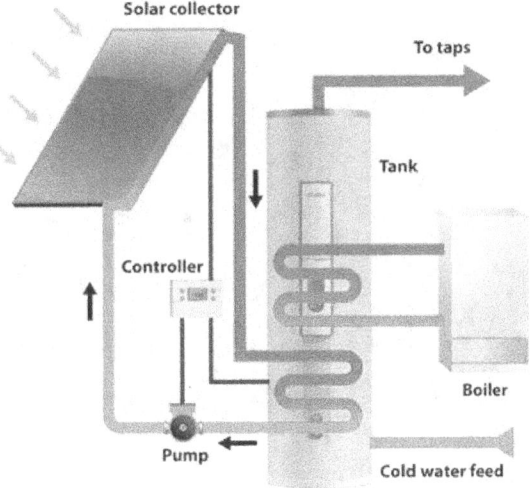

The image to the right shows a basic diagram for a solar-thermal heating system. In concept, excess solar energy is used to heat water during the day, which is kept warm in an insulated storage tank. As water holds its temperature better than nearly any other substance, solar-heated water stored in insulated tanks dramatically reduces the need to use oil or natural gas for building heat. Notably, as solar panels are most efficient at colder temperatures,[114] this can save companies significant sums through reduced heating bills.

The benefits of upgrading existing buildings with solar windows and rooftop arrays of solar power will be limited primarily by the construction of the buildings themselves. To truly extend the capabilities of renewables on standing structures, integration could be considered by design at the architectural stage.

The following image shows the circular headquarters for PAF, a gambling management company owned and operated by the Finnish government. The building is wrapped completely in solar panels.

Another example is the Solar Valley Micro-E hotel in Northwest China, designed from the ground-up to maximize solar utility.

The Sanyo/Panasonic Solar Ark is a dual power plant and science museum dedicated to solar energy. Made from factory-rejected solar panels, it generates 500,000 kilowatt-hours of energy per year.[115]

These examples of solar buildings are prototypes, and, with the exception of the PAF headquarters in Finland, are initiatives undertaken by private companies. But these buildings act as solid proofs-of-concept, giving us ideas with which we can extend the integration of renewables within new structures.

Unlike passive solar design – an architectural trend from the 1970's that attempted to design structures that would naturally take advantage of solar energy without the aid of mechanical equipment[116] – the sophistication of today's technology and available commercial products in the renewable sector make "active solar designs" more straightforward. Technology can provide solar shingles, solar windows and solar walls at comparable or lower costs than fossilized energy sources.[117] Such technologies can further integrate through solar-enabled heating and air conditioning systems within a building. Managing a system could be accomplished through something as simple as a smartphone app.

Approaching architectural design with an eye to using active solar saves significant sums of energy and money. According to the U.S. Energy Information Administration, roughly 33% of a given commercial building's total electricity use was devoted to temperature control in 2012,[118] consuming the equivalent of some 410 billion kilowatt-hours of power in aggregate. At a rate of 10 cents per kilowatt-hour, that's equivalent to $41 billion just for commercial HVAC costs – saying nothing of the electricity costs to keep the lights on and run equipment.

Every building that adopts an active solar system avoids unnecessary energy expenses and further presents less demand on the local electrical grid. Each building further helps create a cascading effect of *increasing* the amount of energy generated while *decreasing* total energy consumption.

This cascading effect becomes important when other infrastructure is selected for potential solar energy deployment. Take bridges for example.

The Sunshine Skyway Bridge in Tampa, for example, is 4.14 miles long (roughly 21,900 feet).[119] If we were to assume an average of only 40' of deployment surface on each side of the 430-foot tall structure, the Skyway bridge would feature 1.75 million square feet of surface area that could be mounted with clear, aesthetically unobtrusive solar panels. When we shortly consider the implications of solar road canopies, the 94' wide driving surface would add another 2.05 million square feet of surface area to a total of 3.81 million.

As one square foot of solar panels can reliably generate 30 kilowatt-hours annually in areas with solid sun exposure, the Skyway bridge could passively generate 114 million kilowatt-hours of electricity per year without having to buy a single piece of land – enough to power nearly 11,000 homes.

Another promising example of usable infrastructure is the parking lot. Our nation is littered with parking lots that are flat and open:

As they're dark in color, they absorb lots of heat, and anyone who drives in summer knows full well that they turn cars to ovens – presenting serious safety risks to children or pets left inside. This makes parking lots uniquely well-suited for solar panels, a concept that's been increasingly adopted as solar panel prices have dropped.

Solar parking lots solve several problems at once. They generate electricity that can be used to power the building that owns the lot; charge the electric cars of customers or employees; or sell energy back to a local or municipal electric grid. Solar canopies further provide shade that keeps cars cool. Across an urban landscape, solar parking lots could generate a massive amount of electricity that could further increase our total generating capacity (and thus supply) while reducing demand from external energy sources.

Assuming a parking lot of 150,000 square feet (roughly equivalent to a local Lowe's),[120] sticking to the figure of 30 kilowatt-hours of electricity generated per square foot, per year, that parking lot would generate roughly 4.5 million kilowatt-hours annually. If for sake of argument, the local Target, Home Depot, Walmart, Costco and Sam's all had the same footprint, the combined total would be 27 million kilowatt-hours annually generated. That's enough to power more than 2,500 homes. It's plain, then, to see how these numbers can add up quickly over large areas of urban sprawl. This becomes even more true once road medians and solar road canopies enter the equation.

Road Medians

As one of the greatest publics works projects of the 20th century, highways present uniquely attractive locations to install renewable energy. They're completely cleared of obstructions, and as they're generally flat and straight, they get extensive sun and wind exposure. They're often close to cities, almost always municipally owned, and easily accessible.

While roads themselves can also be extended to integrate solar power via overhead canopies, highways often feature medians – unpaved barriers between lanes of traffic (especially on city outskirts) – which are perfect locations to deploy renewable energy. Like the highway itself, medians are usually municipally owned as well. Yet they serve no functional purpose apart from separating directional traffic. Additionally, they require constant maintenance (mowing, landscaping, etc.), which comes with significant additional costs. So, why not repurpose them for solar panels and wind turbines, solving two problems at once? Japan, South Korea, China and The Netherlands have done exactly that:

In the case of South Korea, their solar road median also doubles as a covered bike path for non-motorized commuters (image on top and on bottom-right).

This approach provides several benefits. First, sun exposure is maximized. Sunlight will reach panels from the moment the sun rises until the moment it sets – and solar still works on cloudy days.[121] Since highways span thousands of miles in aggregate, the surface area presented by employing medians is tremendous – surface area that's within the ideal distance for maximum effectiveness of renewables. This approach is also *easy*. Highway medians are common and simple to access, yet serve little utility in their current form. Repurposing them as solar farms creates massive utility by using publicly-owned land to generate energy – land that would otherwise cost billions of dollars at fair market value.

Wind turbines can also be integrated with solar panels at regular intervals. They cast minimal shadow, present no risk to motorists, and can complement electricity generation twenty-four hours a day. The Enlil Turbine,[122] for example, is designed to be installed on highway medians to serve this exact purpose.

Installing renewables within highway medians could be a simple extension of municipal highway management. All a work crew would need to do to install, repair or replace a deployment section is simply drive up and perform the required work – just as they would with any road maintenance today. Power lines could be run in modular conduit channels and connect to municipal electric grids. The scale of implementation could be increased piecemeal, as funding allowed, to be extended indefinitely so long as there is available median surface.

Just as importantly, as this energy is generated by the municipality, it can be sold by the municipality – costing significantly less than if sold by a for-profit utility company. This further provides an extra revenue stream which cities can use to extend their service offerings or fund other initiatives outside of standard municipal budgets.

These side benefits, however, aren't the central focus. A one mile-stretch of solar arrays (at ten feet wide) would comprise 52,800 square feet of surface, which at 30 kilowatt-hours generated per square foot annually would arrive at a combined total of 1.58 million kilowatt-hours – enough to power 150 homes. Any energy generated by wind turbines would be supplementary. **There are approximately 4 million miles of public roads in the United States.**[123] Yet while not all road surfaces boat medians (or proximity to power consuming locations), it does present attractive opportunities to install overhead solar canopies to generate energy – especially within cities.

Solar Road Canopies

At concept, a solar road canopy is simply a rectangular canopy over a road surface that's built on pylons. As roads are generally flat, straight and cleared at the sides from overhead obstructions, they're another easy way to deploy millions of solar panels. The four million miles of American road surfaces – some 8.3 million lane-miles[124] – are among the most comprehensive in the world; they connect every city in the nation together and cover thousands of square miles – some 18,860 if we were to assume each lane of road had the national average width of 12 feet.[125]

Extrapolated into square feet, that's a cool 528.88 billion. At 30 kilowatt-hours per square foot, per year, even one-tenth of that area would output 1.58 trillion kilowatt-hours – roughly *half* our national energy consumption. Further, roads also remain municipally owned, which like highway medians and public infrastructure allows for large-scale deployment while avoiding the need to purchase expensive land. Considering the map below, how many other public locations do we have available – at such size or interconnectivity – to deploy a nationwide network of anything to a single standard? Roads are the zenith for integrating renewables within large-scale infrastructure.

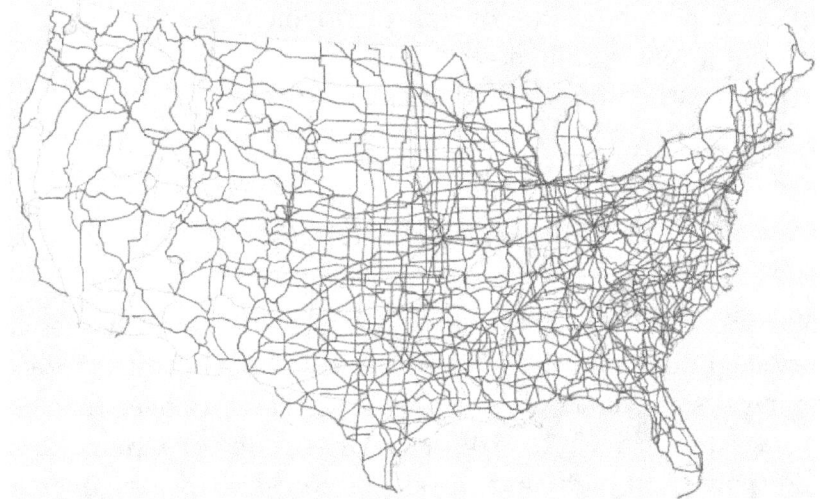

However, power is not the only benefit presented by solar road canopies. Heat management is another major aspect – especially within cities. As summer months continually increase in temperature through the encroaching impact of climate change,[126] cities are finding it ever-more challenging to deal with conditions hot enough to melt asphalt.[127] Placing solar canopies over roads would perform the same effect as placing solar canopies over parking lots – providing shade, keeping the surface below cool and reflecting heat upward instead of having it absorbed into the ground.

Another important aspect is the potential for stormwater management. Water management on roads is a major consideration of both safety[128] and environmental impact.[129] By integrating gutter systems into the solar canopies, such canopy arrays can divert water into nearby water management systems instead of letting it fall on road surfaces. In the context of drier climates with limited rainfall, this can present a highly attractive way to replenish municipal water supplies – especially if solar canopies are deployed over a larger scale. In the context of the National Aqueduct, this presents an especially effective method to deliver auxiliary water as well as store solar energy generated during the day – helping keep roads drier and safer during storms.

While solar road canopies would generate immense energy over distance, further benefits could include integration with road technology aids. Heating elements to keep panels clear of snow, municipal WiFi networks, wireless charging for electric vehicles, directional GPS guidance and auxiliary wireless guidance for autonomous vehicles are all examples in this respect.

As all canopies would require in abstract are metal pylons, a cross-lattice to install panels, wiring and plumbing, the overall deployment expenses and logistical challenges are significantly limited – all the more since they would be installed on municipal property and thus spared of land costs.

What About Storage?

Applying solar and wind power to standing structures, highway medians, and road canopies integrates renewable energy into municipal infrastructure, creating a self-reinforcing system that's easy to scale and maintain over time. The next question is storage: how can we keep the energy generated by renewables for later use? Solar doesn't work at night and works less efficiently during overcast days. Wind power only works when the ambient environment is windy. Renewables, therefore, will only be as effective as their ability to pair with storage mediums. This has proven significantly challenging – and expensive – for large-scale renewable deployment in the past, and has presented risks to energy availability at times of unexpected demand.[130]

The National Aqueduct (which we'll go over in Chapter Seven) is the primary method Universal Energy employs to solve this problem within the framework.

But there are several other promising energy storage mediums that can work on a smaller scale when needed – especially in remote locations.

One option comes from Tesla. Their flagship battery products: the Tesla Powerwall (home use) and Powerpack (commercial applications), provides piecemeal solar storage solutions at relatively modest costs.

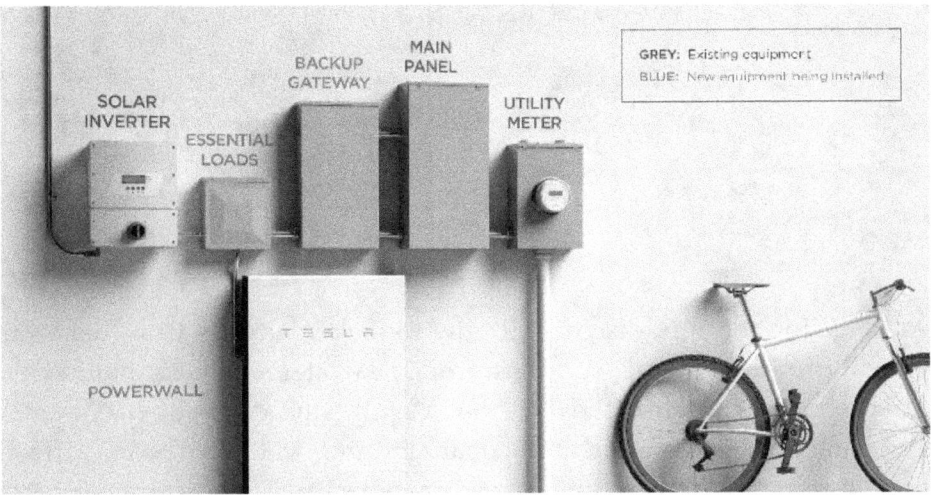

On a commercial scale, Powerpack units can work together, in parallel, to maximize energy storage capacity. As each unit has a storage capacity of up to 200 kilowatt-hours, an array of hundreds could easily store the energy needed for most municipal applications.

Liquid-metal batteries, such as the ones sold by Ambri corporation, provide another option for mass power storage. Unlike lithium-ion batteries used by Tesla and most commercial electronics, liquid-metal batteries use a combination of magnesium and antimony suspended within a liquid salt electrolyte to store electricity.[131] These materials are highly abundant and accessible, and modular battery designs allow for straightforward manufacturing. As their battery arrays are intended to be installed within standard-sized shipping containers, they enable flexible deployment strategies to provide instant-on capability to any electric grid in the world.

Ambri System
500 kW / 1 MWh
2 Ambri Cores, 500 kWh each
Dimensions (excluding power electronics):
L: 20 ft x W: 10 ft x H: 8 ft.

power electronics

500 kWh Ambri Cores

"Beta Core" lab-based system

Scientists in Beijing recently discovered how to make batteries from potassium-ion,[132] which provides ample supplies from seawater.[133] These batteries are especially promising because they do not require the often environmentally destructive practices of lithium and cobalt mining, yet also boast attractive features within energy density and charging capacity. The current prototypes retain 90% of their energy storage capacity after 10,000 charging cycles, which shows significant potential for a technology in its infancy.[134]

A fourth viable option is heat. As opposed to storing electricity directly, thermal energy storage systems capture heat that can be converted into electricity via thermoelectric features. Some of the most prominent options on the market today are molten salt batteries that store heat energy in a salt which doubles as an electrolyte, efficiently enabling high-energy output.[135] Solar-thermal troughs and hydro-solar thermal arrays use solar energy to generate excess heat for future conversion into electricity[136] – by itself a core function of the National Aqueduct.

These advances in commercial battery technology are effective options designed for use at scale. Implemented as such, they can help sustain critical city-wide systems or provide backup functionality for municipal power grids in the event of blackouts. As their scale increases over time, they can provide expanded storage functionality for renewable energy until the point where the municipality becomes self-sustaining. This becomes all the more possible once cities are connected to the National Aqueduct.

It's important to reemphasize that these technologies and the commercial products they reflect are still in their infancy, and have barely scratched the surface of their future potential. As with the first car, aircraft or smartphone, new technologies advance over time through increased economies of scale, future investments in research and development, manufacturing efficiency, and consumer feedback. Their cost and performance today are not what it could be tomorrow, next year, or in the next decade.

But even in their infancy, these renewable energy sources, when integrated into large-scale infrastructure and combined with next-generation commercial energy storage, complete a circle. Deploying renewables in cities transforms their capabilities by maximizing their scale while minimizing costs. Once complete, this circle sets the stage for an evolution in city design that dramatically expands the potential of the contemporary and future metropolis. Cities, now net energy consumers, can one day become net energy *producers* with increased potential to improve quality of life for their inhabitants. This change is the key to the cities of tomorrow, built by our hand today.

A city is not gauged by its length and width, but by the broadness of its vision and the height of its dreams.

- Herb Caen

Chapter Three: A Tale of New Cities

As a framework, Universal Energy places a primary emphasis on modularity and standardization because these concepts enable flexible – and innovative – deployments of new technologies. That's why Universal Energy looks to cities for large-scale renewable integration. While cities have the highest population density, and thus demand the lion's share of both energy and resources, they also allow us to integrate renewables on a large scale as a byproduct of municipal infrastructure budgets. Deploying renewables within cities in a standardized, modular capacity affords Universal Energy the flexibility to serve important secondary purposes and solve future problems.

To see how, consider the following six images of civil infrastructure. Imagine, as we did last chapter, that they were completely integrated with renewables. Every window and roof of every building. Every road and off-ramp. Every bridge. Every highway median. *Everything*.

This scale of renewable integration within any municipality will generate tremendous energy. But that's only a feature of the strategy; it's not the end goal. The goal is to not just remove the net energy *demand* cities place on our national electric grids, but also to further turn that into net energy *production* – functionally transforming cities into power plants.

To see how this possible, let's run some numbers. Miami-Dade County in South Florida spans an area of 2,431 square miles, 1,898 of which is land.[137] It assessed a net energy consumption of 132.13 billion kilowatt-hours in 2018.[138] We saw earlier how one square mile of solar panel surface can reliably generate 836.4 million kilowatt-hours per year. That means Miami-Dade County would only have to integrate renewables with 159 square miles of its infrastructure – less than 8% – to become resource independent. At 10%, the county generates more power than it consumes. At 16%, it's a power plant. One might be forgiven for wondering what possibilities could arise should that integration reach 30%, 60%, 90% - all the more so as renewable technologies advance over time.

We see how this approach can remove the energy cities consume on external power networks and transforms cities from net energy consumers to net energy producers as a standard function of municipal operation. But from there, several key side benefits can help cities take full advantage of their next-generation infrastructure. Of them, standouts include smart grids, intermittency avoidance and centralized resource production.

Smart Grids

At sufficient scale, municipal deployment of renewables enables cities to "detach" from external energy-generating infrastructure by zeroing out their demand on regional electric grids. Once this deployment of renewables expands to the point where cities can reliably generate and store more energy than they consume, they can function as power-generating entities, like nodes on a network.

But it's important to note that this power generation doesn't necessarily have to plug into *existing* grids. Instead, they can also create their own separate electric grids that operate in parallel. Multiple grids could serve the same area, which makes them highly reliable and capable of operating intelligently to respond deftly to spikes in energy demand. To explain what I mean by that, it's useful to see how our electric grid works today and why it's not up to par to meet the challenges it will likely face in the future.

As of 2017, our national electric grid is comprised of 7,600+ decentralized power plants[139] that are owned by 3,200+ competing utility companies[140] that transmit electricity through 450,000+ miles of high voltage power lines, relay stations and transformers.[141] In other words, it's a total mess.

Whenever a power line goes down (storms, transformer overload, accidents), any location in the service area will go dark and will remain so until new power lines are constructed or a workaround is built. (See: Puerto Rico after Hurricane Maria).

Due to the difficulty of preventing these disruptions, electrical outages leave an average of 500,000 Americans without power for two hours or more on any given day.[142] This is a costly problem. The National University System Institute for Policy Research concluded one 2011 blackout in San Diego cost the city between $97-$118 million.[143] That's for one non-disaster-related blackout in one metropolitan area of one state. Nationwide, the Lawrence Berkeley National Laboratory suggests that power outages cost the U.S. economy some $80 billion each year.[144] Worse, in times of extreme heat or cold, blackouts can present a risk to public safety. In the past two decades, power outages have been blamed for hundreds of deaths (141 in California during a single July heatwave in 2006, for example).[145]

Addressing these problems with current methods will be both challenging and expensive, akin to making a computer built in the 1980s compete with today's latest models. By some estimates, it will cost upwards of $1 trillion to improve U.S. electric grids to simply meet demand by 2025.[146] Municipal deployment of renewables completely changes that. A stretch of renewable-integrated highway median or solar road canopy does not need power lines – *they are the power line.*

They are an electric grid. A business district with buildings covered in solar windows serves the same purpose.

Within a city, integrated renewables provide their own electric networks that operate on top of present infrastructure. Connected redundantly to both each other and existing electric grids, they can route and reroute energy as necessary. Paired alongside the National Aqueduct or other storage mediums, they can function as independent nodes to deliver power where it is needed most.

This provides some important side benefits. For starters, it increases the security of our electric grid multifold. To see how, take a look at the image on the following page that represents today's electric grid at the regional scale. In this setup, if a central power line or substation were to be disabled, the entire section of the grid they serve would go dark. Our electric grid is then vulnerable to environmental disasters, terrorist attacks or freak accidents.

By integrating renewable energy within infrastructure on a large scale, our grid becomes substantially more reliable. Wide swaths of it would have to be destroyed in order for it to stop functioning completely – making it far more resilient than our current approach to electricity transmission.

Moreover, a modular, redundant electric grid provided by integrated renewables allows for improved power management by municipal utilities, affording secondary methods that can be engaged as needed based on spikes in demand or blackouts. Municipal authorities would also have an ample supply of usage data that can help create predictive models for intelligent system design. Paired with today's sophisticated computing and accompanying software, this allows energy management to become more automated and efficient. It allows for the system as a whole to be upgraded more effectively, as the specific information the system provides can help determine the areas to best concentrate on for improvement – allowing energy networks to organically, and intelligently, evolve.

Overview of standard (current) electric grid:

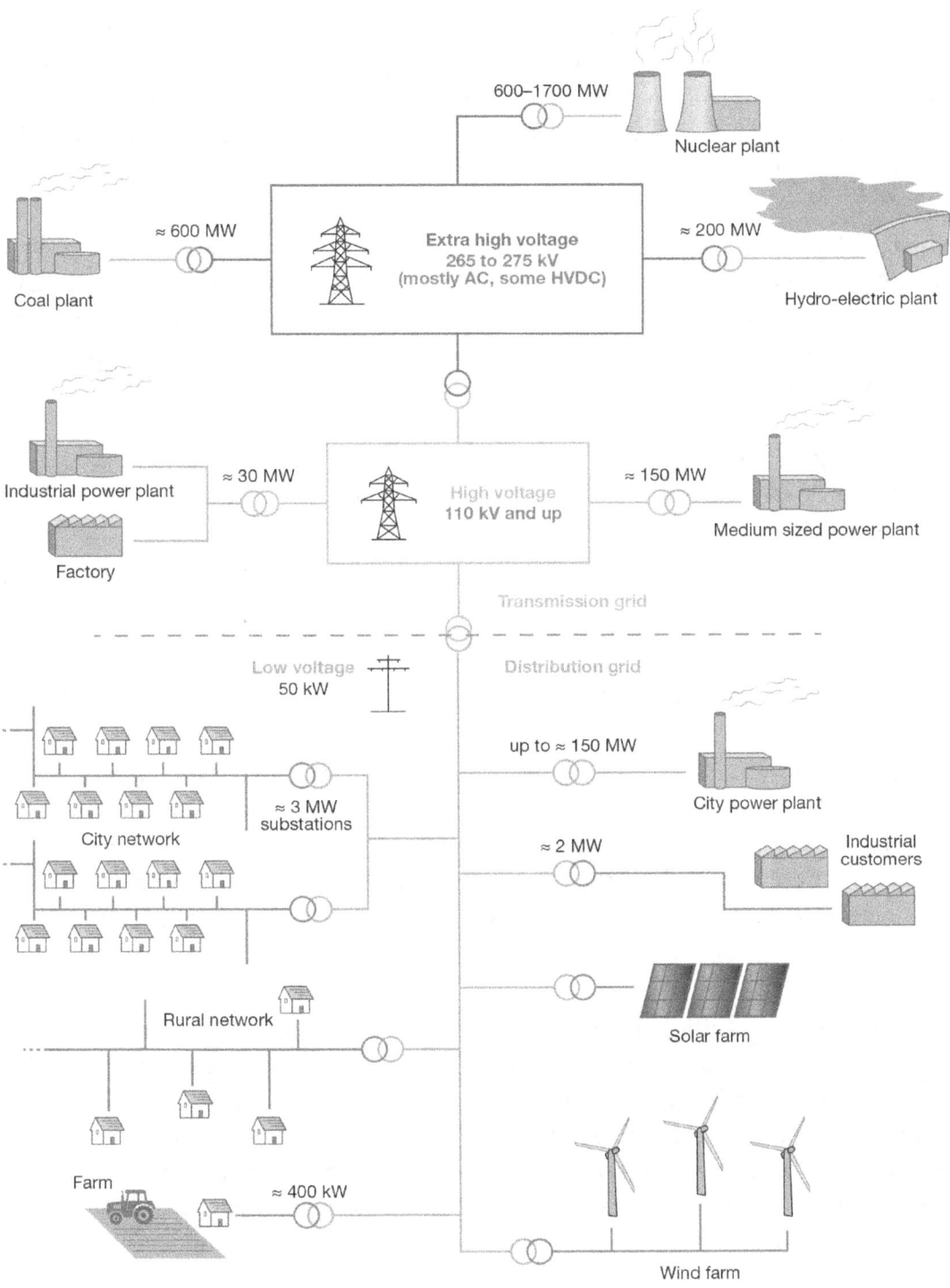

Reducing Intermittency

As touched on earlier, a primary obstacle to renewable power today is the question of intermittency – referring to an energy source that's not continuously *and instantly* available for conversion into electricity. Solar and wind power might be able to generate lots of energy, but if demand spikes at a time when they're not functioning (at night or on non-windy days), their utility wanes significantly.

A particular focus of Universal Energy is integrating these smart, responsive electric grids within the National Aqueduct as a parallel energy source that can engage on demand. It would also be backed by a baseload power network of thorium reactors (which we'll go over next chapter), which by themselves present tremendous energy for nominal use and external resource production. The combined result is a saturation effect, as each power source stacks on the other – and an integrated backup – to each generate energy from each other in parallel. Intermittency is thus removed as a primary obstacle, as the entire system is "instant-on" from at least two of three energy sources.

Centralized Resource Production

Universal Energy seeks to make resource scarcity irrelevant by making energy scarcity irrelevant, as the transformation of cities into energy-generating centers allows for the synthetic production of nigh-unlimited resources in locations close to consumption. As we'll review later, this enables metropolitan regions to do things like extract fresh water and hydrogen fuel from seawater, grow food indoors and synthetically manufacture building materials – all to scale.

This re-imagining of cities has such potential to improve how we generate energy and acquire resources that it's easy to focus on the technical benefits of such capabilities – power, infrastructure, resources, utility management. Yet the social benefits hold equal potential. When every aspect of a city is overhauled in such circumstances, each new upgrade acts as a basis for further improvement that can be reinvested to increase a collective quality of life. It helps reinforce optimism of what's actually possible when we stop and realize that investing in our future best interests is *actually* in our future best interests, and removes psychological barriers to initiatives seeking to continue this trend over the long term. We believe, fundamentally, in what we can build *when we actually build it.*

The benefits of this investment also does not limit itself to the commercial, tourism, and cultural hubs that define our national image. There is no shortage of places outside of America's elite coasts that could use a facelift. Centering municipalities within a national energy framework and investing in next-generation infrastructure is perhaps the fastest way to see that result delivered. Reaching that end for all American cities is a central goal of Universal Energy, one we'll focus on specifically within Chapter Twelve.

Once cities become central to generating electricity, next-generation manufacturing, and resource production, they can provide for themselves and their surrounding regions, markedly increasing the quality of life for millions. Side benefits of this approach naturally include expanded job opportunities, stimulated economy and reduced crime. At the same time, national electricity demand is proportionally reduced as a result of integrated renewables within municipal infrastructure. From there it sets up the next stage of the framework, which is to leverage next-generation nuclear to overhaul our resource production schema, and scale our energy generation capability to uncharted heights.

In the years since man unlocked the power stored within the atom, the world has made progress, halting, but effective, toward bringing that power under human control. The challenge may be our salvation. As we begin to master the destructive potentialities of modern science, we move toward a new era in which science can fulfill its creative promise and help bring into existence the happiest society the world has ever known.

- John F. Kennedy

Chapter Four: The Thorium Backbone

Using integrated renewables to transform cities into power-generating centers is key to Universal Energy because it helps reduce and eventually remove the demand municipalities place on regional electric grids. As urban regions eventually become net energy producers, we can generate more than they consume, which helps contribute to a nationwide abundance of inexpensive energy that can be devoted to the indefinite production of critical resources. The next step in the framework is to increase our base load power infrastructure, and scale a national energy abundance, to a dramatically higher tier.

"Base load" refers to the minimum amount of power that needs to be generated for a given region over time.[147] Today, this is met through larger "base load" power stations that are supplemented by smaller plants that engage when demand spikes. As base load stations are designed to be constantly operational and generate a lot of electricity, they are more expensive to construct and maintain, which encourages the use of cheaper fuels to power them.

Accordingly, most of our base load infrastructure is presently powered by fossil fuels (coal and natural gas), followed by enriched uranium and hydroelectric.[148] While far superior to environmentally toxic coal, hydroelectric and natural gas present their own ecological drawbacks.[149] Hydroelectric can only be deployed in limited locations, and enriched uranium is both limited in quantity and primarily deployed in reactor designs that present concerns of both weaponization and risks of catastrophic failure. To make matters worse, the majority of our base load infrastructure is decades old.[150] More than half of our base load power infrastructure was built before 1980, and 75% of our coal-fueled power plants are at least thirty years old with an average expected lifespan of forty years.[151]

Universal Energy seeks to solve these problems by replacing our base load infrastructure with next-generation technology that's designed to work alongside other power systems intelligently, employing the same concepts of standardization and modularity that's applied to integrated renewables. The technology it looks to for this role is a clean, safe and highly efficient form of atomic energy that comes from the element thorium – not enriched uranium – to provide an immense source of base load power for our national energy grids.

In saying this, it's important to mention that nuclear power can be a polarizing subject – and for good reason. Atomic energy can be dangerous. It can make weapons of mass destruction, cause regionally-devastating meltdowns and produce toxic waste that lasts for millennia. Risks aside, certain types of nuclear power can also be incredibly expensive, leading many to doubt its long-term economic viability. For these reasons, nuclear has become politically controversial in much of the world, especially within the United States.

Yet while many of these concerns are conceptually valid, nearly every single one centers on the consequences of nuclear reactors that run on a combination of *enriched uranium* and *pressurized water*. And the reason most reactors have worked this way in the past is because atomic energy as we know it was born from initiatives designed to produce *both* nuclear weapons and civilian power as directed from national leadership during the Cold War.[152] Consequently, there are few ways to decouple "traditional" nuclear reactors from nuclear weapons development. Any attempt to do so with certainty quickly reaches into the billions of dollars, to say nothing of the ecological risks and their accompanying expenses.

But thorium is not "traditional" nuclear. Reactors fueled by thorium don't use water and aren't pressurized – the main issue behind reactor "meltdowns." Thorium reactors don't use solid fuel, either – the entire reactor core is liquid. It's physically impossible for one to "melt down" like their pressurized-water counterparts and they can't wreak serious environmental havoc if sabotaged. Thorium reactors can consume *both* nuclear waste and weapons-grade nuclear material as fuel, but at the same time are difficult to use to build nuclear weapons. Their waste has a minimal environmental footprint as well, and that waste *becomes safe over decades* as opposed to millennia.[153]

Thorium reactors further fit the requirements of Universal Energy. They operate at high temperature and offer plenty of excess energy for supplemental resource production. They can be built to a single standard, small in size and modular in function that can be mass produced and deployed anywhere in the world. The designs being proposed already include the cogenerative features that Universal Energy seeks for a dynamic energy framework. As we'll see later in this chapter, the technology has been proven to work impressively and has further seen financial investments well into the billions from seven countries (as of this writing) – including the United States.[154]

But in advocating their benefits, it's important to also note that thorium reactors have their criticisms. Some come from people simply opposed to nuclear power as a concept, favoring exclusive use of renewables. Others come from nuclear engineers, cautioning against discarding traditional reactor designs that, while riskier, have seen the lion's share of research and development with regards to atomic energy. Others still come from people who doubt the viability of a science that – to be fair – has a contingent of enthusiastic backers who at times oversell thorium's benefits without recognizing the challenges, however solvable, to deploying the technology on a large scale.

These criticisms are taken seriously by this writing and will be addressed directly in this chapter, situated fairly and factually within the context of the resource, climate and energy challenges humanity will be facing in the future. This discussion will also draw a noteworthy distinction between renewables and thorium, and directly address why we even need both in the first place.

Good point. Renewables Are Awesome. So Why Do We Need Thorium?

As a framework, Universal Energy functions on the recognition that until we develop true fusion energy[155] there will be no one singular technology that is capable of meeting humanity's energy and resource requirements. Renewables are essential to city-level energy reduction and eventual independence, all the more so as their integration increases in scale. But by themselves, renewables fall far short of the threshold needed to reliably meet the demands of base load power nationwide, which itself is a **far** lower bar than the levels of energy we require to solve resource scarcity and climate change.

Further, even if renewables *could* generate enough energy to solve these problems, the carbon emissions behind their base material extraction, manufacture, transport and installation at sufficient scale would undermine the endeavor from the start. Solar and wind power, as we'll review later in this chapter, requires lots of raw materials to construct – between 10,000-17,000 metric tons to generate a single terawatt.[156] That says nothing of transmission or storage – nor the energy needed to source, process and integrate materials for those functions. It also says nothing of the considerable difficulties presented by their end-state disposal.

That's not a problem if the energy used in every step the renewable manufacturing chain is carbon-neutral, but only clean nuclear is capable of generating that level of carbon-free energy as a modular standard. Further, only clean nuclear is capable of generating enough carbon-free energy to power auxiliary functions of synthetic resource production. To see why this is the case, let's compare nuclear and solar at scale:

One of the largest solar power stations in the world is the Topaz Solar Farm in southern California. At a cost of $2.5 billion and spanning 7.3 square miles, the Topaz Solar Farm deploys nine million solar modules to generate an aggregate of 1,270 gigawatt-hours annually.[157] That's certainly impressive. But it's dwarfed when compared to the generating capacity of base load nuclear.

The Limerick nuclear power plant in southeast Pennsylvania, for comparison, has a generating capacity of 2,270 megawatts and annually outputs 19,000 gigawatt-hours.[158] That plant is barely half the size of the Palo Verde nuclear plant in Arizona, which has a generating capacity of 3,942 megawatts and annually outputs 32,840 gigawatt-hours of energy.[159]

Even as one of the largest solar power stations in existence – located in one of the most solar-effective areas on the planet – the Topaz Solar Farm generates less than 5% of the output of a large base load nuclear power station.

Across a city – or many of them – that capacity definitely matters. Renewables serve the purpose of rapid installation, flexibility of deployment and integration within municipal infrastructure, uniquely suiting them for supplementing national energy generation and reducing regional energy demand – all the more so once integrated into the National Aqueduct. But it would take thousands of square miles at a cost of many trillions of dollars to meet our energy demands in full through renewables alone,[160] a threshold that modern nuclear reactors can meet at a fraction of the physical, material and economic footprint.

That's where thorium comes in.

Like renewables, thorium's role in the framework is to provide a nigh-unlimited source of clean electricity. Yet thorium can do so at a degree and to a density that presents an unrivaled capability to not only exceed our current base load infrastructure, but also present such an abundance of energy that it causes the price of electricity to plummet. Most critically, thorium is capable of this *while also generating enough residual heat energy to power inexpensive resource production.*

That's the essential capability that only clean nuclear can meet on the scale we need to solve resource scarcity and climate change. We'll devote the rest of this chapter to see how thorium reactors can serve as a vital component of Universal Energy, bridging the divide between nuclear and renewables while expanding our energy production capabilities and transforming our resource supply chain.

Credit: XKCD (modified with love)

In doing so, we'll be focusing on five key points:

- A brief overview of thorium and nuclear power. (Page 87)
- Why we don't use thorium today. (Page 90)
- How thorium works differently than "traditional" nuclear. (Page 97)
- How we know thorium reactors are a feasible and economical source of power. (Page 107)
- Why criticisms of thorium are wrong on the facts (Page 115)

A Quick Note:

While *The Next Giant Leap* takes care to explain all technologies behind Universal Energy in detail, thorium reactors (and the nuclear science that makes them possible) are the most sophisticated and technically complex systems in the framework. Further, in the interests of intellectual honesty, this chapter also takes care to fairly summarize and address criticisms to thorium + nuclear power and explain why they are wrong on the facts.

As such, while this chapter is written in accessible language that's easy to understand even if you don't have a technical background, it's still the longest and most detailed in this book. If you're feeling overwhelmed, please feel free to jump to Chapter Five: Water and Hydrogen on page 133 and come back at any time to finish reviewing how thorium is key to a clean energy future.

So, What Is A Thorium Reactor?

When people think of nuclear power, they commonly think of a large facility with tall steam stacks, perhaps also containing potentially dangerous materials that could cause calamity under the wrong circumstances. That's the classic "Pressurized Water Reactor" which works via a combination of pressurized water and enriched uranium. In contrast, Liquid Fluoride Thorium Reactors (LFTRs) use thorium within a high-temperature liquid moderator – no pressure needed – and they work in a way that avoids atomic energy's most serious problems.[161] Here's a short list of their highlights:

- LFTRs are highly efficient – hundreds of times more so than Pressurized Water Reactors.[162]

- LFTRs are extremely safe. Because their fuel and reactant are liquid and not under extreme pressure (unlike traditional reactors), it is physically impossible for them to "melt down."[163]

- Thorium is more stable than other radioactive elements and is safe to handle in raw form unless ingested or inhaled. Additionally, it does not require additional enrichment to power a reactor.[164]

- LFTRs produce far less waste than Pressurized Water Reactors and can also consume both nuclear waste and weapons-grade nuclear material as fuel.[165] Of what small amounts of waste remain, it takes only decades for it to become safe as opposed to millennia with reactors powered by enriched uranium.[166]

- The LFTR's thorium fuel supply is highly abundant – thorium is about as common as lead – making it thousands of times more plentiful than fuel-grade uranium (only about 0.7% of all uranium in Earth's known land reserves).[167]

- The thorium fuel in LFTRs is difficult to weaponize. While theoretically possible, the weapon would be unstable, far weaker than traditional nuclear weapons, and would be significantly less practical for use in conflict.[168]

- As a result of their efficiency and safety, LFTRs can be **much smaller** than Pressurized Water Reactors. Where Pressurized Water Reactors often sit on multi-acre compounds and require large buffer zones in case of emergencies, LFTRs can be around the size of a house or even smaller.[169]

- LFTRs are significantly less expensive to build than Pressurized Water Reactors, and their small size allows them to be mass-produced on assembly lines in a standardized and modular capacity. That means nuclear reactors can become iterations of a product model as opposed to custom-built facilities. The cost savings presented by this capability are immense.[170]

- Although recent thorium designs are experimental, the technology is proven to work both reliably and impressively – an emphasis that will be elaborated upon within a later section of this chapter.

LFTRs are superior to today's Pressurized Water Reactors in nearly every way possible, and their capabilities have been known to science since the 1960's.[171] But that prompts an important question: why aren't we using them today?

To answer that, we'll need to cover some background that's easier to understand by first reviewing a few terms surrounding atomic energy. What follows is a quick refresher from science class, or a primer if you're not familiar with how nuclear power works. (Feel free to skim it, or to skip it now and refer to it as necessary.)

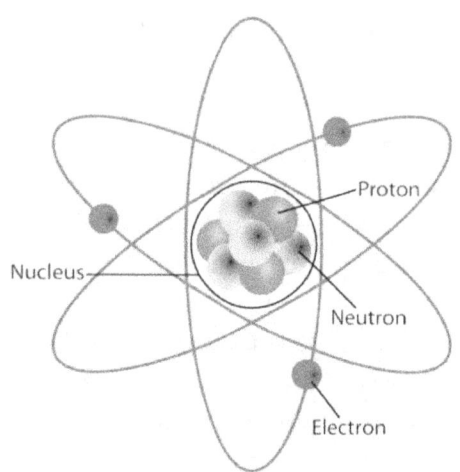

Atom: the building block of matter, composing everything we see and touch. Atoms generally have three types of particles within them. The center of the atom houses the **nucleus**, which is comprised of a given number of positively charged **protons** and neutrally charged **neutrons**. The nucleus is orbited by negatively charged **electrons**.

The different elements in the world are made up of atoms, and each element has a specific atomic arrangement of these particles as shown in the periodic table of elements. Elements and the nature of their atomic composition are the basis of all chemistry and nuclear science.

Radioactive decay: the process in which an unstable atom spontaneously emits radiation in the form of atomic particles or energy. Any substance that naturally undergoes radioactive decay is considered to be "radioactive."

Isotope: an unstable variant of an element, usually as a result of radioactive decay and/or something called transmutation (explained next). Isotopes have numerical designations reflective of their atomic composition. For example: uranium-233 and uranium-235 are isotopes of the element uranium.

Transmutation: the process in which one isotope of an element becomes an isotope of another element through nuclear means (like absorbing a neutron).

Fission: the splitting of an atom's nucleus, releasing tremendous energy and "fission products" (usually radiation + isotopes of other elements). For example: reactors fueled with enriched uranium work by using a neutron to split the nucleus of uranium-235 into kryptonium-92 and barium-141.[172]

Fusion: the joining of atomic nuclei together to form a new element, releasing more energy than even fission. For example: fusing tritium and deuterium (isotopes of hydrogen) into helium, which is how our sun works.[173]

Fissile fuel: an isotope of an element that can undergo fission directly inside a reactor. Uranium-233 and uranium-235 are *fissile fuels*.

Fertile fuel: an isotope of an element that can't undergo fission directly, but can if transmuted into a fissile fuel. Thorium is a *fertile fuel*.

Enrichment: the process of adding greater levels of a radioactive isotope within a nuclear fuel supply. For example: Light Water Reactors use "enriched uranium," which involves adding more uranium-235 to a fuel supply to sustain fission. Nuclear weapons use "highly enriched" nuclear material to sustain a faster chain reaction. Thorium reactors (especially LFTRs) **do not** require enrichment.[174]

Breeding: a process in certain reactor designs that employ transmutation to transform a fertile fuel into a fissile fuel. Any reactor that undergoes a breeding process is considered a "breeder reactor." LFTRs are breeder reactors.[175]

Pressurized Water Reactor (PWR): 1950's-era reactor designs that use highly pressurized water to help regulate and make possible a fission reaction inside a reactor core. Pressurized Water Reactors use solid fuel and are the most common nuclear reactors operating today.[176]

Light Water Reactor (LWR): a type of Pressurized Water Reactor that uses enriched uranium to generate electricity. Most Pressurized Water Reactors take this form.[177]

Heavy Water Reactor (HWR): a type of Pressurized Water Reactor that does not use enriched uranium, but rather uses a type of water with an extra neutron, known as deuterium oxide or "heavy water," to sustain fission. These reactors are less common, but still present proliferation risks – especially for weapons-grade plutonium.[178]

Molten Salt Reactor (MSR): a type of advanced reactor design that uses a special type of non-radioactive salt that becomes liquid at high temperatures to act as both a moderator for the reactor and a carrier mechanism for nuclear fuel. They operate at standard atmospheric pressure and have a liquid fuel supply. LFTRs are a highly efficient form of Molten Salt Reactors that also undergo breeding.[179]

With these terms defined, we'll take a minute to review a bit of our history with atomic energy – specifically addressing why thorium isn't the primary source of nuclear fuel today.

The Unholy Alliance: Electricity and Bombs

Most nuclear reactors, including those within the United States, are fueled by uranium-235, an isotope representing less than 0.7% of all naturally existing uranium on Earth.[180] Uranium-235 is a fissile fuel, meaning that the possibility exists for its atomic nucleus to split into isotopes of other elements if hit by a fast-moving neutron, releasing levels of energy that are millions of times greater than any known chemical fuel source. For reference: burning a single molecule of methane releases 9.6 eV (electron volts) of energy.[181] Fissioning a single uranium-235 atom releases 200 MeV (million electron volts) of energy.[182] That's a huge difference.

For nuclear fission to work for power generation, it involves a concept known as "criticality": a threshold, or "critical mass," where there is enough fissionable material present for the reaction to sustain itself. As it exists in nature, elemental uranium is not capable of doing this. Yet the isotopes uranium-233 and uranium-235 are. If these isotopes are extracted and placed in a controlled environment (or if enriched into a fuel supply) the fission reaction becomes sustainable over long periods of time. Within Pressurized Water Reactors, this reaction efficiently produces heat, which boils water into steam that turns a turbine and generates electricity. In concept, traditional nuclear reactors are just a very efficient and sophisticated implementation of steam power.

But for several reasons, this past approach is less than ideal. For starters, the uranium-235 fuel cycle is more reactive than thorium and harder to control once it reaches criticality. Further, its sustainability as a fuel source is limited, and spent fuel rods must be replaced (along with the reactor core) every 18-24 months[183] – requiring reactor shutdown. As spent fuel is highly radioactive and contaminates anything it comes in contact with (including disposal equipment), this process creates an effectively endless supply of radioactive waste. If that wasn't enough, Pressurized Water Reactors can present extremely dangerous conditions if any part of the reaction became unstable or uncontrollable.

Pressurized Water Reactors were invented in the 1950's and their designs have remained conceptually consistent since then. Most work just fine. Yet should key systems fail, the uranium-235 fuel supply (which is normally placed in extractible rods)[184] could remain stuck inside the reactor. Should this happen, the reaction would continue unrestrained, generating enough heat to melt the fuel supply and cause it to pool at the bottom of the reactor's pressurized water core. In this circumstance, the reaction would accelerate exponentially to amplify heat and water pressure until it eventually caused a steam explosion – resulting in the spread of highly radioactive material over a region. That event is called a "meltdown" and is effectively what happened in Chernobyl.[185] Needless to say, such events are catastrophic beyond hyperbole.

Because of the risk of meltdowns, however unlikely, Pressurized Water Reactors must be built with extensive safety features: containment domes of steel-reinforced concrete that are several feet thick, massive cooling and pressurization apparatuses, and redundant mechanisms that engage in case any systems were to fail. Pressurized Water Reactors also must be built in sparsely populated areas with large buffer zones in case the surrounding region needed to be evacuated in an emergency. The expenses and security concerns of this reality prompt an important question, though: why in the heck are we using uranium-235 within Pressurized Water Reactors, even though better alternatives exist?

Simply stated? Because at the end of the day, **the uranium-235 fuel cycle is mankind's best-known pathway to building nuclear weapons.**

During World War II, scientists working for the U.S. government (and Third Reich)[186] discovered that certain fissile isotopes had a unique property: if enriched highly enough, they could reach a super-critical state. And if, in this super-critical state, they were rapidly bombarded with neutrons, it could create a nuclear detonation – resulting in the most powerful man-made force in existence.

As it was two of those detonations that ended World War II, the significance of atomic weaponry could not be downplayed, especially once the Cold War unfolded. Thus, as civilian nuclear power developed as an energy source, so did the development of nuclear arms and their delivery mechanisms. These two sectors converged to ensure our continued use of uranium-235 as fuel.[187] But not just because its highly enriched forms are far more powerful than conventional explosives in weapons of war. The other, more essential reason is that the uranium-235 fuel cycle can be leveraged to artificially create plutonium-239 – a far more potent weapons fuel that does not naturally exist on Earth in any significant quantity.[188] **And you need plutonium-239 to build hydrogen bombs.**

Nuclear bombs come in varied shapes and sizes. With the right materials, building a basic nuclear weapon is, in theory, relatively straightforward. The general idea is to:

1. Find a way to rapidly combine highly enriched fissile material together into a critical mass (explosives usually do the trick),

2. Introduce a high-intensity neutron source to spark a fast-fissile chain-reaction, and

3. Tada! – you've got yourself a nuclear bomb. (Please don't actually try this at home).

The following image shows a nuclear weapon similar to the description above – a gun-type assembly weapon, which is the bomb the United States dropped on Hiroshima. It's fairly simple, conceptually speaking.

Gun-type assembly weapon.

Once the bullet joins with the target, a critical mass is reached. A neutron initiator is engaged, starting a fast fission reaction that results in a nuclear detonation.

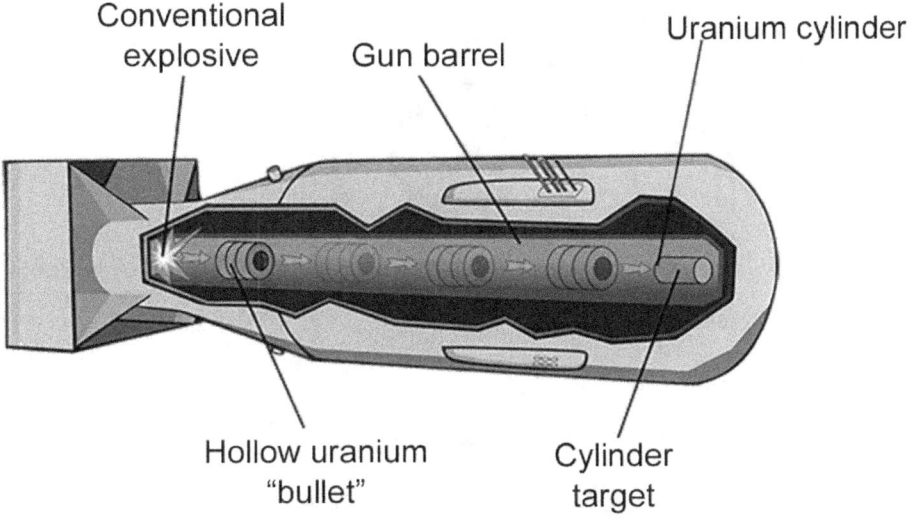

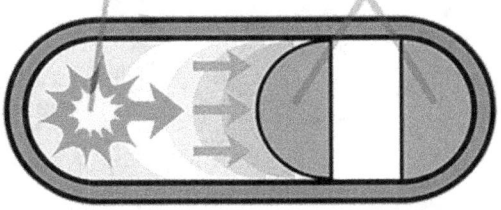

Gun-type assembly method

Image Source.[189]

However, building a powerful bomb that's still compact enough to function as a missile's warhead requires an *implosion-type* assembly. This method delivers a critical mass through implosion – compressing a larger sphere of fissile fuel into a smaller sphere by means of explosives – a highly sophisticated and difficult process. Implosion-type devices can not only be significantly smaller than gun-type devices, they are also far more efficient and thus far more destructive.

Implosion-type assembly weapon

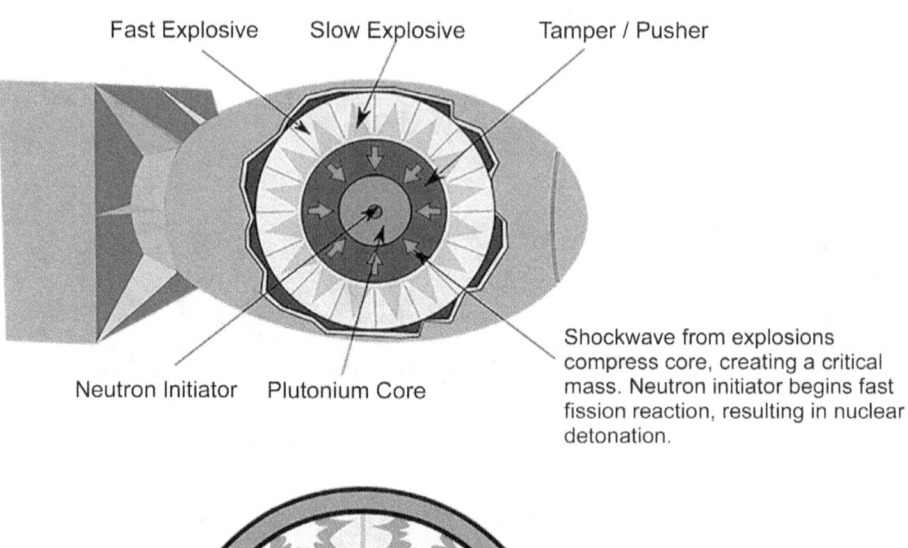

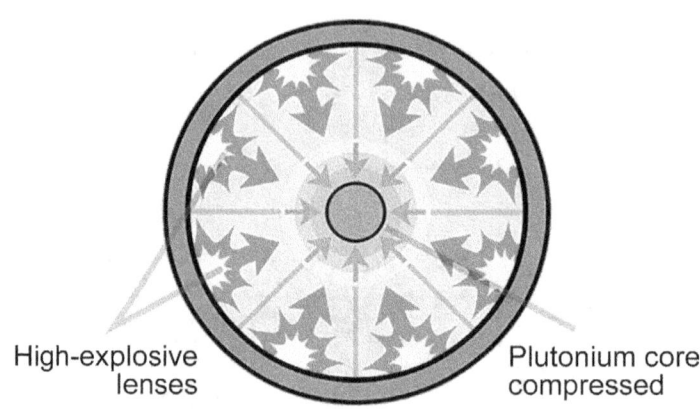

Image Source.[190]

Uranium-235 isn't very effective in implosion-type devices, as it has a high critical mass requirement – complicating its deployability in military conflict.[191] Plutonium-239, on the other hand, has a much lower critical mass requirement,[192] but only exists in trace amounts on Earth.

Yet the fuel and reprocessing cycles within Light and Heavy Water Reactors provided a convenient method to source plutonium-239 for implosion devices.[193] As these devices can be built small in size (basketball or smaller), this led to future weapons designs that leveraged their immense energy to fuse isotopes of hydrogen together, providing a dramatically more powerful explosion. Thus, the hydrogen bomb was born,[194] as was our reliance on uranium-235 to source the plutonium that makes them possible.

The following image shows the basic stages of the thermonuclear detonation of a hydrogen bomb, employing a Teller-Ulam design:[195]

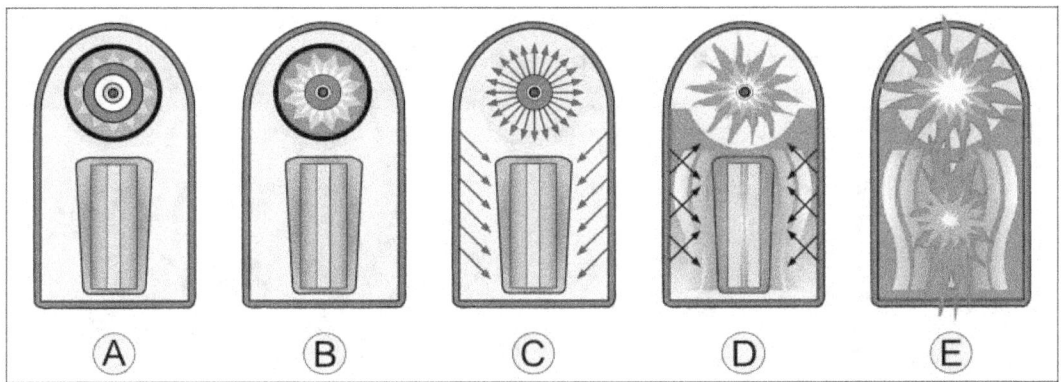

Image source.[196]

A): The warhead, in its inactive state. The "primary" – the implosion-type fission bomb – is on top. The "secondary" – the cylinder-shaped object below – is *fusion fuel*. It's comprised of something called lithium-6 deuteride that's wrapped around a "sparkplug" of plutonium-239, both of which are suspended in polystyrene plastic foam.

B): The bomb is triggered. The high explosive charges of the primary activate and compress the plutonium core into a super-critical state. A fission detonation occurs.

C). Within the first nanoseconds of detonation, the fission primary emits high levels of X-ray energy that reflect inside of the bomb case. This irradiates the polystyrene foam.

D). The heat and X-ray irradiation cause the polystyrene foam to turn into superheated plasma, which expands massively and compresses the secondary. As this occurs, the excess neutrons from the primary's fission reaction cause the plutonium in the secondary's sparkplug to undergo fission, creating even more heat, pressure and radiation.

E). At such extreme heat and pressure, the lithium-6 deuteride separates into tritium and deuterium, which are isotopes of hydrogen. Under these circumstances, these isotopes fuse together to form helium in a reaction thousands of times more powerful than the fission detonations of atomic bombs.

In short: by harnessing the heat, radiation and pressure of an implosion-type fission bomb, thermonuclear weapons fuse isotopes of hydrogen together to create helium, essentially forming a second sun when detonated. With potential yields in the megatons, we can now build weapons that make the bombs dropped on Japan seem like firecrackers in comparison.

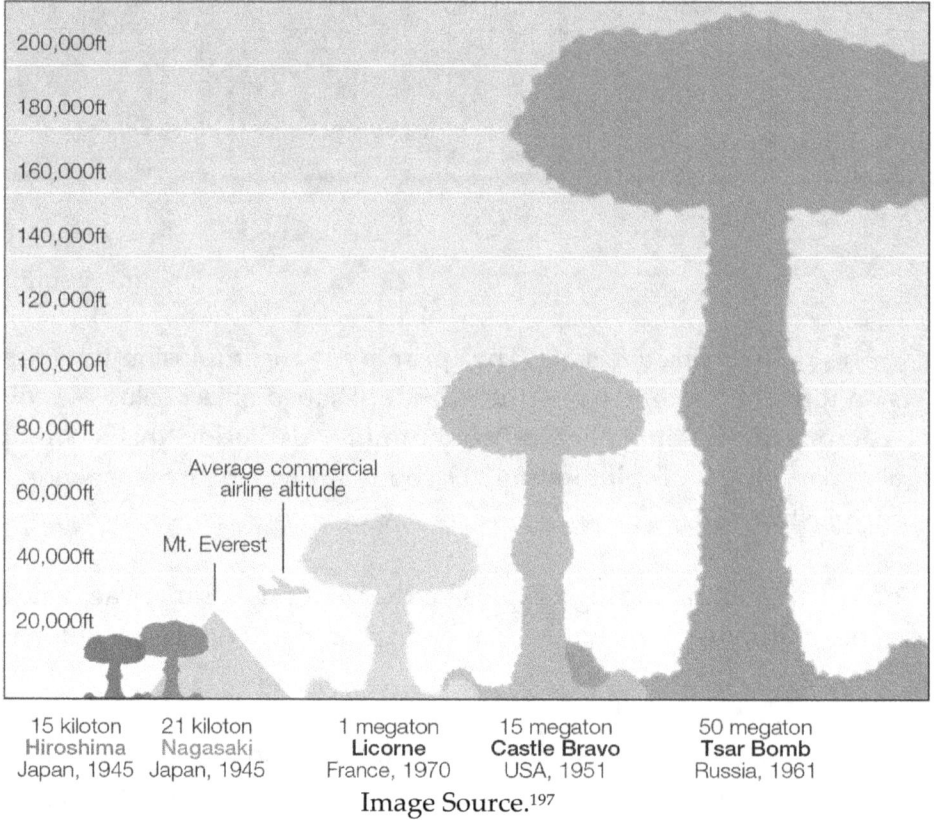

Image Source.[197]

But without plutonium-239 to facilitate a thermonuclear detonation, none of the thousands of hydrogen bombs in the world could exist. And without using Pressurized Water Reactors, there would not be a straightforward way to efficiently produce plutonium-239.

This is why we fuel our power plants today with nuclear dynamite that creates waste products that last for thousands of years and rank among the most toxic substances in existence: to build and maintain nuclear arsenals. That's the dirty secret behind our approach to nuclear energy. We have corrupted the most powerful energy source that we have ever discovered in the name of an arms race, and at the cost of a world placed in perpetual jeopardy.

If we're willing to contend with the 15,000 nuclear weapons in global arsenals as enough,[198] and finally forsake the drive to intertwine their production with civilian nuclear energy, thorium gives us another option. One that can avoid nearly every complication with past approaches to atomic power, as well as usher in a clean energy future that exceeds every threshold of our present limitations.

Thorium to The Rescue

Although the name "thorium" comes from the Norse god of thunder, thorium isn't as reactive as its namesake suggests, ranking among the least reactive radioactive elements.[199] It is safe to handle in its raw form so long as it's not ingested, and by itself isn't particularly remarkable. This lack of natural reactivity and radioactivity, however, is what makes it an ideal fuel in next-generation reactor designs.

As a LFTR is a type of Molten Salt Reactor (MSR), it powers nuclear fission at **normal atmospheric pressure** through a wholly liquid core that is self-regulating – a completely different setup from the solid fuel rods and pressurized water cores used in traditional nuclear reactors. As discussed previously, meltdowns are problems with solid fuel reactors because a runaway reaction can't be controlled, which leads to catastrophic results because the solid fuel melts and creates ever-greater pressure until it eventually causes a steam explosion.[200] But an MSR is designed to operate in "meltdown" conditions naturally. In the case of a LFTR, it's one of the few circumstances in which thorium is sufficiently reactive – and even then, it's a slow and steady reaction.

Compare LFTRs and Pressurized Water Reactors to the fable of the tortoise and the hare, and you're on the right track.

Liquid-fluoride thorium reactor

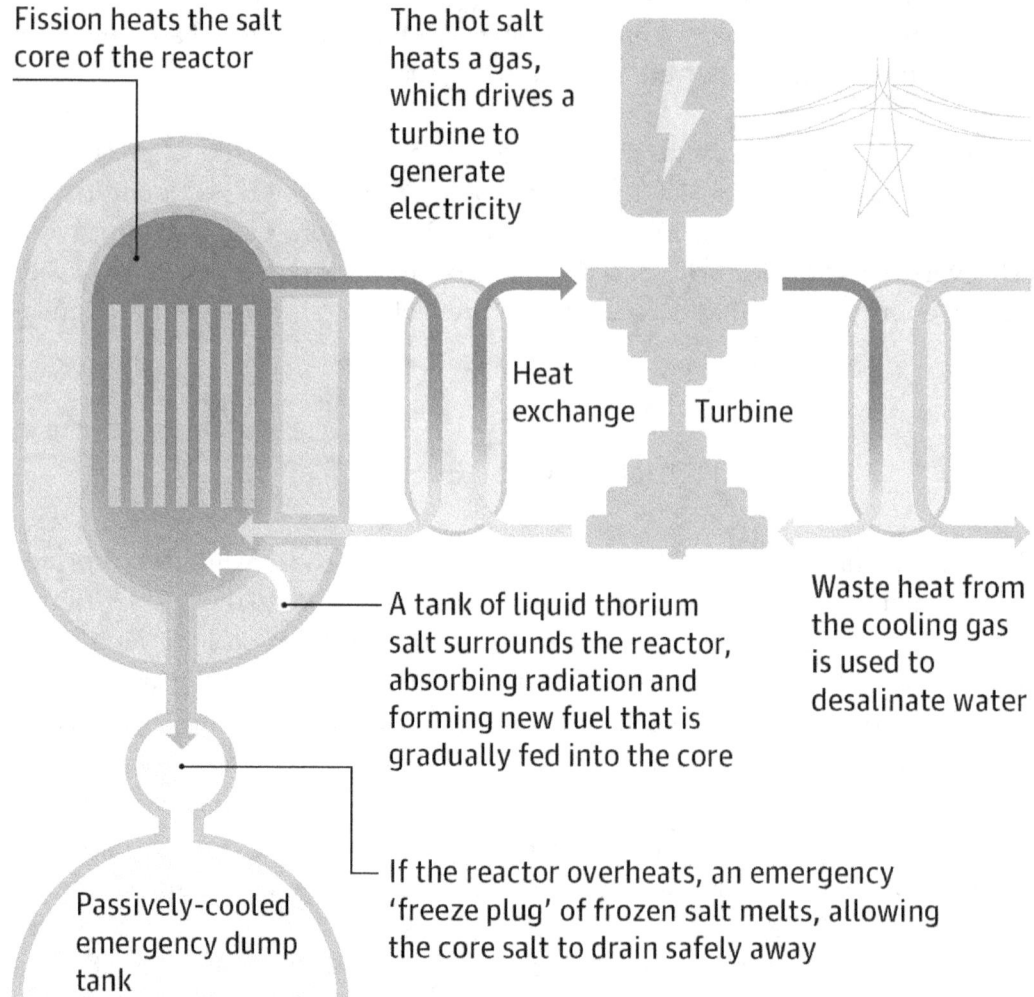

Image source. [201]

The reaction works like this: thorium-232 and uranium-233 – the kind of uranium that's difficult to use in bombs – are dissolved into molten salts (usually lithium fluoride – but could be any number of alternatives)[202] and fed into the reactor. The molten salts act as a carrier for the thorium fuel and as a catalyst for the reaction, which keeps the fuel supply at high temperature and at the same time helps refuel the reactor over time through breeding.[203]

More technically, a LFTR's core fissions transmuted uranium-233, releasing heat, energy and excess neutrons that combine with the *fertile* thorium-232 in the liquid molten salt to form more *fissile* uranium-233 through transmutation. Then, the newly-transmuted uranium-233 is fed back into the reactor core, making for sustained, efficient, and self-regulating fission.[204] This fission reaction, through a series of heat exchangers, then heats an inert gas that is sent through turbines to generate electricity.[205] As this reaction generates lots of heat, there remains plenty of leftover energy that can also be used for on-site water desalination and hydrogen production, which we'll discuss in the next few chapters.

As LFTRs can reprocess and resupply their own fuel from the waste products of the original fission reaction, in addition to a hefty supply of fertile thorium (known as the "blanket"), LFTRs essentially refuel themselves for long periods of time at high efficiencies. Just how efficient are LFTRs? One ton of thorium-232 in a LFTR gives us the energy equivalent of 250 tons of uranium-235 in a traditional Light Water Reactor, or 4.16 million tons of coal in a coal-power plant.[206]

The breeding process can allow the reactor to continually reprocess and produce its own fuel from thorium for *up to 30 years* without replacement (although this time would be likely be shorter due to the eventual need to replace certain components of the reactor core).[207] For a fuel supply, though, that's still some 15 times longer than traditional Light Water Reactors.[208] Plus, LFTRs are 54% efficient (some 20% higher than most coal plants) and use 99% of their fuel.[209] And as that 46% efficiency loss primarily takes the form of heat, we can re-capture that heat to power auxiliary systems that produce resources.

The use of thorium-232 in LFTRs is superior to the use of uranium-235 in a Pressurized Water Reactor in other ways as well, particularly when looking at safety, sustainability, scalability, security and cost.

Safety. Because LFTRs operate at normal atmospheric pressure, far less can go wrong. And in case of emergencies, the fix is simple and effective: gravity. As the reactant is liquid, it can drain into smaller storage tanks with insufficient critical mass to sustain the reaction, thus it freezes.[210] This makes it *physically impossible* for a LFTR to "melt down" in the traditional sense, even under catastrophic circumstances.[211] If a LFTR was targeted by terrorist attacks and blown up, the liquid reactant would flash-freeze into a solid once exposed to the open air. Additionally, once its fuel cycle has completed, LFTRs produce substantially less radioactive waste than Pressurized Water Reactors.[212] The radioactive waste that remains is also short-lived – staying dangerous only for decades as opposed to millennia.[213]

Thorium is plentiful and sustainable for long-term use. About as common as lead, the global supply of thorium is three to four times greater than all forms of uranium – and only 0.7% of all uranium is fissile.[214] Thorium is also a common byproduct of rare earth metal mining, presenting straightforward opportunities in the short-term for easier acquisition.[215] There is enough thorium in the United States alone to power the country for the next 10,000 years.[216]

LFTRs have a greatly reduced environmental footprint. By nature of their operation, most radioactive waste inside LFTR cores is consumed by the reactor.[217] This enables LFTRs to consume other types of waste as well, including weapons-grade fissile material and even the nuclear waste generated by traditional nuclear reactors.[218] In turn, LFTRs can then act as nuclear garbage disposals that also generate electricity for decades.[219] The physical amount of waste remaining once the reaction consumes all fuel is less than 1/1000th of the waste produced by Light Water Reactors.[220] Additionally, LFTR waste decays quickly, with the most toxic radioactive isotopes having a half-life of only 30.17 years.[221] This means that a supply of radioactive waste from a LFTR would become less radioactive than natural uranium within a period of 300 years or less.[222] More toxic radioactive waste from Light Water Reactors can last for thousands of years.[223]

LFTRs reduce the possibility of proliferation. The weaponization of a nuclear reaction is unique in that only uranium-235 and plutonium-239 have been known to make a militarily effective bomb. However, while not on the same scale it is technically possible to create a rudimentary nuclear device using material produced in LFTRs – namely through uranium-233 and neptunium-237.

Yet doing so is considerably more difficult and less reliable than with traditional nuclear materials and traditional nuclear reactor designs. Further, uranium-233 and neptunium-237 – even if fashioned into a nuclear weapon – would likely make the device ineffective for military purposes. The reasons?

> **Purification difficulties and inherent dangers.** It's been theorized that if an LFTR is using something called a fluorinator, neptunium-237 can be extracted via a chemical process, which has potential to undergo a fast-fissile reaction and enable a nuclear detonation.[224] But the critical mass requirement for neptunium-237 is roughly 60 kilograms, which is higher than even uranium-235 – and emits 2,000% more gamma emissions than plutonium-239.[225] That would make such a weapon far more dangerous to

build and less practical to deploy even if the expertise to weaponize neptunium-237 existed.[226] Further, chemical purification to this degree requires highly expensive and purpose-built infrastructure that can't be obtained by entities other than states with sophisticated science programs.

Moreover, part of the breeding process to transmute thorium-232 into uranium-233 involves the production[227] of an invariable amount of uranium-232 – which, while perfectly safe within a reactor, also emits high levels of both alpha[228] and gamma radiation.[229] For those inclined, ionizing radiation (the potentially harmful kind) is commonly measured in **"rem"** (Roentgen Equivalent Man).[230] In general, the more radiation one absorbs, the more harmful the effects become. At standard levels, the uranium-232 contaminant within a 5kg sphere of uranium-233 would generate up to 38 rem per hour.[231] For use in a weapon, uranium-233 has a minimum critical mass of 16.5 kilograms[232] - presenting an aggregate dose of up to 125 rem/hr. Serious radiation sickness begins with short-term exposure of 150 rem, and anything over that is potentially lethal.[233]

At sufficient mass to build a nuclear weapon, this would make the material too dangerous to handle by human beings and would require the employment of sophisticated (and expensive) remote-assembly robotics – traits not shared by other weapons-grade nuclear material.[234] Additionally, uranium-232's gamma emissions damage sensitive electronics and increase material heat,[235] which can prevent a sophisticated nuclear device from detonating under precise and exact conditions – hard requirements for effective use as a weapon.[236]

These are important distinctions in light of concerns from nuclear agencies that cast aspersions on thorium's proliferation resistance. One notable example of these concerns comes from a 2010-era report by the United Kingdom's National Nuclear Laboratory[237] that states:

> *"Contrary to that which many proponents of thorium claim, U-233 should be regarded as posing a definite proliferation risk. For a thorium fuel cycle which falls short of a breeding cycle, uranium fuel would always be needed to supplement the fissile material…Attempts to lower the fissile content of uranium by adding U-238 are considered to offer only weak protection, as the U-233 could be separated in a centrifuge cascade in the same way that U-235 is separated from U-238 in the standard uranium fuel cycle…The argument that*

the high U-232 content would be self-protecting are considered to be over-stated. NNL's view is that thorium systems are no more proliferation resistant than U-Pu systems though they may offer limited benefits in some circumstances."

As with neptunium-237, "proliferation risk" is contextual – and much of that context derives basis from academic postulation as opposed to tangible capability. Just because a state could *theoretically* make a nuclear device from uranium-233 doesn't mean it can make one *practically* – all the more so since there is negligible data or expertise to aid in the creation of such a device. And even if that effort was successful, none of that says such a device would be sufficiently powerful for use in a conflict – or even if it can be effectively deployed in the first place.

Once those factors enter the equation, the optics change significantly.

The UK National Nuclear Laboratory is, of course, correct. Both uranium-233 and neptunium-237 are fissile fuels, and both have potential to undergo nuclear detonations.[238] Further, the uranium-232 contaminant within uranium-233 probably wouldn't stop a crude bomb from detonating crudely – even if it killed its makers.

But it would stop a bomb that relies on sophisticated technology to implode in the precise detonations required for miniaturization to a warhead-scale. It also ignores that LFTRs can be designed to minimize the risk of weaponization barring major infrastructural investments that would draw the attention of international atomic energy monitors. And even if a state had enough neptunium-237, the critical mass requirements would hinder the ability to deploy a weapon of sufficient yield unless that weapon was carried by aircraft or large, long-range missile. The presence of both of these factors would remove either isotope from consideration as a primary charge for a thermonuclear device, and would further preclude both from use as a first-strike weapon.[239]

Further, even if we did grant credence to a rouge state's ability to invest the time, effort, and risk to either purify uranium-233 or build a weapon with neptunium-237, it's important to emphasize just how difficult this is to do – all the more so to do so *quietly*. If any rogue nation tried to build such infrastructure, any intelligence agency with a satellite would know exactly what was going on in short time (which is how we know Iran, North Korea, etc., have nuclear weapons programs).

Nuclear weapons design isn't secret anymore – no 75-year old technology is. The hydrogen bomb, even, is old enough to collect social security. But reaching certain milestones towards making one are only possible with highly expensive and sophisticated systems *built specifically for that purpose*. They're not the sort of thing one picks up at Walmart, and their procurement would certainly raise flags among international monitors and foreign intelligence services – especially since there's only a few entities in the world that manufacture them. If that wasn't enough, they remain among the most controlled machines on the planet.

If your life's ever lacking excitement, try wiring a few million dollars to an offshore bank account for an order of krytron tubes, ultra-fast relay switches, large gas centrifuges and a hefty supply of lithium-6. At the very least you'll see a whole lot of government property and personnel you didn't know existed appear *awfully fast*. Should any state try the same, that property and personnel usually manifests in the form of an airstrike to destroy such a program in its infancy. That's usually long-before said state has even conducted the multitude of tests needed to see if their bomb design even works, as (likely) happened in Syria in 2007.[240]

At the levels of sophistication and expertise required to covertly obtain the necessary materials and successfully make a bomb out of uranium-233 or neptunium-237, building a bomb with traditional nuclear materials sourced from the ground or ocean[241] becomes an easier prospect.

Even so, commercial LFTR designs would need to be required to intentionally contaminate the reactant with materials that would make weaponization harder from the start.[242] This wouldn't permanently remove the risk, but it would make it much more difficult for all but the most dedicated actors. In those cases, if a state is advanced enough to make a nuclear weapon from thorium, they don't need thorium to make one in the first place.

And regardless, it's still poor bomb fuel. Even if they could be efficiently extracted, uranium-233 and neptunium-237 are ineffective fast-reacting fissile fuels compared to highly enriched uranium-235 and plutonium-

239. There are only two known nuclear weapon tests that have ever used uranium-233 – none have ever used neptunium-237. Both of the uranium-233 devices were largely considered failures due to weaker-than-intended explosive yields, respectively at 22 kilotons (U.S. – 1955) and 0.2 kilotons (India, 1998) – relative pittances compared to modern nuclear weapons.[243] In the first device, the uranium-233 was chemically purified (which, again, is highly difficult to do) and was significantly complemented by plutonium-239 to increase yield.[244] As a consequence of those lackluster tests, there exists little research or expertise to weaponize uranium-233 or neptunium-237, nor avoid the inherent dangers of doing so.[245]

And even if there was research or expertise, what's the endgame? To bring dynamite to a thermonuclear missile fight? In a *realpolitik* sense, the leverage gained by building a nuclear weapon is only as valuable as its effective usability in a conflict – or a hedge against the same. A single U.S. Navy *Ohio*-class submarine can launch 24 missiles – *each* armed with up to 12 thermonuclear warheads that *each* yield 475 kilotons – and the U.S. Navy has fourteen of such submarines. Nothing from thorium is ever going to produce anything that can hold a candle to that.

This is why every state with nuclear ambitions has instead invested in uranium-235 and plutonium-239, for their use is easier and safer than hijacking the thorium fuel cycle to produce weapons-grade material. While again, this does not totally alleviate concerns of proliferation through thorium, it does reduce them to a significance on par with a state making an in-house weapons program of their own volition – and that's becoming increasingly more plausible as technology advances globally. With these considerations in mind, the clean energy benefits that thorium and LFTRs bring simply outweigh the theoretical risks of either being used by a dedicated actor for nefarious purposes.

LFTRs are simpler, smaller and less expensive than Pressurized Water Reactors. As traditional nuclear reactors have to be pressurized to 160 atmospheres just to function – pressure equal to a mile below the ocean's surface[246] – they require redundant processes and complex systems to manage the reaction and ensure nothing goes wrong. Additionally, as Pressurized Water Reactors present the potential for catastrophic environmental damage should a reactor melt down or be destroyed through sabotage, they further require extensive security infrastructure. Combined, these factors cause such reactors to rank among the most expensive and over-engineered systems on the planet:

Light Water Reactor fueled by uranium-235	Liquid Fluoride Thorium Molten Salt Reactor (LFTR)
Fuel: Uranium-dioxide solid fuel rods	**Fuel:** Uranium-233 and thorium-232 in a solution of molten lithium-fluoride salts
Fuel lifetime: Approximately two years. Requires reactor shutdown to replace.	**Fuel lifetime:** 30 years without replacement. Current reactor core lifetime is in excess of six years
Fuel input per gigawatt output: 250 tons uranium-235	**Fuel input per gigawatt output:** 1 ton thorium-232. *250 times more efficient*
Annual fuel cost for 1-GW reactor: $60 million	**Annual fuel cost for 1-GW reactor:** $10,000 (estimated)
Total unit construction cost: $7.0 billion	**Total unit construction cost:** $1.0 billion* (1-GW reactor)
Coolant: Highly pressurized water with a graphite moderator	**Coolant:** Self-regulating with passive gravity emergency shutdown
Weaponization potential: High	**Weaponization potential:** Low
Physical footprint: 300,000 square feet + large buffer zone	**Physical footprint:** 2,000-3,000 square feet (size of a house). No buffer zone required

Table Sources: [247]

*Unit cost is expected to reduce over time due to scaling the learning curve of manufacturing if constructing standardized systems.[248]

As LFTRs are spared the size, expense and security requirements of Light Water Reactors, they can be built much smaller and less expensively. They can also be built closer to population centers (as opposed to Pressurized Water Reactors that need to be geographically isolated), considerably reducing the infrastructural requirements to transmit power to electric grids.

LFTRs can be built in a modular, prefabricated capacity. Today's nuclear reactors are designed as unique, custom systems that are each made to order – significantly increasing their total cost. Yet recent improvements in manufacturing today allow LFTRs to be built on assembly lines as iterations of product models in the form of small modular reactors.

This provides two main benefits:

First, efficiencies inherent in modern manufacturing enable us to reduce construction costs over time as more identical units are produced. This is often referred to as "the learning curve," or "learning ratio" – the reduction in manufacturing cost every time the number of produced units doubles.[249]

In computing, Moore's law has shown that computer processing power at a given price doubles every two years.[250] In aerospace manufacturing, the reduction in per-unit cost has been roughly 20% every time the number of produced units has doubled.[251] As applicable to the manufacturing of Light Water Reactors, the University of Chicago estimates a learning ratio of 10% in their 2004 study *The Economic Future of Nuclear Power*.[252] As LFTRs can be built on assembly lines, that percentage would likely be higher, expected to be on the order of aerospace-grade manufacturing.

But even at 10%, this would mean that by the time the 1,000th LFTR was constructed it would cost around 40% of the first commercially produced unit. This means that if the estimated price tag for a LFTR stands at $200 million currently, as more units were produced that cost would fall over time – making them increasingly more affordable and economically viable. The following except is from *Thorium: Energy Cheaper than Coal*, written by Robert Hargreaves, PhD, Professor of Nuclear Physics at Dartmouth University:

> "Boeing, capable of manufacturing $200 million units daily, is a model for LFTR production. Airplane manufacturing has many of the same critical issues as manufacturing nuclear reactors: safety, reliability, strength of materials, corrosion, regulatory compliance, design control, supply chain management and cost, for example. Reactors of 100 [megawatt] in size costing $200 million can similarly be factory produced. Manufacturing more, smaller reactors traverses the learning ratio more rapidly. Producing one per day for 3 years creates 1,095 production experiences, reducing costs by 65%."

The second main benefit of manufacturing LFTRs on an assembly line is standardization, and standardization provides modularity. This becomes important when building small modular reactors because not only are smaller, modular and standardized reactors considerably less expensive to construct, they are also easier to deploy.

If you recall, a core requirement of Universal Energy is widespread deployment, as many regions that suffer from the consequences of resource scarcity are geographically remote and/or feature terrain that's hostile to the construction of something as large as a power plant. A smaller LFTR manufactured on an assembly line can be built rapidly and plugged into any grid in a relatively short time period.

So if, for example, a region needed to quintuple its electricity generation capacity in a matter of weeks, small modular LFTRs make this possible – and they make this possible effectively anywhere. This also would pay dividends toward disaster-relief efforts, peacekeeping missions, ocean trash cleanup and possibly even space exploration.

How We Know It Works

The thorium fuel cycle has been known to science from the start of the atomic era, and reactor designs associated with that cycle have been around since the 1950s. The first successful use of thorium came from the Department of Energy's MSRE experiment, a 1960's-era project from the Oak Ridge National Laboratory working from prior research to build a molten salt reactor for aircraft propulsion.[253] The 7.4 Megawatt reactor went online in 1965 and worked successfully for four years until the experiment was cancelled in 1969 in favor of Light Water Reactors.[254] Light Water Reactors eventually became the national standard largely because they could produce both energy and weapons-grade nuclear material.[255]

While that unfortunate result came to pass, the results of the MSRE experiment nonetheless conclusively showed that the reactor concept was viable,[256] as have other tests since. The MSRE experiment confirmed predictions and expectations, showing the safety, efficiency and heat transfer potential for LFTRs was present using 1965-era capabilities.[257] Several other countries and companies have since made progress on LFTR technology. Notable high-profile projects include:

China: The Chinese government has invested $3.3 Billion into molten salt reactors in Gansu province under the name "Thorium-Breeding Molten Salt Reactor (TMSR)."[258] These reactors are being built underground, and are intended to generate up to 100 megawatts of power. Their reactor models heavily leverage cogenerative design, intending to use excess energy to power other resource-producing systems including fresh water, hydrogen and hydrocarbon fuels.[259]

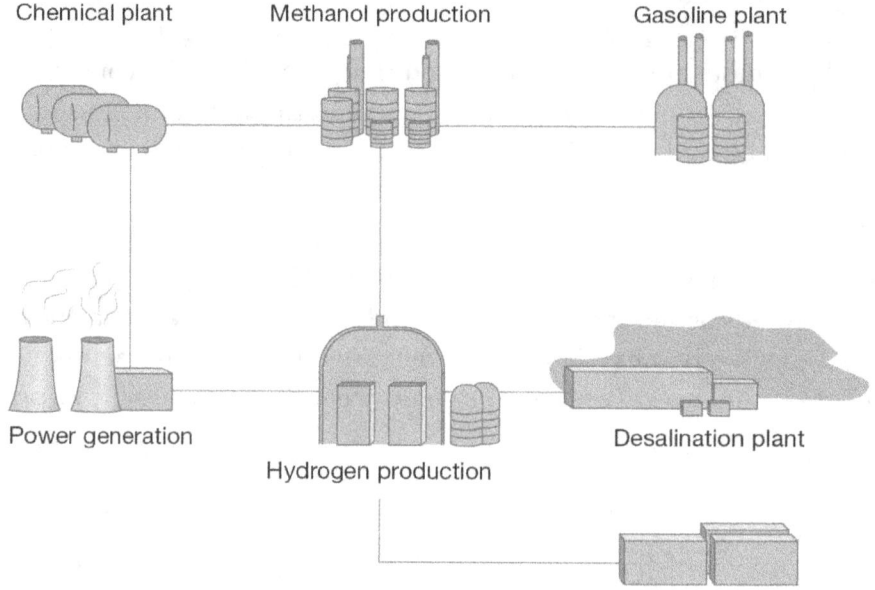

Image source: [260]

Although a large focus of this project is electricity for civilian usage, the Chinese government hopes to apply the results of this project to military applications like drones and future fast aircraft carriers.[261] As reactor miniaturization would be required for placement within something as small as a drone or warship, such an advance would present significant implications for NATO states in both civilian and military sectors – making a matching investment in LFTR technology all the more pressing. As of this writing, the Chinese program intends to have a functional reactor prototype by 2025 with large-scale commercialization by the early 2030's.[262]

The Netherlands: The Dutch NRG (Nuclear Research and Consultancy Group) is one of the leading European nuclear service providers. They have constructed a prototype Molten Salt Reactor that began fluoride salt irradiation on August 10th, 2018.[263] Instead of turning a live reactor "critical" for sustained civilian power, the NRG intends to conduct a series of experiments (referred to as SALIENT) to reinforce the validity of thorium energy and use the experiments as templates for future LFTR development.[264]

Unique among other thorium R&D efforts outside of G8 states is the NRG's installation in Petten, which has all of the decontamination, cleaning, salt production, radiation shielding and fine element analysis equipment to build a nuclear reactor in-house.[265] As a semi-private venture, the NRG can now compete with state-level actors to fine-tune the necessary manufacturing requirements to achieve LFTR viability.

India: A nuclear power since the 1950's, India is no stranger to the promise, challenges, and risks inherent to atomic energy. Yet among members of the "nuclear club," India is unique in that it has the largest thorium reserves of any sovereign nation – some 11.5 million metric tons.[266] This has led India to accelerate research and development on thorium-powered molten salt reactors as a part of its three-stage nuclear program.[267]

India's latest effort is the Kalpakkam prototype fast breeder reactor, designed to generate 500 megawatts of electricity. The Kalpakkam prototype is expected to reach criticality by the early 2020s.[268]

Explaining the benefits of the reactor model to the *Times of India*, the Director General of the International Atomic Energy Agency noted that *"fast reactors can help extract up to 70 percent more energy than traditional reactors and are safer than traditional reactors while reducing long-lived radioactive waste by several fold."* It's worth noting for our purposes that while fast breeder reactors are akin to LFTRs in both theory and function, and present promising results even in initial prototype stages, the designs have had stability issues in the past and present varied engineering challenges to long-term stability.[269]

For this reason, India has been running a forerunner reactor to the prototype they're building in Kalpakkam under the Fast-Breeder Test Reactor program. This smaller reactor has had its own technical challenges, yet has reliably produced impressive amounts of energy even while operating at significantly less than total capacity.[270] In doing so, it has provided Indian nuclear scientists with the data needed to complete the larger Kalpakkam reactor – which itself can be used as a stepping stone to further advancement in breeder reactors that leverage the thorium fuel cycle. The third stage of India's nuclear program is designed to use thorium exclusively.[271]

Russia: while India has nearly completed their prototype fast breeder reactor, Russia has successfully deployed the technology since the early 1980s at the Beloyarsk Nuclear Power Station. The station currently operates two fast breeder reactors – the only two in the world that are currently operational – respectively generating 600 and 885 megawatts each. However, unlike Indian variants, these Russian fast breeder reactors are fueled by enriched uranium, due to plentiful Russian reserves and the security "benefit" of uranium's dual support of civilian energy and nuclear armament.

Things are changing, however, as Russia is currently developing a high-temperature breeder reactor fueled by thorium, which, like China's variant, is expected to divert waste heat energy to desalinate seawater and extract hydrogen.[272] This Russian reactor will also be partially fueled by discarded weapons-grade material – paying homage to the ability of Molten Salt Reactors to safety generate electricity as a byproduct of armament reduction initiatives.[273] Professor Sergey Bedenko from the School of Nuclear Science & Engineering at Tomsk Polytechnic University and a co-author of a paper[274] on the project, noted:

"Current reprocessing and recycling technologies still results in radioactive waste that contains plutonium...Our technology tackles this problem as it allows 97% of weapons-grade plutonium to be [consumed]."

The main advantage of such plants will be their multi-functionality...Firstly, we efficiently dispose one of the most dangerous radioactive fuels in thorium reactors, secondly, we generate power and heat, thirdly, with its help, it will be possible to develop industrial hydrogen production."

This project also has the benefit of state backing. As of 2016, President Vladimir Putin has directed Russia's state energy institutes, Rosatom and Kurchatov, to deliver a proposal on how to leverage thorium for next-generation reactors while improving thorium procurement through rare-earth metal extraction.[275] As reactor technology improves with future research and development, we may see more sophisticated Russian LFTRs that can be manufactured at scale.

Germany: As of this writing, Germany has transitioned from nuclear power to a more renewable-focused approach in response to anti-nuclear political pressure.[276] The results have been mixed at best, and Germany still relies heavily on coal to complement the intermittency and unreliability of using renewables for baseload power.[277] Germany once had functional thorium reactors that operated at high efficiencies and output. Although the design wasn't a Molten Salt Reactor, their THTR-300 high-temperature thorium reactor worked successfully between 1985 and 1989,[278] but was decommissioned in favor of light-water reactors. Despite its higher costs and unique engineering requirements, the experimental THTR-300 reactor proved thorium's viability as a fuel and presented a rich supply of test data for future high-temperature reactors.

The United States: as the world's first nuclear power, the United States has extensive experience with atomic energy. Nearly all nuclear engineering today is derived from American designs – including reactors fueled by thorium.

Yet most of these reactors were designed with the intent of providing an ample supply of weapons-grade nuclear material alongside a civilian power program, and the United States remains the only nation in the world to deploy a nuclear weapon in an armed conflict. For these reasons, alongside debates over how to dispose of the nuclear waste associated with Pressurized Water Reactors, atomic energy in America faces significant political resistance – even though it generated 60% of our emissions-free power in 2016.[279]

The political mood is changing, however. 2016 saw the first new American nuclear reactor to come online in decades,[280] and that reactor now generates enough energy to power 650,000 homes.[281] Myriad companies, from startups to long-established nuclear engineering firms, are now exploring advances in thorium reactor technology. Several are even investing in "microreactors," scaled-down modular reactor designs that can be mass-produced on assembly lines. Although some would still use uranium-235, the mass-produced approach is still important for several reasons:

First, it leverages one of the attributes that makes America the world's wealthiest economy: its ability to build sophisticated systems on a large scale. When it comes to mass-producing cutting-edge technology with minimal room for error, America's manufacturing prowess shines brightest. This gives the United States perhaps the best advantage when mass-manufacturing modular reactors on assembly lines to a single standard.

Second, the advances made in miniaturizing reactor technology – even if still fueled by uranium-235 – can be extended to miniaturizing LFTR technology in the future, lowering costs, barriers to entry, and barriers to scale.

Third, microreactors can be built small enough to deliver via train, ship, or even truck or aircraft. That enables them to be transported and deployed effectively anywhere – a key goal of Universal Energy.

Several companies are proving to be pioneers in this future frontier of nuclear energy, both within and outside of thorium:

- **Westinghouse's** eVinci's microreactor design is factory built, fueled and assembled. It's also small enough to transport on a truck, and boasts a 0.06 acre footprint with less than 30 days onsite installation.[282] It can match the energy output of up to 380 acres of wind turbines and 79 acres of solar panels.[283] And at zero emissions, the equivalent energy output with diesel would produce 230 million pounds of CO_2. It was partially envisioned to power military bases and research stations in frigid climates where wind turbines freeze, sun is scarce and diesel fuel is the only viable source of energy. Westinghouse's current designs – planned for release in 2024 – estimate constant operation for upwards of ten years without refueling.[284]

- Corvallis, Oregon-based **NuScale Power** has its own microreactor designs. Their Small Modular Reactor is designed to provide scalable power generation up to 720 megawatts. While based on the light-water model, their modular design eliminates two-thirds of the internal parts of traditional Pressurized Water Reactors and also incorporates a passive auto-shutdown that doesn't require external power, additional water or operator action. These distinctions present critical advantages over previous reactor designs, not only in terms of safety but also because they avoid the need for the redundant and expensive containment systems.

    At a deployment area of 15x82 feet, the containment vessel and reactor core are roughly 5% of the size of a traditional nuclear power plant[285] - small enough to be delivered by rail, barge or truck. To date, the company has secured more than $300 million in funding from the Department of Energy.[286] They plan to construct a 12-module Small Modular Reactor plant at the Idaho National Laboratory by 2026,[287] that would provide up to 720 megawatts of emissions-free power.[288]

- **General Atomics** is a defense contractor specializing in aerospace and nuclear engineering. As a former subsidiary of General Dynamics, it's been on the front lines of nuclear advancement since its commercialization, with a proven track record of building reliable high-performing reactors. Their latest reactor concept is billed as an "Energy Multiplier Module," which is a series of modular microreactors that can be deployed together as a unit and buried below ground.[289]

Image Source: General Atomics

General Atomics reactor designs mimic the benefits of LFTRs in function – breeding, automatic operation for up to 30 years, passive non-mechanical safety measures, the ability to consume both nuclear waste and weapons-grade material as fuel and a high-temperature loop that can be used for supplemental resource production. Although the design is still in concept stages, they have received more than $60 million in funding thus far from the Department of Energy, and continue to join other companies domestically and abroad in developing Small Modular Reactors.[290]

- **Terrestrial Energy** is a joint Canadian-U.S. startup specializing in thorium-fueled molten salt reactors. It has been working alongside the United States Department of Energy and Oak Ridge National Laboratory to bring smaller-scale LFTRs to market.[291] Larger than a microreactor yet significantly smaller than a traditional nuclear power plant, their patented Integral Molten Salt Reactor (IMSR) is modular in design and scalable from 80 to 600 Megawatts on 17 acres or less.[292] As with any LFTR, Terrestrial Energy's design includes secondary and tertiary heat loops for supplemental resource production. The company is expecting to start their first reactors by the 2020's with larger-scale commercial viability thereafter.[293]

As designed, each of these modular reactors can be assembled in groups to meet the output of base load power.

The litany of investments in LFTR and other small modular reactor technologies are made because we know these technologies will work, as prototype after prototype have proven it so. We know reactor miniaturization works because we've designed and built miniature reactors after thousands of iterations of modeling and tests. We know that standardization and modularity in the design of advanced systems gives way to greater scalability and flexibility in deployment, because we've seen these concepts produce this exact effect in every other industry they've been employed.

The developed world isn't investing many billions of dollars into next-generation nuclear technology on a whim, nor would it do so if the science was in doubt. It's perhaps for this reason why market growth in microreactor technology is increasing at an annual rate of 19%[294] - the investment follows the data, and the data supports the investment.

What About Costs and Criticisms?

It must be mentioned that atomic energy has critics who doubt the technology's viability on grounds ranging from safety, waste, and proliferation to economic and logistical feasibility. Some critics see nuclear, in and of itself, as a fundamentally unredeemable technology. Others bring up points that are valid in abstract, yet are used to cast inappropriately generalized aspersions on nuclear's promise. Thorium, of course, is no exception.

While there are myriad sources that might articulate these criticisms, I've chosen three articles for their accessible language and range of arguments. The first two argue against nuclear as a concept, and the third argues against thorium specifically.

I invite you to review them individually, in full, so that you understand where they're coming from in their own words.

The first article is a February 2019 piece in *ThinkProgress* entitled *"Taxpayers should not fund Bill Gates' nuclear albatross,"* which casts doubt on Bill Gates' effort to address climate change by increasing nuclear's American market share.[295] The author, Joe Romm, writes:

> *"The reality is that nuclear power is so uneconomical that existing U.S. nuclear power plants are bleeding cash — and in many places it's now cheaper to build and run new wind or solar farms than to simply run an existing nuclear power plant. Saving the existing unprofitable nuclear plants would require a subsidy of at least $5 billion a year, according to an analysis last July by the Brattle Group. So, given existing plants are so uneconomic, it's no shock that building and financing an entire new fleet of nuclear plants is wildly unaffordable — especially since a new nuclear plant can cost $10 billion or more."*

The second article is from *Popular Mechanics*, entitled *"The Alexandria Ocasio-Cortez 'Green New Deal' Wants to Get Rid of Nuclear Power. That's a Great Idea."*[296] The author, Avery Thompson, had this to say:

> *"…nuclear simply has too many downsides to ever be a viable way to produce electricity in the U.S. Primarily, it's just too damn hard and expensive to build new nuclear capacity in 21st century America. To see why, take a look at the Watts Bar 2 nuclear*

power plant. Watts Bar 2 started supplying power in June of 2016, becoming the first nuclear power plant to be built in the U.S. in two decades. Overall, Watts Bar 2 was the result of 37 years of construction. For most of that period the reactor simply languished in a sort of economic limbo. Construction was halted in 1985 due to low energy prices and only resumed in 2007. At that point, Watts Bar 2 faced another decade of construction delays and cost overruns. The reactor was initially scheduled to be completed in 2013; delays pushed back the start date to 2015 and again to 2016. The initial cost was estimated to be $2.5 billion; the final cost, after a series of unanticipated hurdles, was $4.7 billion...Reactors are gigantic beasts, and sustaining a nuclear reaction while drawing power from it requires an absurd level of engineering. Reactors are expensive, bulky, and complicated, to say nothing of the waste products they produce or the fear of a catastrophe like Chernobyl or Fukushima."

Lastly, there's the thorium-specific criticism written in The Guardian nearly a decade ago.[297] In June 2011, Eifion Rees, an ecologist who advocates for a post-industrial "ecolocracy" (his words), wrote a piece titled *"Don't believe the spin on thorium being a greener nuclear option."* In summary, Rees makes the case that thorium is merely a way to deflect attention from the dangers of the uranium fuel cycle within Pressurized Water Reactors that would still be used until thorium had proven large-scale commercial viability.

He states that the nuclear industry itself remains skeptical of thorium and cites the 2010 paper from the UK's National Nuclear Laboratory[298] mentioned earlier in support. Rees continues that that even if LFTR waste is much shorter lived, it will still be toxic and emit harmful radiation. Rees's final point is that the effectiveness of renewables is rapidly improving, so even if thorium proves itself in the next 30 years (by Rees' assessed timeline), it will arrive to solve a problem that's no longer present. He concludes that the combination of these factors make thorium too new, too untrusted, too potentially dangerous, and too expensive when we already have an energy solution – renewables – in hand.

These articles and mindsets were cited verbatim because I believe in intellectual honesty. They were also cited because they're wrong. Not just on the facts when it comes to 2020-era technology, but also just as importantly on the role of nuclear in the context of the energy, resource and ecological requirements of our time.

To explain why, we'll boil down the primary arguments of these three pieces (and the NNL assessment[299] cited in The Guardian) into three overarching claims that we'll review in order.

- Claim One: Nuclear energy is too expensive in terms of cost and safety, and always will be, thus we should focus exclusively on renewables.

- Claim Two: Thorium's an unproven technology with unproven viability, thus we should focus exclusively on renewables.

- Claim Three: Thorium advocates overstate its waste and proliferation benefits, thus we should focus exclusively on renewables.

Claim: Too Expensive

Nuclear power is sophisticated technology, and sophisticated technology costs money. This feeds into a common narrative from nuclear detractors that atomic energy is cost-prohibitive. If waste and proliferation concerns weren't enough, they say, there's simply too many expenses behind nuclear engineering, deployment, operation and safety to make the technology viable.

Some of these concerns have merit in abstract. But they don't necessarily apply to nuclear as a technology any more than concerns about burning zeppelins apply to aircraft as a technology. Further, nearly all focus on Pressurized Water Reactors and point to their drawbacks as cause to paint the entire field of atomic science with a wide brush.

The reason why this chapter focused on smaller LFTRs and made repeated mention of small modular reactors (even if powered by uranium-235) is because they avoid these problems, particularly cost criticisms, by design. Large-scale Pressurized Water Reactors further present major safety hazards if things go wrong, especially as they increase in size. They need massive containment and cooling apparatuses, redundant safety mechanisms, security features, buffer zones and control systems. These highly expensive components are not necessary for LFTR and newer small modular reactor designs. That alone is a differentiating cost factor on the scale of billions of dollars.

Then there's the fact that most every power plant in America today – nuclear especially – is built as a unique entity, designed and deployed **to order.** The plant's concrete, metal, wiring, HVAC, control systems, thermal management, waste processing, plumbing, walkways, stairwells, doors and all points in between are *designed, architected and engineered from the ground up **each and every time*** – as is the regulatory approval for each specific building site.

For every nuclear power plant thus far built, a company gathered an architect, structural engineer, nuclear engineer, electrical engineer, environmental engineer, fluid engineer and mechanical engineer to design the plant. Then they hired the contractors to build it, the banks to fund it, the insurance companies to underwrite it, the bond companies to back it, the lawyers to sign off on it and the regulators to approve it. And if they wanted to build another, they then had to wipe the slate clean and do it all over again from scratch because the math and materials that worked for plant A didn't work for plant B.

It's no wonder, then, how a nuclear plant can cost billions to construct today. Any system of any level of sophistication would cost the same under those circumstances. If your car had to be built this way, it would cost millions of dollars. If a commercial aircraft had to be built this way, it also would cost billions of dollars. Automated manufacturing changes this, as it does in any other sophisticated industry. It started with Ford's assembly line and now works to build jetliners. Once we apply automated manufacturing to nuclear, the same cost effectiveness will be achieved as it has in literally every other industry automated manufacturing has been employed.

Another major additional cost contributor is inappropriate regulation. Nuclear power carries a requisite responsibility of standards and demonstrated operational expertise to develop at scale, and regulations are of course critical to appropriately determine safety and proper function. However, many of these regulations are antiquated and more appropriately geared to deal with Pressurized Water Reactors on the scale of a base load power stations designed and built using 1970's-era technology. When it comes to the construction of molten salt reactors or small modular reactors in power modules, these regulations present obstacles that are disproportionate for the level of technical deployment.

Take paperwork, for example. While paperwork is an unavoidable reality of regulatory compliance, American nuclear plants today annually spend between $7 million - $16 million just to document such compliance.[300] As the annual regulatory liability of a nuclear plant today hovers around $60 million, those cost figures add up.[301] And that's just the costs for running a nuclear power plant. Construction of new plants can only begin after waiting nearly a decade for regulatory approval – a hard sell to any entity that must bleed cash for fees and loan interest before even breaking ground.³⁰²

This most certainly doesn't mean that regulations in and of themselves are a bad thing – especially for an industry as sophisticated as atomic energy. But the goal of regulation should be to ensure safety and ideal standards of operation. They shouldn't hinder the nuclear industry by requiring 2019-era technology to comply with regulations geared to 1970's-era reactor designs.³⁰³ Nor should they make nuclear technology jump through unnecessary financial hoops, the costs of which are then pointed at to feed a narrative of financial unviability in a self-fulfilling prophecy. We regulate plenty of other sophisticated industries: commercial aerospace, private space ventures, submersible crafts and industrial chemistry without burdening them to insolvency. We can and should do the same with nuclear.

When we summarize the costs behind past implementations of nuclear, we see many of them – non-standardized and ad-hoc construction, outdated regulatory schemas and inappropriately expensive compliance requirements, focus on large size as opposed to modular scalability – are untrimmed fat. Once we start mass-producing small modular reactors for flexible deployment to a single standard, manufacturing costs plummet. As regulations shift towards next-generation nuclear built on that standard, costs of regulatory compliance, research and development, maintenance, and scalability will concordantly follow suit.

Further, in touting renewables as an immediate alternative, advocates correctly celebrate their benefits yet also ignore the costs of renewable integration on a comprehensive or even base load scale. Universal Energy's model seeks to minimize these costs by way of municipal infrastructure. Yet barring a hyperbolically expansive scale of deployment, there's no way to generate true base load energy on the scale of Universal Energy's target with renewables alone without prohibitively expensive and logistically daunting land purchases. And that's without looking at the cost of the renewable technology itself, either levelized, maintenance, replacement or end-of-life processing.

It will cost money no matter what energy source we employ to solve future energy needs – and costs will vary depending on how they're integrated. Both renewables and clean nuclear can and should be employed to their greatest strengths and cost efficiencies. But suggesting that nuclear is too expensive while touting renewables as an inexpensive alternative makes hefty omissions about the ultimate costs of end-to-end renewable implementation. It also makes hefty omissions towards manufacturing, transporting and installing said renewables in a carbon-emitting supply chain – saying nothing of the immense material extraction required to do so. Once those factors are included in the analysis (as we'll see shortly), the circumstances change quickly.

Claim: Unproven

In a vacuum, the critics are right: thorium-fueled LFTRs, and for that matter, uranium-fueled Small Modular Reactors, have yet to prove themselves as large-scale viable technologies. But no emergent technology in history has ever arrived to market with "proven viability." In 1890, the car had unproven viability; the same goes for aircraft in 1910. The computer had unproven viability in the 1970's. The touch smartphone had unproven viability in the early 2000's – Microsoft CEO Steve Ballmer famously said the iPhone would "never gain market share."[304] The internet, even, had unproven viability – one of the foremost computer scientists at the time dismissed it as a fantasy in a February 1995 article in *Newsweek* entitled "Why The Web Won't Be Nirvana." The author, Clifford Stoll, PhD, wrote:

> *"After two decades online, I'm perplexed. It's not that I haven't had a gas of a good time on the Internet...But today, I'm uneasy about this most trendy and oversold community. Visionaries see a future of telecommuting workers, interactive libraries and multimedia classrooms. They speak of electronic town meetings and virtual communities. Commerce and business will shift from offices and malls to networks and modems. And the freedom of digital networks will make government more democratic. Baloney! Do our computer pundits lack all common sense? The truth is no online database will replace your daily newspaper, no CD-ROM can take the place of a competent teacher and no computer network will change the way government works."*

It could be that Mr. Stoll or Mr. Ballmer might find comradery with Mr. Romm, Mr. Thompson, Mr. Rees, and other critics today. Expertise in a specialized focus can lend itself to tunnel vision that hinders views of the larger picture. Emergent technologies always need refinement. That's how technology works. That's how it has always worked.

That's why data storage went from a million dollars per gigabyte in the 1980s to less than ten cents per gigabyte today.[305] That's why your smartphone, laptop or flat screen television doesn't cost millions of dollars today. That's why you can send a video to any continent on the planet within seconds, whereas back when JFK was President an intercontinental call could cost a small fortune.

Technology emerges, designs improve, innovations are incorporated – and the market responds. In an era where we can manufacture error-unacceptable systems on assembly lines in a matter of days, it's actually quite possible to mass-produce a technology that avoids the drawbacks of designs chosen in the 1960s specifically because they helped make hydrogen bombs. Now that it's 2020 and nations worldwide are pouring billions into thorium, there's no reason to wager they'll fail, particularly since:

- The 1960's MSR experiment was successful and proved thorium workable.

- Germany's THTR-300 proved the thorium fuel cycle workable.

- The Chinese TMSR has proven thorium breeder reactors workable.

- India's fast breeder reactor has proven the science behind thorium.

- Russia - the only nuclear power on par with the United States - has proven fast breeder reactors workable, and are currently investing in advanced thorium reactors.

- North American companies, including Westinghouse, NuScale Power, Terrestrial Energy, and General Atomics have proven that LFTRs, Small Modular Reactors, and Molten Salt Reactors are all workable.

None of these countries or companies deal in fantasy – and they wouldn't collectively devote tens of billions to fantasy, either. The thousands of highly educated women and men working with them, their peer-reviewed studies, the prototypes they've built and the agencies that issued them grants are all behind the future viability of both thorium and Small Modular Reactor designs.

That we need to iron out some present challenges to realizing large-scale commercial deployment isn't an indictment of the technology – it's a reflection of opportunity. Every globalized technology in history has risen to meet these challenges, and the international efforts and investments behind thorium stand to repeat the same result.

Claim: Weapons and Waste

We spoke to weaponization earlier in the chapter. LFTRs and the thorium fuel cycle can be weaponized in theory, but not easily and not practically by military standards. That's the important distinction: it's *theoretically possible* for a state to make a crude bomb with uranium-233 or neptunium-237. But that says nothing about miniaturizing the bomb into a missile-driven warhead or keeping the mechanisms stable for long-term storage, nor the ability to do these things in secret – all of which are required capabilities to present even a moderate threat. Most nuclear powers today have thermonuclear capabilities that uranium-233 or neptunium-237 can't meet in a best-case scenario, and no state's going to raise the scales in a conflict to a nuclear fight they're certain to lose badly.

When considering terrorism, terrorist organizations fundamentally lack the material procurement power and expertise needed to manufacture such a weapon – let alone source the tools needed to do so. And when an inexpensive home gene editing kit[306] or a basic knowledge of industrial chemistry[307] could wreak as much havoc as a dirty bomb, denying humanity the most powerful source of baseload-scale power we have in the name of thwarting terrorism seems foolish.

The waste angle, too, is highly contextual. Yes, nuclear waste is dangerous and toxic – including waste from thorium. But thorium's waste lasts for only 300 years compared to the millennia from waste from the uranium-235 fuel cycle.[308] And most nuclear waste being produced today still has potential to function as fuel within Molten Salt Reactors[309] – which by itself only needs one ton of thorium to generate one gigawatt of energy.[310]

Of the waste that one ton produces, 83% becomes stable in 18 years, with the remaining 17% reaching stability ~282 years later.[311] It's well within the realm of feasibility for us to hold that waste in secure underground facilities for that timeline, especially since the most danger occurs only during the first few decades of storage, with the material becomes progressively safer over time.

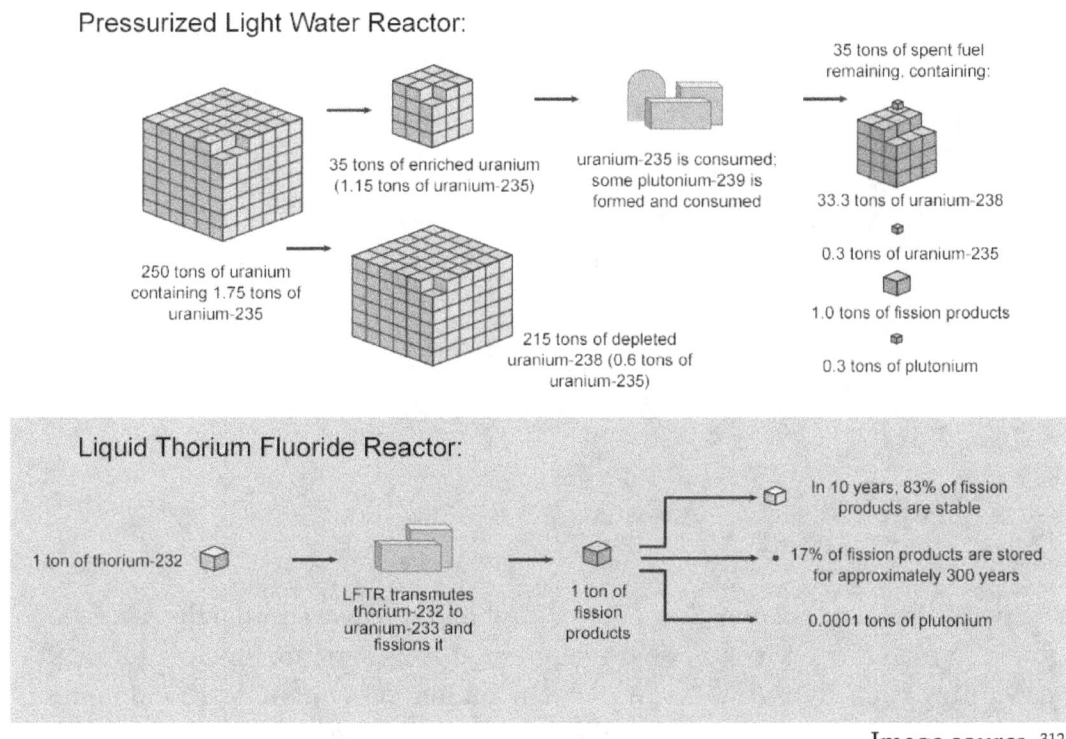

Image source.[312]

Nuclear critics say that's still unacceptable, and that we're still better off sticking with renewables. Yet even though renewables are essential for a clean energy future, they're far from waste free – and it's highly disingenuous to suggest otherwise. Wind turbines are difficult to recycle and often end up in landfills (the bulldozer in the below-left image is the size of a school bus). Solar panels produce gallium-arsenide (arsenic) and chemical batteries contain strong acids and heavy metals like mercury, lead and cadmium.[313] These are highly toxic substances *that don't lose their toxicity over time as radioactive materials do.* Once they leech into the environment, they become a permanent part of it.

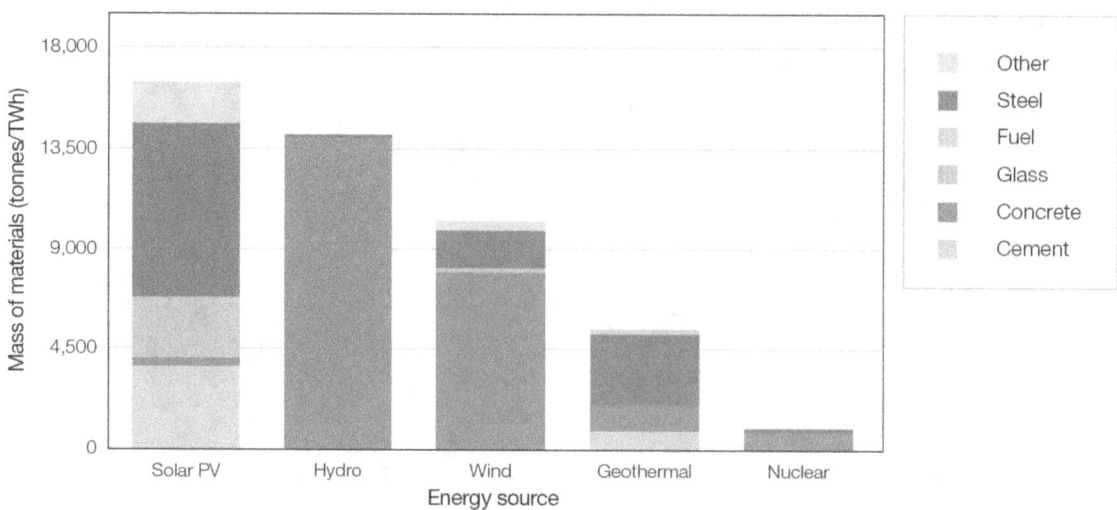

Sources: DOE Quadrennial Technology Review, Table 10.
Murray, R.L. and Holbert, K.E. 2015. Nuclear energy: an introduction to the concepts, systems, and applications of nuclear processes (7th ed.). Elsevier.

Then there's the material question. From the table above, we see that the materials required to manufacture solar panels weigh in at 16,447 metric tons per terawatt of generating capacity[314] (one terawatt = one billion kilowatts). At 10,260 metric tons, wind comes in third.[315] Mindful of Universal Energy's annual generation target of 12.5 trillion kilowatt-hours (12,500 terawatt-hours), this would require a daily power generation capacity of at least 34.25 terawatts. That translates to a material throughput of 534,530 metric tons of material for solar and some 351,400 metric tons for wind. And that's *just* for the United States, some 4.3% of humanity. The rest of the world's material requirements to exclusively leverage renewables would represent a higher order of magnitude.

Such figures don't incorporate the carbon emissions or waste footprint inherent to sourcing these materials in the first place – nor the emissions produced in the manufacture of billions of solar panels or wind turbines. Nor do they incorporate the carbon emissions and cost externalities – both financial and ecological – presented by transporting, installing and wiring billions of solar panels and wind turbines, which, in such implementations, would certainly require vast purchases of expensive land.[317] Think of how many copper or rare earth mines we'd need. How much diesel fuel we'd consume. How many non-recyclable materials or toxic chemicals we'd produce or dispose of at any step of the supply chain. The waste implications involved with such considerations are enormous, as are the environmental impacts.

For example, the following image shows the Escondida copper mine in Chile:

This image shows the Mountain Pass rare earth mine in California, which sources the lithium-ion needed for advanced batteries.

This image shows the Katanga copper-cobalt mine in Congo-Kinshasa.

Those examples are only *three mines of thousands worldwide*. Also not shown are the processing plants, smelting facilities, or human toils required in any of the extractive efforts required to rely on a renewable-exclusive portfolio for our future energy needs – nor the ultimate end-state waste created once these systems are ultimately discarded. These considerations, even in a vacuum, carry immense humanitarian and ecological consequences. The following images show child workers mining cobalt and lithium for rechargeable batteries:

These images show large volumes of discarded electronics (e-waste):

Should our civilization move to a renewable-exclusive energy portfolio, these images would represent barely a fraction of the environmental pillaging and humanitarian maladies that would consequently arise. That's the uncompromising reality of facts as they are.

In this mention, it's important to re-emphasize that use of lithium-ion, cobalt, copper or rare earth metals isn't inherently destructive in and of itself – nor are the laudable efforts to switch to either renewables or next-generation battery technology to help wean the global supply chain off of fossil fuels. But it's nonetheless vital to recognize that *exclusive* reliance on these technologies in a carbon-emitting manufacturing chain comes with massive ecological and humanitarian consequences that still manifest even if out of sight and mind of the developed-world consumer. The inclusion of such technologies remain central in the Universal Energy framework because the negative aspects associated with them would drastically reduce (if not vanish) should they become sourced, processed and recycled under a clean, carbon-neutral nuclear power schema.

Simply stated: clean nuclear is the key to making renewables truly clean.

Most importantly, the Universal Energy framework is made possible only by the cogenerative benefits of high-temperature reactors. Without that energy source, we can't extract fresh water and hydrogen fuel from billions of gallons of seawater, nor can we initially keep that water hot to store energy generated from renewables on a nationwide scale. We're left with chemical batteries that carry an intensive material throughput themselves – now on a massive scale – yet without the energy abundance needed to solve resource scarcity. All that effort, time and material invested – along with what would certainly be trillions of dollars – and we'd still be facing the same malady that's dogged us from the dawn of time.

In that context, storing a few hundred barrels of radioactive material that becomes inert in 300 years is the better option – all the more so since Universal Energy still employs renewables to their greatest strengths within municipal infrastructure. But as we're only able to manufacture renewables in a carbon-free capacity with next-generation nuclear, the numbers behind renewables as a singular solution don't add up *even if* we had the capital and material basis to pull off such an endeavor. As we don't, the waste implications of attempting to do so far eclipse anything nuclear brings to bear – and no amount of selective criticism or ideological advocacy is going to change that.

Moving Forward

Building a clean energy future requires investment into technologies capable of delivering such ends. As circumstances stand today, only thorium and renewables, working together, are capable of meeting this challenge. Yet deploying thorium as a base load solution requires investment and additional research that necessitates an appropriate social focus.

Yet there are outside forces slowing down this process – some of which we have already familiarized ourselves with. Because of the social hesitancy around nuclear power, politicians are often less eager to embrace advances in the technology. It is, of course, politically safer to stage photo-ops in front of wind turbines and solar panels while quietly pivoting towards coal or natural gas for baseload power when the cameras are turned off. This is effectively what Germany has done.[318]

But that's not good enough anymore. The future risks to humanity's long-term resource supplies, the state of Earth's warming climate, the state of future population growth, and the corresponding state of future ecological collapse rank among the most serious problems our species has ever faced. And in the short-term future we will be facing all of them, more or less simultaneously.

Depending on renewables alone will not generate the immense energy required to provide for our needs and long-term growth estimates. Depending on renewables alone will not bring about an abundance of raw, inexpensive energy required to synthetically produce resources to an effectively unlimited scale. Depending on renewables alone will not extend baseload energy redundancy and reliability to provide all this and more for the indefinite future. There must be a political and social inflection point to recognize this reality.

By themselves, neither renewables nor nuclear can meet this challenge. Proponents of both technologies must realize that meeting future energy and resource challenges will require a joint venture between both – at maximal investment and output – working together in synchronicity. Nothing short of that end will get us to where we need to be.[319]

When the political and social will becomes present – a likelier "when" than "if" due to the realities of future energy and resource needs[320] – politicians need to consider regulatory streamlining. As we reviewed earlier, new nuclear initiatives face challenges that are as much regulatory as they are technical,[321] and that's a problem that needs to change.[322] Regulation is a necessary – yet significant – component of the cost of implementing atomic energy, and regulators need both the freedom and impetus to craft regulations for 2019-era technology.

From there, we can begin to delve into the engineering nuances required to develop modular thorium reactors that can be safely mass-produced to a single standard. The aerospace industry is a perfect model for this goal, as it has automated the fast-turnaround construction of highly-sophisticated systems with extremely tight tolerances and even lower margins for error. If Boeing can mass-manufacture a commercial jetliner in nine days under these requirements,[323] we can do the same for modular reactors. Even so, some of the open considerations involving these engineering nuances include:

Design and process optimization. Molten salt reactors may not be a new technology in concept, but they have neither the research behind them nor the operational hours of Pressurized Water Reactors fueled by uranium-235. This has left several outstanding technical challenges. Among them:

- Whether the reactor design should use a one or two-fluid exchanger, both of which present engineering tradeoffs.[324]

- How to minimize corrosion of the reactor moderator (which used to be graphite, but would now be replaced with molybdenum alloys that have far greater resistance to fluoride salts and damage from fast neutrons).[325]

- How to prevent salt freezing if the reactor heat gets too low.[326]

- How to remove excess beryllium from the fluid exchanger.[327]

- How to contain gamma emissions from the uranium-232 within the reactor.[328]

Meeting these challenges is absolutely feasible – as we've solved harder problems at larger scales – but they still must be addressed.

Reactor ignition. Currently, there isn't a standardized method to turn a Molten Salt Reactor "critical" and keep it operating for sustained power. Thorium is an excellent fuel source within breeder reactors, but that reaction has to be started somehow.[329] New reactor designs (of any type) are frequently ignited by an array of neutron sources.[330] But some of the most portable include beryllium and americium,[331] which are safer and less expensive than either radium or plutonium that could also serve in this function.[332] Science and industry will need to determine if these methods work for thorium, and, if not, to identify a method that minimizes safety risk and doesn't come with strict security controls.

Extra proliferation prevention. We've touched on the difficulties inherent to using the thorium fuel cycle to make a nuclear weapon, but any standardized design brought to market should include internal mechanisms to make that even harder. Deliberate material contamination, integrated process software to verify the presence of automated control systems (hash comparison),[333] and "call home" features that remotely alert the manufacturer should unauthorized tampering occur are all available options that can perform this function. Any new nuclear standard must include the most effective anti-proliferation mechanisms as a function of domestic and international law.

Export designs, controls and marketing. Universal Energy and the technology it proposes seeks to create a new market for large-scale energy generation, one with nigh-limitless economic potential in developing and modernizing nations. How we market such technologies, and what controls we place on them, are issues that must be negotiated between private industry, global regulators, and diplomatic and security services. Once such technologies reach greater degrees of maturity, haste becomes critical, as we never want to be playing catch up in a global market against foreign competitors who got the jump on American innovation.

It's well within our capabilities to answer these questions and address these challenges, and do so in a way that revolutionizes our approach to the most powerful source of energy we have ever discovered. If we were to truly invest in Molten Salt Reactor and microreactor technologies, it would lay the foundation for a clean, sustainable, affordable, and rapidly-deployable means of baseload power generation nationwide. We could advance our world and support the American economy for generations in one stroke. And while this would present yet another tool, alongside renewables, to dramatically multiply our potential for energy generation, it also would serve a more direct and even more important function: excess energy for fresh water and hydrogen fuel.

I believe that water will one day be employed as fuel, that hydrogen and oxygen which constitute it, used singly or together, will furnish an inexhaustible source of heat and light.

- Jules Verne

Chapter Five: Water and Hydrogen

Of the resources facing impending scarcity, water and fuel are especially important as they respectively make life possible and power vital aspects of our advanced economy. They are also critical to growing, processing and delivering food. Consequently, these resources are the most likely to drag us into conflict when they run low – which is unsurprisingly why we've been fighting over them throughout much of history. Universal Energy seeks to solve this problem by generating enough energy to synthesize these resources, further enabling us to produce synthetic building materials on an effectively unlimited scale.

With renewables and thorium in place, the next focus is seawater.

It's important to note in this context that humanity isn't facing a water crisis in abstract; we're facing, specifically, a freshwater crisis. 71% of the planet's surface is covered by water, yet less than 2% of that water is fresh – and 80% of that freshwater is locked in polar ice.[334] For consumption purposes, that last 20% of freshwater - 0.4% of all water on our planet – is the only percentage that has historically mattered. Thanks to modern seawater desalination technologies, that is no longer true today.

The desalination of seawater is a well-proven concept.[335] The same is true with extracting hydrogen fuel from water via electrolysis, as running an electric current through water chemically separates it into hydrogen and oxygen.[336] Yet both processes require lots of energy, which has traditionally made them expensive. The energy generated by LFTRs and municipally integrated renewables removes energy cost as a major factor, allowing us to extract fresh water and hydrogen fuel from seawater on a massive scale.

This begins with a system known as a Multi-Stage Flash Distillation Chamber (MSFD),[337] as seen in this diagram:

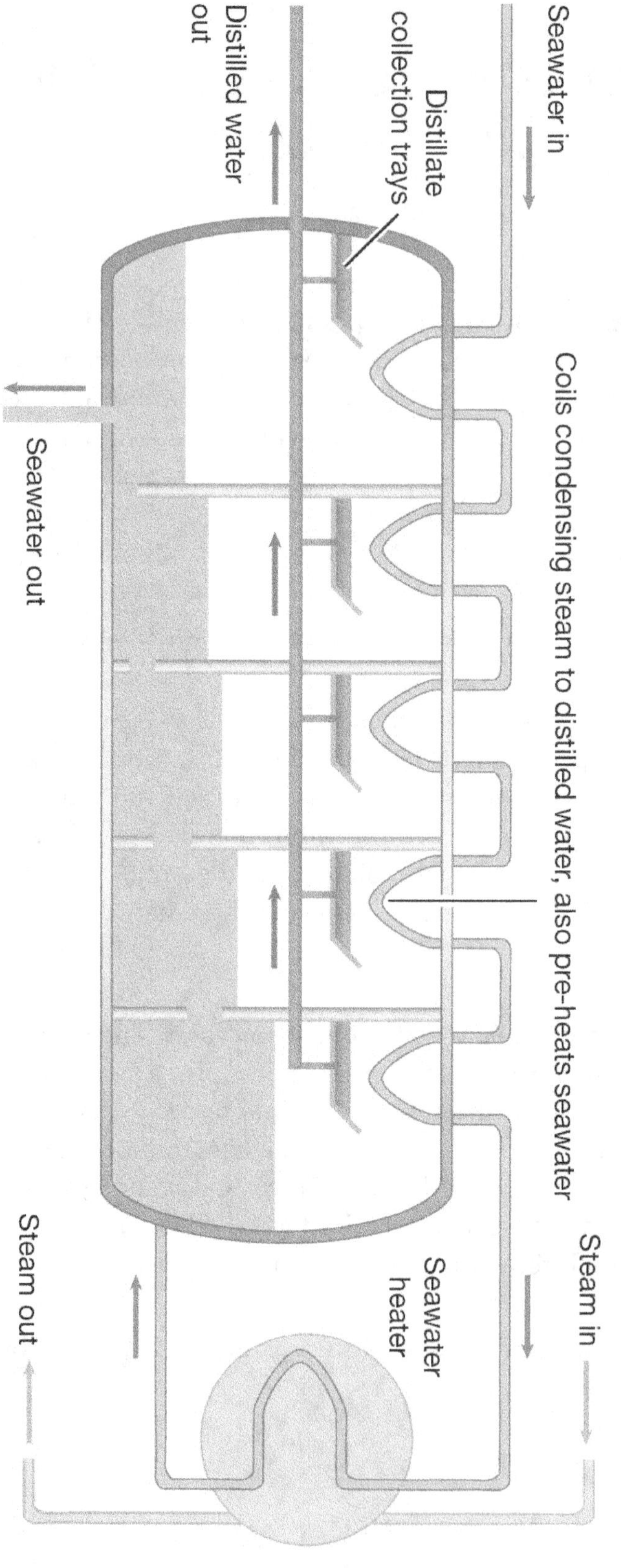
Seawater in
Distillate
collection trays
Distilled water
out
Seawater out
Coils condensing steam to distilled water, also pre-heats seawater
Steam in
Seawater
heater
Steam out

Flash distillation for desalination of seawater [338]

An MSFD facility features a series of interconnected chambers (referred to as "stages") set at varied temperatures and pressures relating to the boiling point of water. Seawater is pumped in through one end and heated to reach a certain temperature. Once at the right temperature, it's then pumped into subsequent stages, each of which has a different temperature and pressure. This process forces seawater to instantly flash-turn to steam, which is then collected via a condenser and turned into liquid fresh water.

From there, the remaining hot brine is pumped back into the system to counterflow with the influx of cold seawater, helping to heat new seawater and recycle a majority of heat energy in the process.[339] What waste remains is essentially very salty water, which can be evaporated to leave only salt.

Diagram of counterflow heat exchanger:

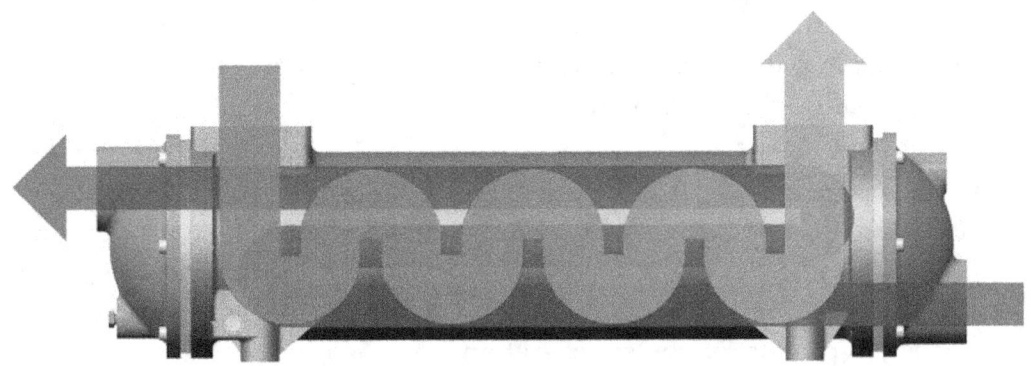

Image source.[340]

MSFD is common today; 60% of all desalinated seawater in the world[341] is produced through this method, and more than 18,000 MSFD facilities exist globally.[342] However, MSFD is an energy-intensive process with high operating expenses, making it more difficult to justify at larger scales. Universal Energy seeks to lower the cost of multi-stage flash distillation, giving us the ability to desalinate unlimited amounts of fresh water as a function of the framework.

The use of the word "unlimited" here bears special mention. There is a vastness to the oceans that "71% of Earth's surface" does not give justice to. Only 0.4% of water on Earth is both fresh and accessible, and that's been enough for humanity's entire existence until now. With that in mind, we would have to increase our water consumption thousands of times for desalination to even measurably impact sea levels – especially since the water cycle would eventually return all desalinated water to the ocean. And even if we did somehow lower sea levels, it would be to our benefit anyway, as sea levels are rising due to climate change.[343]

Valid questions exist as to the environmental impact of MSFD, both on local ecosystems and on the ocean as a whole.

Conceptually, MSFD doesn't do anything to the environment except place a pipe in the ocean and suck in seawater. Rather than one large pipe that might risk capturing marine life, a smaller series of filtered pipes designed to reduce ecological impact can be used.

As these intake systems can operate twenty-four hours a day, a large volume of water can be secured through a slow yet steady flow – meaning it does not need to be strong enough to measurably interfere with the local ecosystem. The greatest environmental impact of MSFD plants today usually involves the dumping of waste brine back into the ocean with chemicals[344] – steps that need not be taken with modern facilities for two primary reasons:

1. Chemical pretreatment of seawater is not as necessary in modern MSFD plants. Older models have sometimes introduced chemicals to "soften up" water, making it less corrosive and easier to heat, but modern polymers can resist corrosion[345] and Universal Energy provides ample inexpensive heat energy as a byproduct of power generation.

2. Currently, some MSFD facilities pump waste brine back into the ocean, which raises local salinity levels and can cause environmental damage. With Universal Energy, we have plenty of excess energy to boil off waste brine and leave only salt as a byproduct.

That latter point presents an important question, though: if we were to desalinate seawater on a large scale, how do we deal with all the leftover salt? The answer? Simply sell it.

Let's say our implementation of Universal Energy desalinated a total of 500 billion gallons of water annually. Each gallon of seawater contains roughly 4.5 ounces of salt.[346] Therefore, 500 billion gallons of seawater would contain 2.25 trillion ounces of salt, or 140.63 billion pounds. That's a lot of salt – but our national salt consumption is equally high.

According to the U.S. Geological Survey, the United States consumed 69,500 thousand metric tons of salt in 2015 for all purposes.[347] At 69.5 million metric tons, that translates to 153.2 billion pounds of salt. This means a 500-billion-gallon annual desalination effort would yield around 91% of our annual salt consumption. At an estimated price of $40-$50 per long tonne (2,204lbs) 140.63 billion pounds would yield roughly $2.5 billion in profits from annual salt sales (assuming $40 per tonne).[348]

With these concerns addressed, MSFD technologies can be harnessed to produce unlimited amounts of fresh water for any use, with negligible fiscal and environmental costs. This would effectively end water scarcity as a concept. And we can then do the same for fuel.

Hydrogen Fuel

Hydrogen is the most abundant element in the universe.[349] It's light, clean, and highly combustible, with an energy-per-mass ratio greater than any known fossil energy source.[350] This makes it a flexible alternative to petroleum if we go about sourcing and storing it in the right way.

Hydrogen production is currently a $100+ billion industry,[351] yet most current methods of hydrogen production involve extracting the element from oil or coal through high-temperature steam reformation[352] – a process that is both environmentally destructive and will likely prove untenable once fossil fuels eventually become more scarce.[353] In a world with effectively unlimited cheap energy, electrolysis becomes a significantly more attractive method.

Electrolysis is a process that introduces an electrolyte and an electric current strong enough to break molecular bonds of water, chemically separating it into oxygen and hydrogen gas. Like Multi-Stage Flash Distillation, it's not a new concept. Electrolysis has been in use since the 1700s to extract various substances, hydrogen among them.[354] Nor is it particularly complex; you could set up a simple facility in your garage, if you wanted to (just don't smoke).

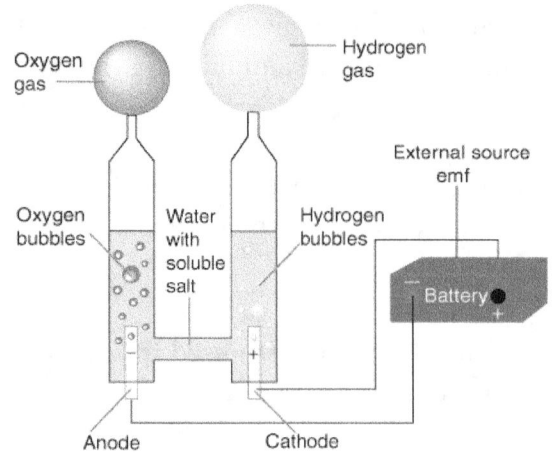

But to produce enough hydrogen for use as a viable fuel on a nationwide or global scale, an industrial setting would be necessary. Commercial hydrogen extraction through electrolysis has traditionally proven expensive,[355] but Universal Energy mitigates this cost factor as a byproduct of generating heat energy from thorium, making the production of hydrogen through electrolysis perfectly viable. Once extracted from water, hydrogen can be harnessed to power an array of systems and processes, to be discussed throughout the rest of this writing.

But even so, challenges to using hydrogen remain as production is only one half of the equation. The other is how to contain, transport and stabilize it – considerations of no small significance. Because hydrogen is highly reactive, it is easily contaminated as it naturally bonds to other substances.[356] And due to its volatility, it has usually required storage in containment tanks at high pressures. While metal tanks work in a laboratory or industrial setting, the weight of these tanks and the safety risks presented by the explosive nature of compressed hydrogen have made this approach questionable for civilian use. Thankfully, recent advancements have given us new alternatives, such as:

Graphene storage: Storing and transporting hydrogen in compressed form requires immense pressure, on the order of 482–690 bar (7,000-10,000 PSI).[357] Currently, this is only possible through metal tanks that have limited utility due to increased bulk and weight. Through Universal Energy, we'll have better materials.

Although we'll be reviewing materials further in Chapter Ten, one of the most noteworthy in the context of hydrogen is graphene,[358] which serves several important roles in the Universal Energy framework. Conceptually, graphene is a one-atom-thick sheet of carbon that is structured in a way that is both ultra-strong and ultra-conductive.[359] This allows graphene to *both* function as an efficient battery[360] and also a structural material – one that is 200 times stronger and six times lighter than steel.[361] As it can be made paper-thin while remaining flexible, graphene is well-suited to make storage tanks for hydrogen in vehicles and other machinery. Just as importantly, these storage mediums can be amorphously shaped, providing greater flexibility in how they integrate with a fuel supply.[362]

Synthetic oil: Universal Energy's approach to solving resource scarcity is based in large part on replacing oil as a fuel source, due both to its finite supply as a fossilized product and its contributions to climate change. But oil has other important uses: it's essential for making plastics and synthetic materials, and it's a critical ingredient for chemical engineering.[363] Oil is type of chemical known as a hydrocarbon, and hydrocarbons are useful for both organic chemistry and fuel for combustion. Oil is the abundant hydrocarbon of our time, so it's what we use. But that doesn't *have* to be the case, especially as oil eventually becomes scarce and thus expensive in the future.[364]

With an abundant supply of hydrogen, we can use it to manufacture synthetic hydrocarbons for lubricants and chemical stabilizers[365] as well as specialized fuels for sophisticated applications like aircraft and rockets.

We can also use synthetic hydrocarbons for long-term hydrogen storage. Once processed into a solution that stays liquid at normal pressure, we can store an effectively unlimited amount of hydrogen in tanks that don't require compression and would work similarly to how we transport liquid fuels today. And unlike the environmental damage caused by crude oil drilling, refining and transport, synthetic oil can be processed at facilities that remove ecologically hazardous steps from the manufacturing chain. This becomes all the more important when using hydrogen as a medium to create fuel from atmospheric carbon capture. (We'll go over those details next chapter).

Fuel Cells: A hydrogen fuel cell is a means of generating electricity from a chemical fuel, in this case, hydrogen. In practice, this allows fuel cells to function as emission-free batteries. Fuel cell technology has been around since the 1950s and has steadily grown since then into a billion-dollar market,[366] with several proven designs powering myriad industrial applications.

In a dynamic with reduced energy costs and improved manufacturing processes, fuel cells become less expensive, easier to build and easier to expand into varied sectors of our economy.

HOW DO HYDROGEN FUEL CELLS WORK?

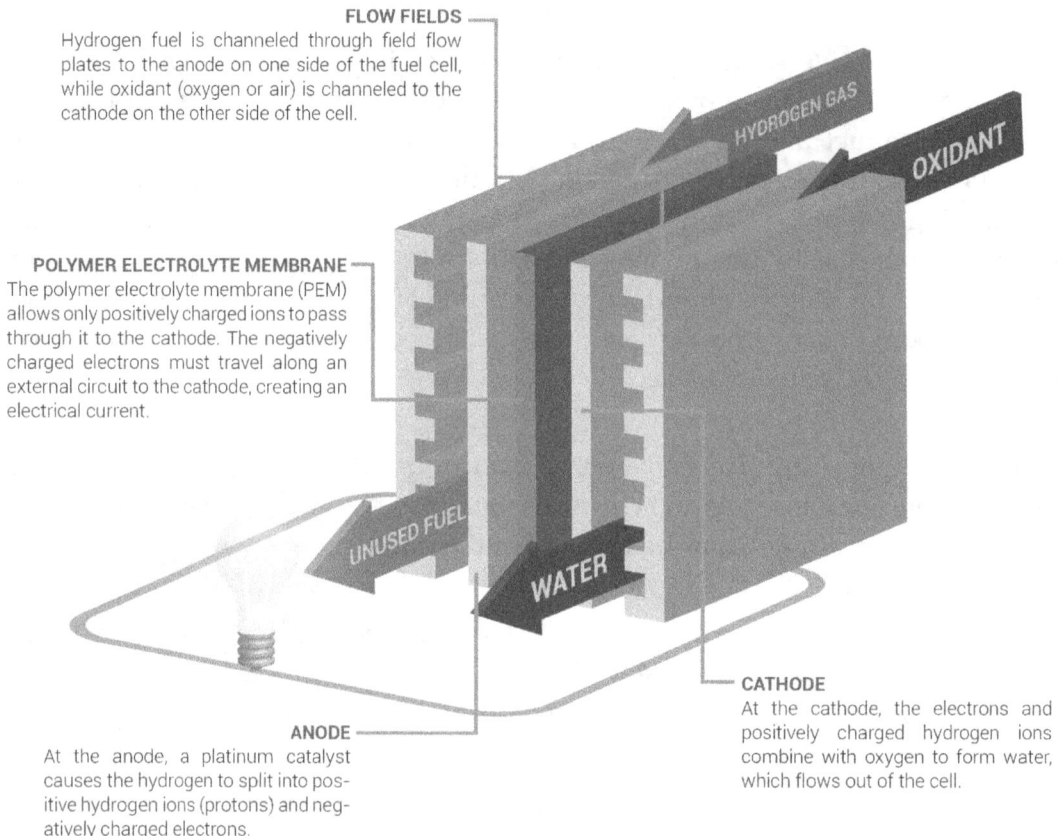

Image source.[367]

Although hydrogen fuel cells are often looked to as a replacement for oil, they also have potential to power remote areas that are environmentally hostile to power generation. There are several circumstances where it's not feasible to rely on local power systems and where solar isn't possible (war/disaster zones, remote research facilities, long-voyage ships, space travel, etc.), yet fuel cells can provide energy as long as a supply of hydrogen exists. Future advances in graphene battery technology can also complement this possibility, allowing for robust energy storage even when far away from civilization's amenities.

Adding Things Together

With municipally integrated renewables doing their part to power cities and reduce the energy demand they place on regional electric grids, employing the baseload electricity generation and excess heat energy of LFTRs to both desalinate seawater and extract hydrogen becomes a routine deliverable of the framework. Yet that merely reflects only the *products* of this approach – its underlying strategic value is itself a step further. As we've discussed previously, a key component of Universal Energy is the intention to deploy energy technologies strategically so that they can *work together* as a team to become greater than the sum of their parts in operation.

The next stage of this goal comes through conceptual "CHP Plants" – (Combined Heat and Power) – which are modular arrays of the technologies thus-far discussed that can be installed rapidly in varied configurations for specific applications. As we'll see later in this writing, these applications might include atmospheric scrubbing of greenhouse gasses, waste processing and recycling, large-scale ocean cleanup or supplemental energy and resource production. Yet they each would deliver their intended goals as a system that's standardized and cogenerative in design, enabling turnkey deployment of energy generation and resource production to solve most any problem caused by the absence of either.

The strength of a structure does not stem from a single beam or fastener – it stems from the sum of its components in perpetual reinforcement.

This mindset is the basis for everything lasting that humans have ever built.

In turn, it will be the mindset for everything lasting that we will ever build.

Chapter Six: Cogeneration

Universal Energy's technologies generate a lot of energy, but the driving mindset behind their deployment is their ability to cooperate *by design*. This concept – the idea of diverting the waste energy of one technology to help power the functions of another – is commonly referred to as "cogeneration," or "Combined Heat and Power" (CHP). But it hasn't really been a central component of past power plant design, and has only recently started to gain prominence.[368] In general, our current power infrastructure just "there," decentralized, ad-hoc systems that are each custom-designed and built to order. They rarely work together.[369] They barely talk to each other.[370] They are built with non-standardized components and powered by non-standardized fuels.[371] Just as importantly: the energy they generate is usually devoted only to powering electric generators, any excess is written off as "waste" instead of used for auxiliary or offsite functions.

This is a squandered opportunity on a monumental scale.

Most power plants today have an efficiency of around 33%.[372] This means 67% of their generated energy is discarded in the form of heat that either dissipates into the surrounding air or is absorbed into the ground. **That's more than twice the energy used to generate electricity in the first place**. If we're going to build an advanced and clean energy future, we must improve the efficiency and utility of our power infrastructure over the long-term.

The problem? Entropy and thermal loss are unavoidable byproducts of energy transfer – which means we're probably never going to be able to build power plants that operate at superb levels of efficiency. But we can harness waste energy to power auxiliary functions at low additional cost. This is something Universal Energy seeks to employ **at the design stage** so that *every aspect* of power-generation is engineered to maximize cogeneration from the ground-up. And not just within a given facility, either, but also externally to other facilities that could be modularly integrated within an indefinitely scalable standard.

This writing commonly refers to these facilities as "CHP Plants" (Combined/Cogenerative Heat and Power), but their defining characteristic is simply a deployment of energy technologies designed to maximize efficiency and operational capability by leveraging waste energy for other useful applications.

While this design is significantly more aggressive than even our most ambitious approaches to energy efficiency, it is not inventive in and of itself, and has several proven demonstrations of its concepts within both power generation and manufacturing.[373] One notable example is the National Renewable Energy Laboratory's REopt™ model (Renewable Energy Integration & Optimization), which is designed to identify renewable and/or energy efficiency opportunities within power infrastructure. One of their flagship projects in Arizona (home to the nation's largest nuclear reactor)[374] involved the integration of light-water nuclear with renewables to efficiently desalinate seawater into hydrogen.[375]

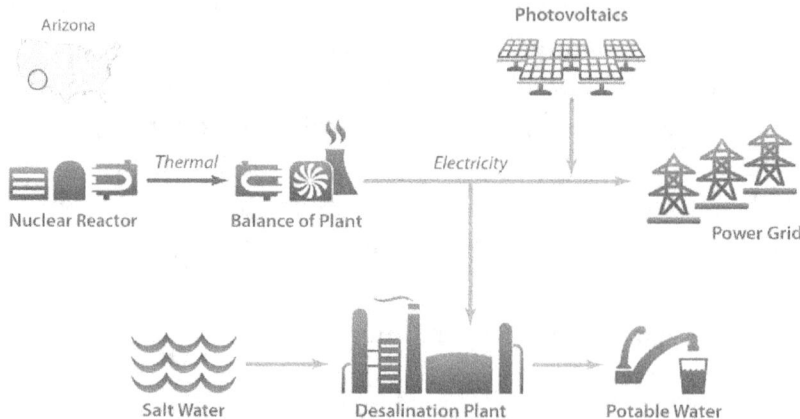

Source: NREL. A desalination nuclear-renewable hybrid energy system uses electricity from a nuclear reactor and solar photovoltaics. Illustration from Mark Ruth, NREL

Another example is China's TSMR project in Gansu province that we reviewed in Chapter Four.[376] Their cogenerative deployment leveraged up to 100 megawatts of clean energy to power resource-producing systems within fresh water, hydrogen fuel and chemical hydrocarbons.[377 / 378]

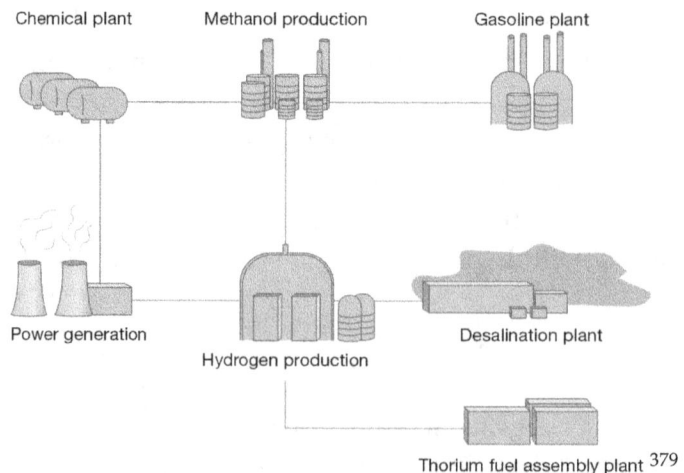

Thorium fuel assembly plant [379]

Russia's thorium project conducted by their School of Nuclear Science & Engineering of Tomsk Polytechnic University further leverages cogeneration to desalinate seawater into fresh water and hydrogen fuel.[380] Cogeneration has also been leveraged on varied scales with other energy technologies[381] and has been committed into official policies of the European Union[382] and the UK[383] that generally require its inclusion to participate in energy tax incentives.[384] The underlying benefit of cogeneration, therefore, is both proven and by itself isn't especially novel as applies to energy generation.

But cogeneration today – even if applied to strong effect – still reflects the same foundational shortcomings of our current energy schema. *Each effort looks for practical – yet piecemeal – improvements within unique, ad-hoc systems that are designed and implemented as such.* Even if NREL's REopt engineers retain the best tools and brightest minds in the world, everything they do for one application in one area will have to be completely re-done from scratch in another because the nature, capabilities, and limitations of the technology deployment hinge on energy systems that are essentially *made to order*.

By leveraging today's manufacturing capabilities to mass-produce identical power-generating systems that are modular and standardized, Universal Energy seeks to promote an energy mindset that's cogenerative-first. This means energy technologies are designed *first* – from blueprint to integration – to work cooperatively with others on a *modular standard*. That, then, makes the framework indefinitely scalable because identical power modules can be extended, integrated and/or swapped on-demand.

This helps change the fundamentals of the energy question from "how can we identify individual efficiency improvements within unique power-generating systems" to questions both more practical and expansive in vision.

For example, let's say we were to ponder the following tasks:

1. What would be required to build a power facility in mid-coastal California that generates 300 megawatts of electricity, desalinates 40 million gallons of seawater and produces two metric tons of hydrogen per day?

2. With that known, how easily could we upgrade that facility's desalination capacity to 60 million gallons, add on another ½ ton of hydrogen production and 100 megawatts more electricity?

3. In case the upgrades in task #2 exceed an allocated budget, what deliverables would respectively meet 35% and 75% of this target?

4. How much land would this require at what size of building envelope and what terrain limitations would apply?

5. All aspects considered, what's the overnight cost of these deployments with a high degree of certainty – plus or minus no more than 5% - all regulatory and permitting aspects considered?

In a world where energy systems are unique and made to order, the presence of these very questions sustains the business models of global consulting conglomerates. Their answers take months to derive and cost millions of dollars.

Yet in a world where capabilities of modular, mass-produced systems are known clearly, and their deployments both integrate and scale with others by design, answering these questions becomes far easier.

The unknowns are removed from the equation because each energy system is model-identical in similar application to a D-cell battery – and can couple, swap or decouple from others just like the very same. Everything from power requirements, performance specifications to product lifetime and physical footprint can be granulized in database tables that could be used to generate dynamic reports with the click of a mouse.

While the engineering complexities of each system of course remain, they're also contained within the module. They do not permeate into the module's capabilities or connection interfaces. Nor do they need to. Few of us know how to design or build a lightbulb, battery, USB hard drive or computer monitor, for instance.

We simply know how it's supposed to work and how to replace it if it doesn't. And, if we want a second one, simply get a second one and plug it in. This is how the world of technology works in nearly every commercial sector. A rare exception is power-generation.

The time has come for that to change.

As applicable to Universal Energy, we'll take a brief overview of how the concept of cogeneration would apply to CHP Plants.

Every technology within the framework deals with electricity, heat, and/or water. When these technologies are deployed in close proximity to one another, they can easily be tied together to harness waste energy.

The central technology behind CHP Plants are Liquid Fluoride Thorium Reactors (LFTRs). As far as power plants go, LFTRs get hot – quite hot – (900 °C / 1,600 °F).[385] As the reactor's heat exchangers are well-separated from anything radioactive and its electric generator is driven by an inert helium loop, waste heat can be harnessed as central energy source for secondary functions.

The first function is seawater desalination.

Normally, a Multi-Stage Flash Distillation (MSFD) facility counterflows waste hot brine with cool seawater to preheat it for desalination, recycling energy in the process. But by directly integrating seawater intake with the heat exchangers of LFTRs beforehand, the seawater can be brought to a boil before it even reaches the facility. This increases efficiency in both systems and translates to cost savings. Further, it sterilizes any microbial life within the seawater – further preventing concerns with the aquatic transfer of invasive species.

The second function is hydrogen production, as there is already an ample supply of heat, electricity, freshwater and electrolyte (ocean salt) on site to extract hydrogen into fuel or base supplies for synthetic materials.

The third function is to use excess energy for atmospheric scrubbing of greenhouse gasses and air pollution, and as we'll review later in Chapter Ten, waste processing via plasma gasification.

This becomes especially important in contexts of addressing climate change, because it is the most direct and effective tool we have available to physically remove greenhouse emissions already present in our atmosphere – and perhaps the most vital method available to reduce the circumstances fueling an inexorably warming planet.[386]

Sample CHP Plant Configuration. Note citation for link to larger version.[387]

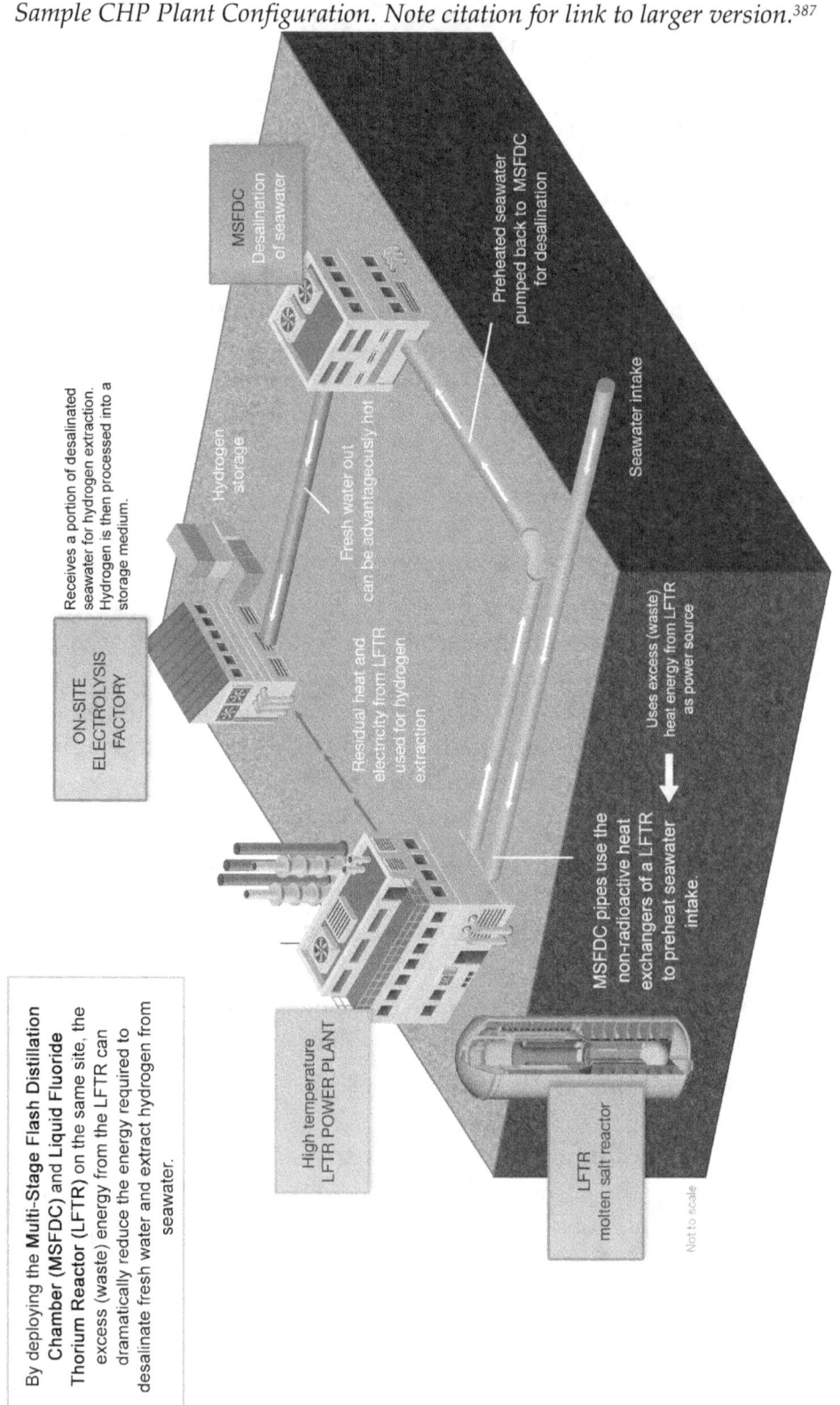

Atmospheric Scrubbing and Waste Processing

Climate change is a consequence of the concentration of greenhouse gases in our atmosphere. When the sun's energy hits Earth, a portion of that energy is absorbed into the ground. Greenhouse gases, like carbon and methane, block more heat from radiating into space than nitrogen, which is the primary gas in earth's atmosphere.[390] In function, this works much like a bedding blanket: the thicker the blanket, the more heat that blanket retains. The more heat it retains, the warmer you become – and this is happening on a planetary scale.[391] This problem adversely affects weather and long-term polar ice melt, but also exacerbates droughts, wildfires, smog and air pollution – saying nothing of mass human migration. Alongside resource scarcity, climate change ranks among the most serious problems of our time, with potentially catastrophic consequences for our civilization should it remain unaddressed.[392]

Worse, solutions to this problem that involve scaling back fossil fuels have proven politically precarious,[393] as the oil and gas industry is a major driver of the global economy and many entrenched power players owe their wealth to its lucrative returns. Even if this wasn't the case, the greenhouse gas emissions from manufacturing, agriculture and global commerce would still persist even if our fuel supply chain was carbon neutral.[394]

Getting ahead of the impacts of climate change are laudable efforts. But they, in all truth, needed to begin decades ago when the first warnings were sounded (and unfortunately ignored).[395] Humanity has already passed an initial carbon tipping point of 400 parts-per-million[396] and global fossil fuel usage and consumption is accelerating[397] – as is our population – meaning that even though we *recognize* the problem of climate change, it's getting continually worse even as we attempt to slow it down.

Switching to a clean, carbon-neutral energy framework like Universal Energy is an essential part of any strategy to avoid the calamitous results of climate change. But even if it was implemented as proposed it wouldn't by itself clean the atmosphere of the greenhouse gasses already present. It can, however, power unique systems designed to accomplish this very task.

An atmospheric scrubber is a machine that strips greenhouse gases from the atmosphere, either by a chemical or mechanical method.[398] A primary example is Direct Air Capture, which blows air through towers containing a solution that reacts with certain compounds, removing them to form a substance that can be further processed into materials – including fuel.[399]

A Vancouver, Canada-based company named "Carbon Engineering" has patented several Direct Air Capture methods to isolate carbon from air through modular fan assemblies that work in unison with each other.[400] Their current pricing models assess a rate of $100 per-ton of CO_2 captured under combined-cycle natural gas, which costs significantly more at present than Universal Energy's target of 2 cents per kilowatt-hour.

Image source: Carbon Engineering.

As with other Direct Air Capture technologies, water and fuel are produced as deliverables alongside renewable electricity through cogenerative functions.

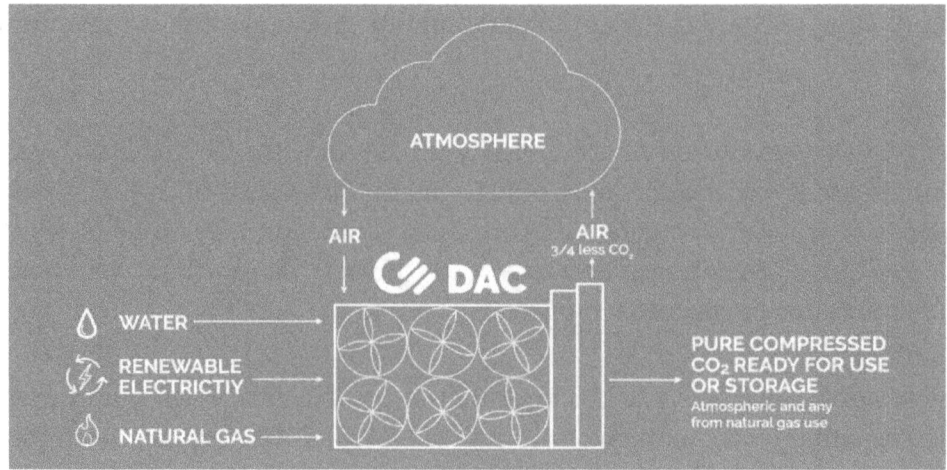

Image source: Carbon Engineering.

Another noteworthy component of this system is use of a chemical cycle that both uses non-toxic materials and is functionally closed-loop,[401] meaning that the chemicals (and thermal energy) used for Direct Air Capture is continually re-used and does not require frequent refueling over time. This makes the method both environmentally friendly and indefinitely scalable.

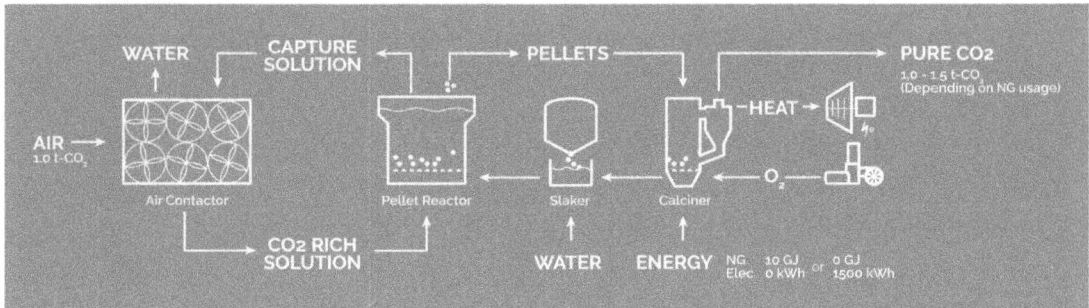

Image source: Carbon Engineering.

Several other ventures have come to market with similar technologies. Iceland's Climeworks' models, for example, are both modular and scalable with their largest units being capable of capturing nearly 5,000 kilograms of carbon per-day.[402] In addition to standard carbon capture, they also are able to condense captured carbon into usable fuels[403] or solid carbon for use in materials.[404]

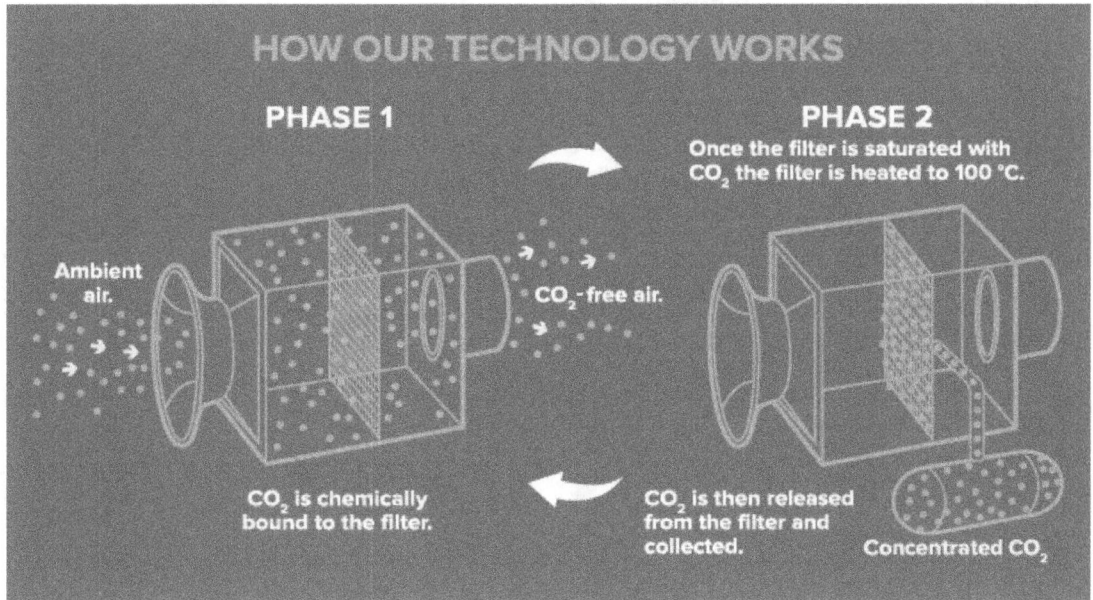

Image source: Climeworks Incorporated.

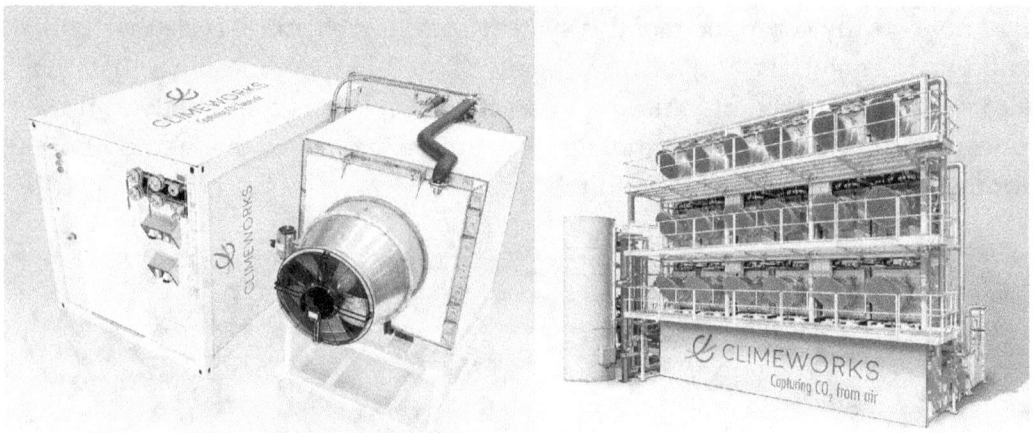

Image source: Climeworks Incorporated.

Other companies, including Silicon Kingdom Holdings in the United Kingdom and Alabama-based Global Thermostat have obtained patents for similar models that perform similar functions in abstract.[405]

The costs of these systems stand to fall significantly over time through continued investment and research, and their operational costs stand to fall even further if implemented within a modular energy framework. And while cost reductions are realized maximally if integrated within CHP Plants, that shouldn't overshadow other highly significant benefits that occur as well.

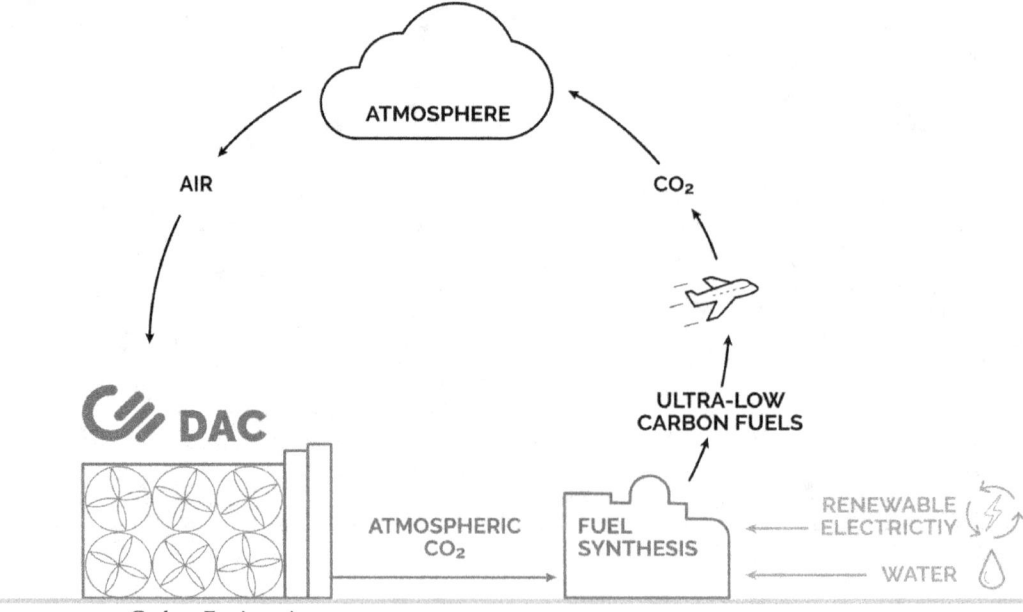

Image source: Carbon Engineering.

Because captured carbon is stored on-site, along with molecular hydrogen from electrolysis facilities, there is ample source material to make unique hydrocarbons with low-carbon emissions.[406] This means that not only are we able to produce hydrogen fuel through CHP Plants, we're also able to further produce specialized hydrocarbons that have applications within chemical engineering, fuels for motor vehicles, heavy machinery, commercial aviation and aerospace.

Further, as any emissions from these applications (especially commercial aviation[407]) can be captured by these facilities and re-converted into usable fuel, we're able to leverage the combined benefits of CHP Plants to not only help clean our atmosphere but present yet another avenue for sustainable fuel production.

Due to the modular nature of this approach, CHP Plants can be deployed both rapidly and at scale – also in a way that is location-agnostic. Building hundreds or thousands of them worldwide is limited only by vision and capital. Considering that their overnight cost is not prohibitively expensive[408] in the context of nationwide power[409] (or avoiding the consequences of resource scarcity and climate change), it's an ideal approach that can cascade in effect to present immense social benefits practically anywhere on the planet.

In an ideal scenario, prefabricated facilities could be delivered turn-key, initiated in a matter of weeks for instant energy and resource production. With an effectively unlimited source of clean energy, problems of nearly any scale – even planetary – become solvable. And the more modular and adaptable these sources of energy become, the quicker they can arrive to deliver solutions that mitigate the impact of energy, climate or resource-driven social problems.

Symbiotic, cogenerative energy deployments are our next stage in the evolution of power generation not just because they solve several problems at once. They also address physical and economic challenges presented by our current ad-hoc approach to power generation:

1. Constructing a single facility that can generate multiple types of energy and resources at one location is significantly less expensive than if these facilities were located far from each other.

2. Consolidating multiple functions into a single facility avoids expenses of transmission and transportation, increasing overall efficiency.

3. Symbiotic design helps establish ideal standards for implementation and operation. This reduces costs and helps encourage greater adoption of the Universal Energy framework.

4. The fresh water produced from CHP Plants can **come out hot** – which will become important as we look at the concept of the National Aqueduct within the next few chapters.

What we can have with such symbiotic deployments of advanced energy technologies – what we can have today – is something that we have never before had in our history: the ability to synthetically, sustainably, and inexpensively produce as much electricity, water and fuel as we could ever need. All this while de-polluting our environment and combating climate change as a dedicated effort. And once we have these functions in hand, we can look beyond them, using any excess waste energy to produce resources to even greater extents.

While we began with electricity, water, and fuel out of necessity, we can extend the framework further in the fields of agriculture, chemical engineering, recycling processes and next-generation building materials. The next step towards that future comes from the National Aqueduct – a vital function of Universal Energy that ties each of its core technologies together into a nationwide network of energy and resource abundance.

Water, the hub of life. Water is its mater and matrix, mother and medium. Water is the most extraordinary substance. Practically all its properties are anomalous, which enabled life to use it as building material for its machinery. Life is water dancing to the tune of solids.

- Albert Szent-Gyorgyi

Chapter Seven: The National Aqueduct

So far, the Universal Energy framework enables production of three out of five critical resources: electricity, water and fuel. But while electricity and fuel can be transported relatively easily, water is a different story. We might be able to leverage Multi-Stage Flash Distillation technology to produce a lot of water at the coasts, but how do we get it inland?

Universal Energy's proposed answer to that question is the National Aqueduct, a system that arguably functions as the heart of the framework. In concept, it's a nationwide array of modular, above-ground pipelines and storage facilities intended to transport billions of gallons of water to any location in the country. In function, it serves four critical roles:

1. Completely solve drought and water scarcity anywhere the network serves.

2. Operate as a fourth energy source alongside renewable-integrated cities, LFTRs and CHP Plants.

3. Serve as a gigantic nationwide battery for renewable energy.

4. Provide endless, sustainable irrigation for agriculture and synthetic material production.

Thanks to three things we've been perfecting for the past 50 years: oil pipelines, high-voltage power lines and interstate highway networks, not only do we have the free space and wherewithal to build this system, we've already built it for other substances – at higher stakes and with higher difficulties. To elaborate, consider a series of three images. First, if you recall from Chapter Two, we see that our nation has a highway system connecting nearly every area of our country:

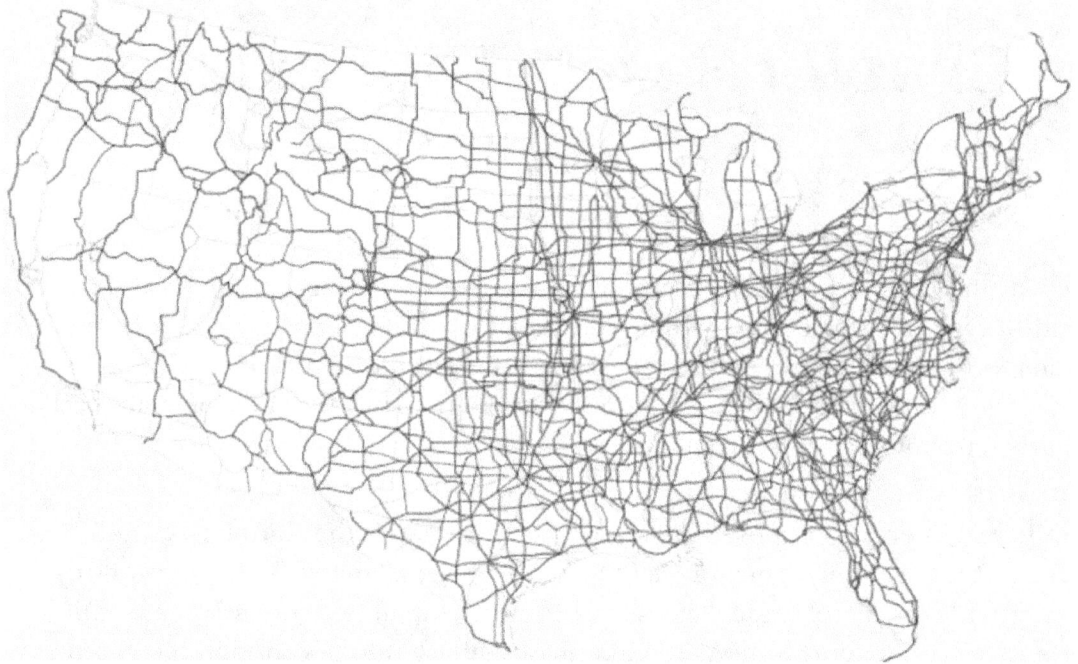

Second, consider a map of nationwide power transmission lines:

Third, consider a map of nationwide fossil fuel pipelines and refinery networks:

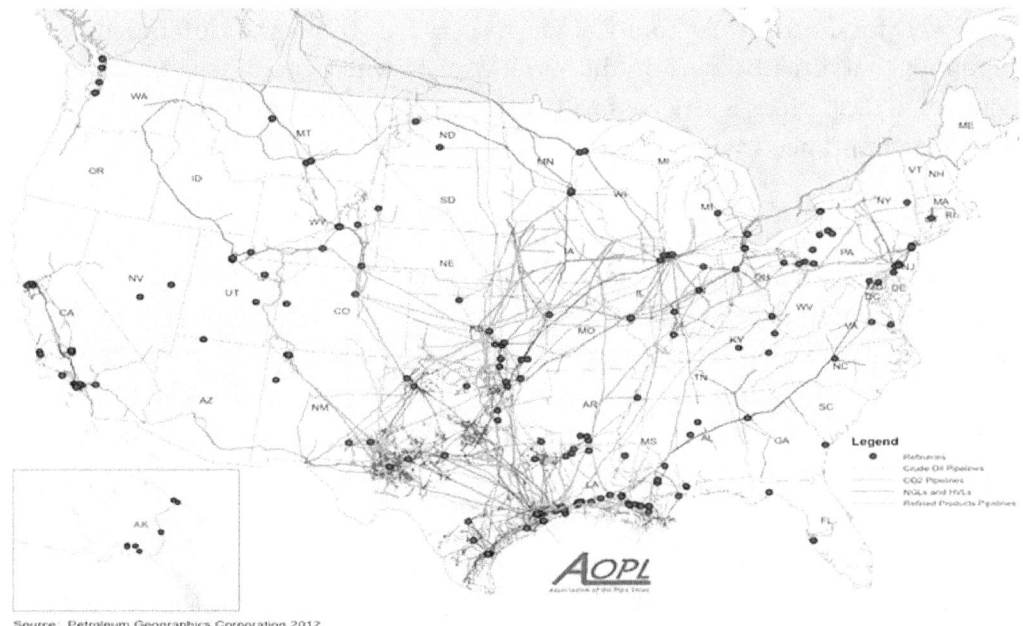

Source: Petroleum Geographics Corporation 2012

From these maps, we can derive two important conclusions:

1. **Highways and high voltage power lines give us plenty of free space to run water pipelines.** Our road networks provide ample open space to install solar panels *while removing the requirement to buy land*, as this land is generally owned by public services that have exclusive authority to build on them.

   Roads and highways also have clearance at each side and tend to be flat and straight – a trait usually shared by high voltage power line networks. This gives us thousands of miles of open, unused space to build a National Aqueduct. As this space has been cleared of potential obstructions beforehand, construction in these locations has far fewer obstacles than commercial land parcels.

   Additionally, their close proximity to power systems (either integrated renewables, LFTRs or CHP Plants) gives the National Aqueduct plenty of energy to power sensors, pump stations, purification mechanisms and heating elements to help **keep water hot**. This hot water functionally acts as a "battery" – further discussed in the next chapter – and the potential energy stored within it can be harnessed alongside pipeline-mounted solar panels and turbines to generate immense electricity.

2. **Running water pipelines is feasible.** We know that the National Aqueduct will work as described because we've already built a similar network of pipelines for fossil fuels today. Our nation has thousands of miles of oil pipelines that already work in the same way as water pipelines would in this context, and oil pipes are necessarily built to a higher environmental standard than we would need with water.

   Water pipelines can come factory prefabricated and be designed for rapid, modular construction. Environmental risks are reduced, as the only substance a leaking water pipeline would spill is fresh water. In tandem, modular deployment and lower environmental risk would allow us to build a water pipeline network at a lower overall cost than with oil pipelines. We can also use the lessons we've learned with oil pipelines to get a head start, as the expertise needed to plan and build such a pipeline network already exists.

Much of the work involved with developing a National Aqueduct has already been done for us in terms of research and development, engineering, and methods of implementation. But to have the National Aqueduct accomplish its intended goals and meet our needs in full, we'll need to establish a few requirements for the system:

Efficiency, reliability and affordability: Humanity has a fresh water requirement of trillions of gallons a year. While Universal Energy is capable of producing that much water in abstract, delivering it with any effectiveness must be efficient and inexpensive. A water delivery system must also be reliable, as any given industry or city can't depend on an external water source if its reliability is questionable.

Scale of delivery: Whatever system we use to transport fresh water must work over thousands of miles, as water must be delivered from the coasts to areas deep inland. And as the land we have to work with isn't flat with consistently warm weather, this system must also be deployable over varied terrain and climates – especially cold climates.

Control of operation: A water delivery system must be locally controlled, as a centralized control structure would be incapable of efficiently managing the water requirements of every agricultural and population center throughout the country. There must also be redundancy in the system – for example, allowing a city in the central United States to receive water from multiple routes in case one becomes incapacitated by some unforeseen event, such as a tornado or earthquake. This calls for a "smart" system approach that would have the ability to both monitor and manage how water is distributed from the point of desalination to the final point of consumption – both locally, regionally and nationwide.

Modular construction and ease of maintenance: Modularity is essential to ideal system design, providing benefits and cost savings in terms of construction time, standardization, reliability and maintenance. Any system to transport water would have to meet these standards by allowing its components to be installed and/or replaced rapidly by design.

To meet all of these requirements, the National Aqueduct would be made of a "smart grid" of above-ground pipelines, storage tanks, and pump stations that would transport desalinated fresh water from the coast to any area inland. These pipelines would feature interior turbines and exterior solar panels that would generate electricity, a portion of which would then be used to keep the water supply hot to thermoelectrically generate power at night. If built, the National Aqueduct would enable us to provide for our national fresh water needs without having to rely on local water sources ever again.

That statement is worth repeating: The National Aqueduct, in conjunction with CHP Plants, would allow us to have unlimited fresh water. And that water would never need to come from the ground, a lake, or a river, unless we wanted it to. This would give natural water sources time to replenish, decreasing much if not all of the drought impact we've been experiencing as of late and benefiting the environment as a whole.

The National Aqueduct's system consists of four primary components: production, transmission, storage and control:

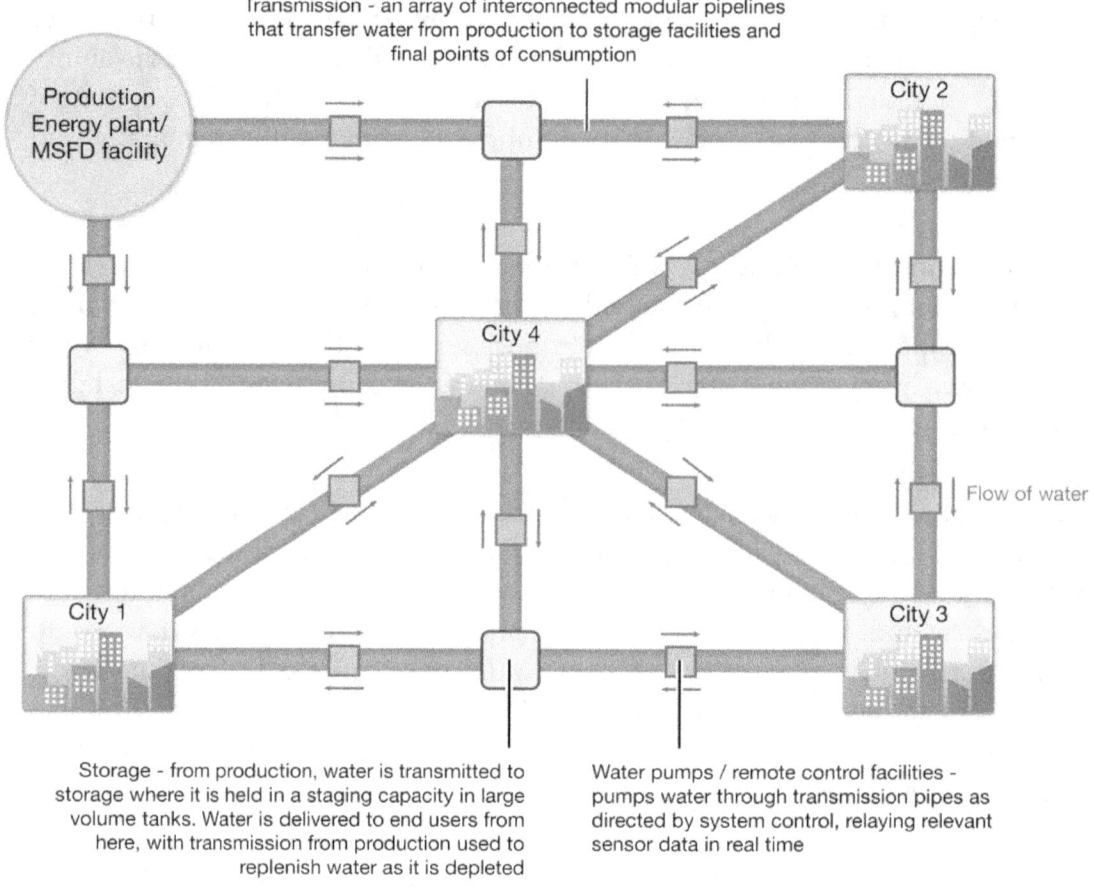

Production

Production is comprised of Multi-Stage Flash Distillation (MSDF) facilities, ideally as part of CHP Plants, as described last chapter. To summarize, these facilities would use the waste energy from LFTRs to power saltwater desalination on our coasts, which apart from the water devoted to producing hydrogen would effectively give us an endless supply of fresh water.

Transmission

The transmission component consists of a series of water pipelines and pumping stations. Instead of building single water pipes as unique entities, the National Aqueduct would instead use factory-prefabricated pipe assemblies that are built to one standard and are designed to couple together modularly. The mindset behind this approach is two-fold: first, it would simplify construction of pipelines over long distances, and second, it provides the ability to rapidly expand transmission capacity with minimal overhead and construction costs, should the need arise.

The pipes themselves would be insulated against environmental elements and could drain or block flow on demand. They could additionally contain a series of sensors that relay relevant data to the system's control component (described shortly). As mentioned, these pipelines would also feature external solar panels and internal turbines, which we'll describe in detail next chapter.

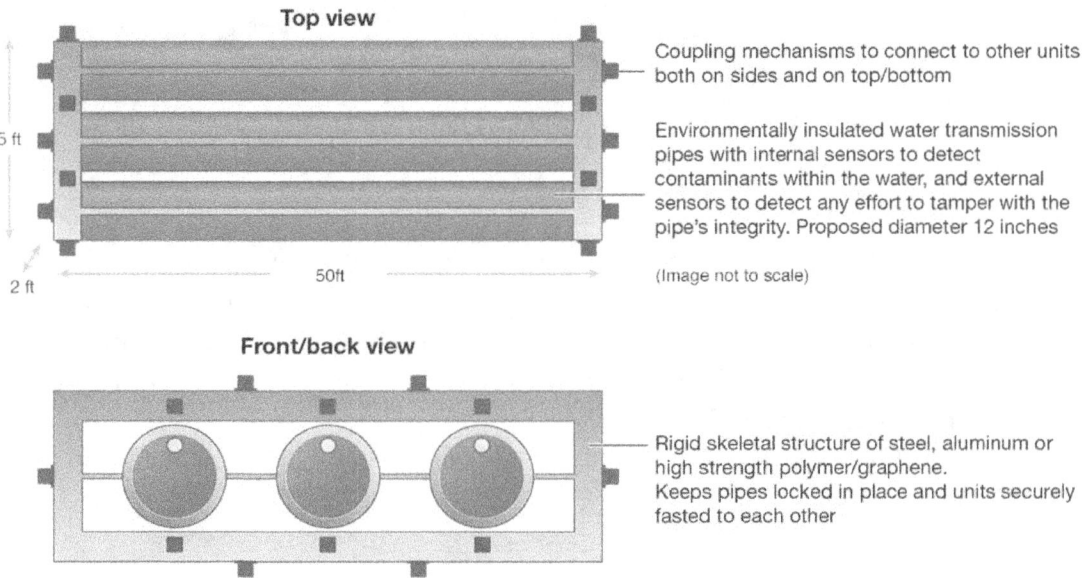

The sensors within the pipeline could serve multiple purposes: they might detect contaminants, determine water quality, or send alerts if they were compromised or modified without authorization. Since each separate pipe within each pipeline could have its own sensors that connect independently to a control network, water quality could be monitored instantly on both local and national levels.

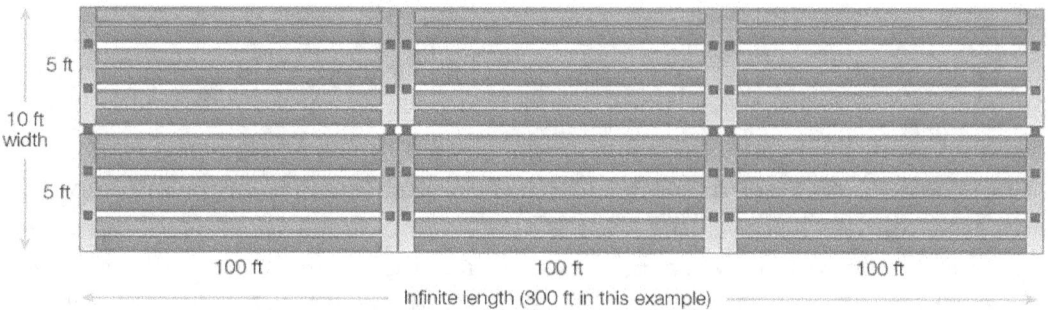

Top view of 4 unit pipe array (not to scale)

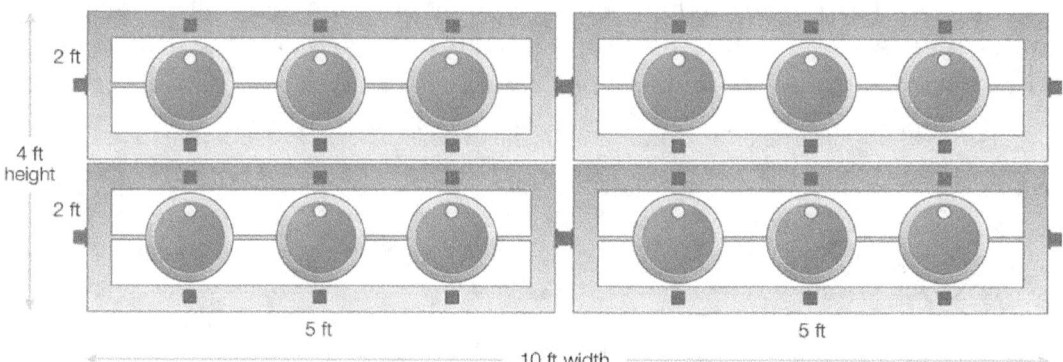

Front view of 4 unit pipe array (not to scale)

As sensor technology has reached levels of sophistication where sensitivity in parts per billion (PPB) is common, the returned data would be useful for water management. This, among other conclusions from sensor data, could influence a range of actions from the control center of this system, ensuring maximum performance, reliability, and security.

Storage

The storage component of the National Aqueduct includes arrays of containment tanks that act as the supply reservoir for a region. Rather than transport water directly from production to areas of consumption (as we largely do with electricity) this system would instead use storage tanks as a staging system. Water from the storage tanks would go directly to cities or other areas of need, and the tanks would be replenished from production facilities as necessary.

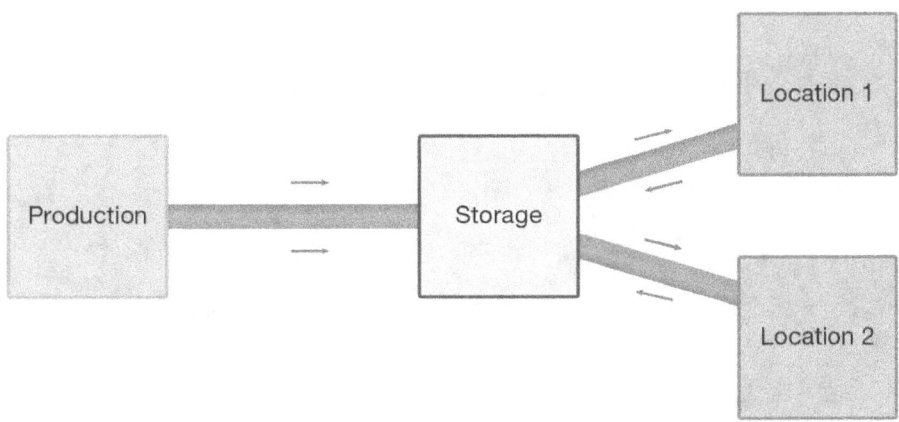

These arrays of storage tanks would contain millions of gallons of water and could be installed as distribution centers throughout the country. Each agricultural and population center would be served redundantly, with each storage center servicing multiple regions. The storage component would additionally provide several important functions:

Staging: it's difficult to guess how much water a region might use with certainty, as external factors such as weather, time of year, and state of economy all impact how much water is consumed. Maintaining supply and pressure over thousands of miles under inconsistent demand would be a logistical nightmare, which rules out any system that delivers water directly.

However, when used as a part of a staging system water storage tanks can contain enough water to supply a region for a certain time period – say a week – which equips them to handle unexpected spikes in demand. In turn, water would be supplied from production facilities to maintain consistent levels.

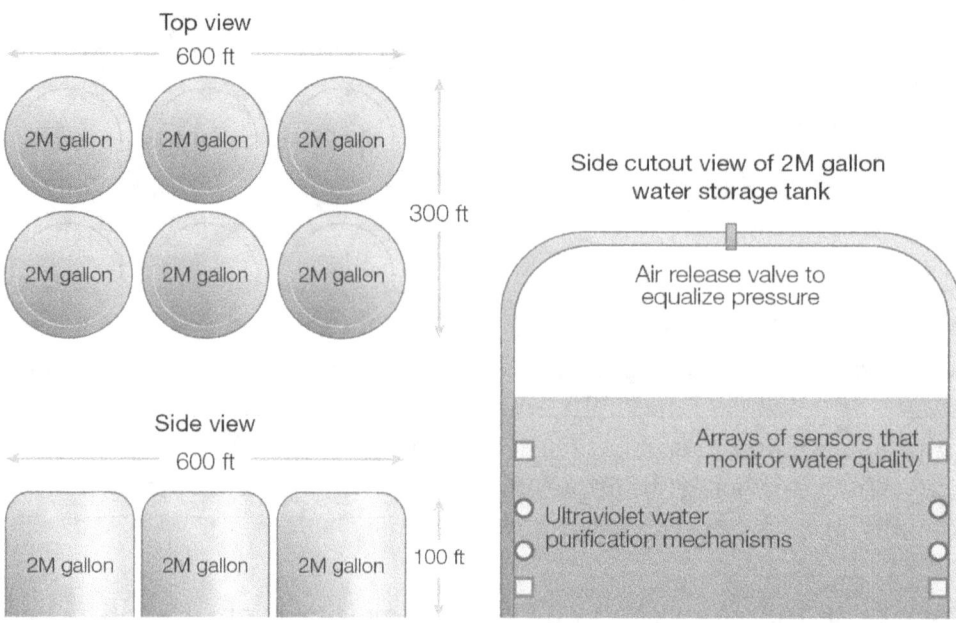

This pays homage to the concept of constant resupply, which bears special mention in this context. The National Aqueduct would be constantly producing and pumping water. This consistent operation is key to meeting our immense water requirements.

For example: a common bathtub faucet has a flow rate of roughly 120 gallons/hour.[410] If that faucet was to never turn off, it would provide 2,880 gallons a day, 86,400 gallons a month and 1.036 million gallons a year – just from your bathtub faucet.

With a water velocity of just 15 miles an hour, a 12" pipe can have a flow rate of 150 gallons *per second*.[411] That translates to 540,000 gallons an hour, 12.96 million gallons a day, 363 million gallons a month and 3.35 billion gallons a year. And that's just from one 12" pipe. Imagine what an array of nine pipes could do, and imagine that pipeline array multiplied by hundreds. A constantly running pipeline network could easily transport hundreds of billions of gallons.

Overflow / active water management. Nodding to the fact that water demand is inconsistent in most areas of the country, there will be times where one region has more water than it needs or needs more water than it has. This requires the system to feature an overflow component.

By design, water storage tanks would only be filled to 70-80% of capacity, which would give them the ability to accept more water from other regions if supply exceeds demand. On the other hand, should a region consume more water than anticipated, overflow in one region can be diverted to another as necessary. With this attention to storage and re-routing, the National Aqueduct can maintain consistent uptime and high degrees of reliability.

Ultraviolet sterilization and filtration. Storing water for weeks on end can lead to circumstances where microbial agents could conceivably cause contamination. Further, there is always the concern of invasive species through any sort of aquatic transfer (think Zebra mussels).[412] In addition to standard filtration mechanisms and high temperature of water, the Aqueduct's storage and transmission component would have a series of UV lights to further sterilize stored water. Ultraviolet light, especially at high intensity, is lethal to any microbial lifeform (bacterial or viral) with high reliability, allowing for long-term water storage without fears of stagnation or outside contamination.

Control

The control component is the brain and central nervous system of the National Aqueduct, yet also works as a decentralized framework allowing any region to control the flow of water in its jurisdiction. This would be accomplished through a series of manned control centers at national, regional, state and local levels, similar to the current management functions of large public utilities.

In this case, the control component would monitor the system to ensure consistency and stability, and further act if it detected a contaminant, outside tampering, or mechanical failure.

In the event of a shortage or an overabundance of water in an area, water could be routed from one storage array to another as necessary; in the event of a problem, the control component could direct the "smart grid" of transmission pipelines and pumping stations to take action. These actions might range from disabling, draining, and isolating a certain section of pipes from transmission lines, to bypassing a whole sector or storage array completely.

The sensor network within the National Aqueduct could further provide data to create models that would be greatly beneficial toward ideal operation, and which could, for example, predict demand over time. A defining feature of the National Aqueduct is that once enough water has been produced, the production component only needs to resupply what has been consumed, allowing desalination facilities to produce fresh water only as needed. Accurate modeling allows water to be resupplied intelligently – knowing when water will be used at what levels based on time of year allows us to address anticipated shortages before they occur.

The National Aqueduct is Universal Energy's means to deliver water anywhere in the country, allowing us to live free of the destructive results of unsustainable water use and natural drought cycles. From here, the secondary component of the National Aqueduct comes into play – in addition to supplying the nation with water, it can store electricity generated by renewables on a massive scale, making it the "world's largest battery." Next chapter, we'll explore why.

A ten-times increase in the weight-oriented density of batteries would enable so many other moonshots, and we will start that moonshot if we can find a great idea.

- Astro Teller

Chapter Eight: The World's Largest Battery

The National Aqueduct is designed to transport an effectively unlimited volume of desalinated water from the coasts to anywhere inland. Powered by Universal Energy, it can deliver on that promise. But like many of the systems included within the framework, the National Aqueduct is not a one-trick pony. It has additional functions to generate electricity and serve as a battery for renewable energy on a massive scale.

To consider a nationwide array of water pipelines a "battery" of any sort may seem strange – especially since "batteries" technically require a chemical reaction to facilitate an electric charge. But the National Aqueduct would have the same effect as a battery in function – far beyond that of the Bath County Pumped Storage Station in Northwest Virginia, a similar concept of pumped-storage hydroelectric power that remains the largest of its kind in the world (currently).[413]

By taking advantage of three aspects of the water transmission system – surface area for solar power, water flow for hydroelectric power, and water heat for thermoelectric power – we have a platform on which to build potent external electricity generation and storage systems.

Best of all, as with CHP Plants, these systems can work together in a cogenerative capacity, presenting additional benefits.

To explain how, we'll go through each of these three aspects.

Surface area for solar panels. Universal Energy intends to deploy National Aqueduct pipelines at the side of highways or under currently existing long-distance power lines. That's because this land is usually state-owned and doesn't need to be purchased, and also because it's pre-cleared of obstructions. If you recall from last chapter, water pipelines are designed to be installed in prefabricated arrays, which boast relatively flat surfaces. So, what if we were to cover their surface area with solar panels? The result, in the following image, is the final intended form of National Aqueduct pipeline arrays:

Top view of 4 unit pipe array (not to scale)

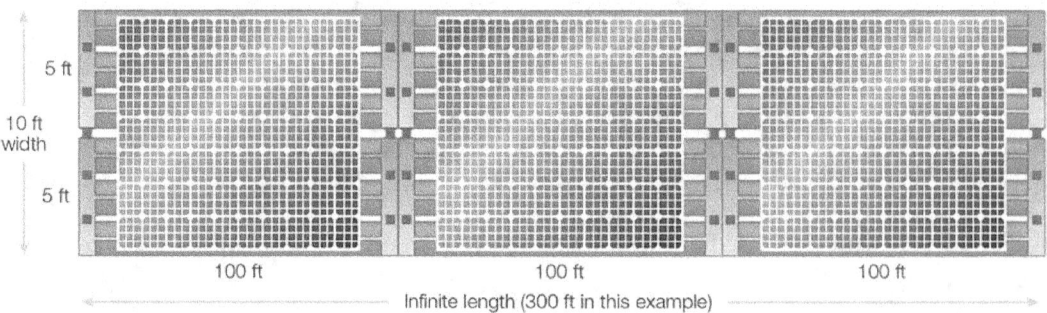

Across the thousands of miles such pipeline arrays would span, integrating them with solar panels stands to generate extremely high levels of electricity – far beyond any single power plant.[414] Alongside solar road networks and other municipally integrated renewables, they can also help reinforce a redundant, smart electric grid.

To see how, let's assume solar-enabled water transmission pipelines had a surface width of 10 feet. Multiplied by a mile, that's 52,800 square feet. Going back to our estimate of 30 kilowatt-hours annually generated per square foot, a one mile stretch of solar-integrated pipelines would generate 1.58 million kilowatt-hours per year – enough to power 150 homes. Keeping in mind that there are four million miles of road surfaces in the United States,[415] even fractional integration would be sufficient to power tens of millions of homes. Total integration, hypothetically, would annually generate 6.32 trillion kilowatt hours – **nearly twice our annual national energy consumption.** And that's just from one of *three* power sources of the National Aqueduct.

Integration with wind turbines. Across the wide, open expanses of rural America, wind power is especially effective. Placement in close proximity with the National Aqueduct would enable a uniquely convenient source of transmission and storage for the electricity they generate, that like other municipal placement, avoids the need to buy expensive land.

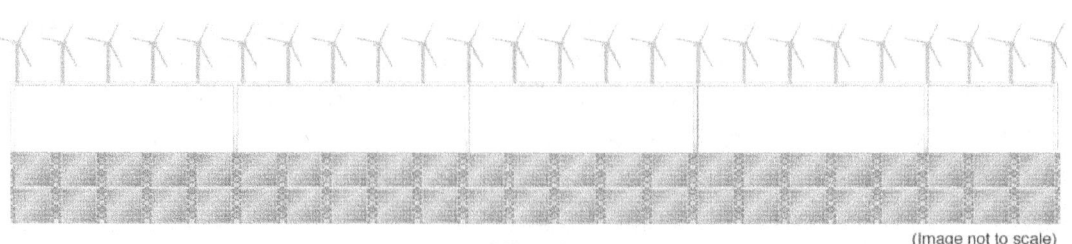

Wind can assist the National Aqueduct in both electricity transmission and thermoelectric battery functions. During peak demand, wind turbines integrated throughout the National Aqueduct could directly power regional areas, as the Aqueduct itself would effectively be a nationwide electric grid that could transmit electricity at high efficiencies. Conversely, the energy they generate could be diverted to water heating elements that keep the Aqueduct's battery function at maximum capacity. As the Aqueduct is a "smart" system that operates within a modular, responsive framework, it would be able to respond to these needs intelligently, making wind integration yet another force multiplier to the National Aqueduct's already multifaceted capabilities.

Internal hydroelectric power. As a part of the energy-generating capabilities of the National Aqueduct, the water pipes themselves could be fitted with internal turbines to generate electricity from the water flow. Hydroelectricity is highly effective as a power source, and by miniaturizing turbines within prefabricated pipeline arrays, nearly every aspect of the water transmission process can be harnessed as power.

A company named Lucid Energy in Oregon is already doing this today through modular turbine assemblies installed within prefabricated water pipes.[416]

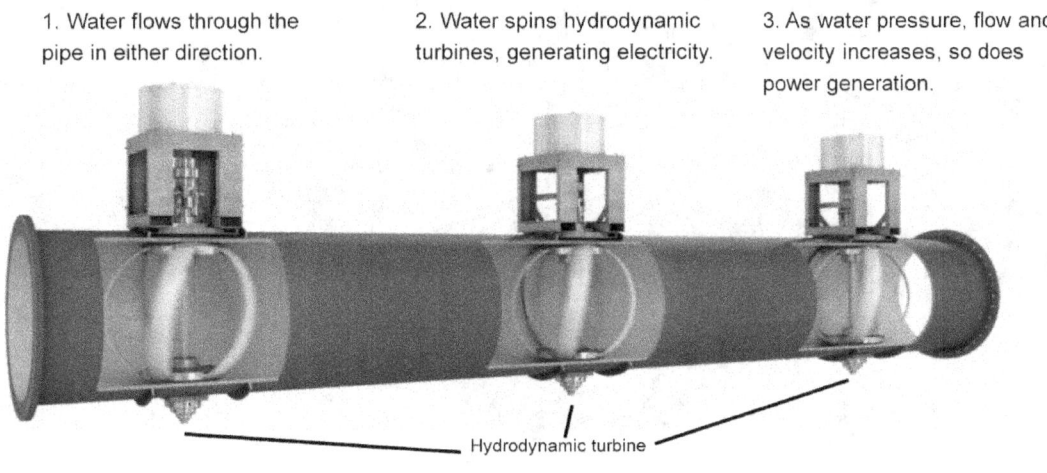

Cutaway view of pipeline

Image source: Lucid Energy

Overview of Lucid Energy Water Turbine:

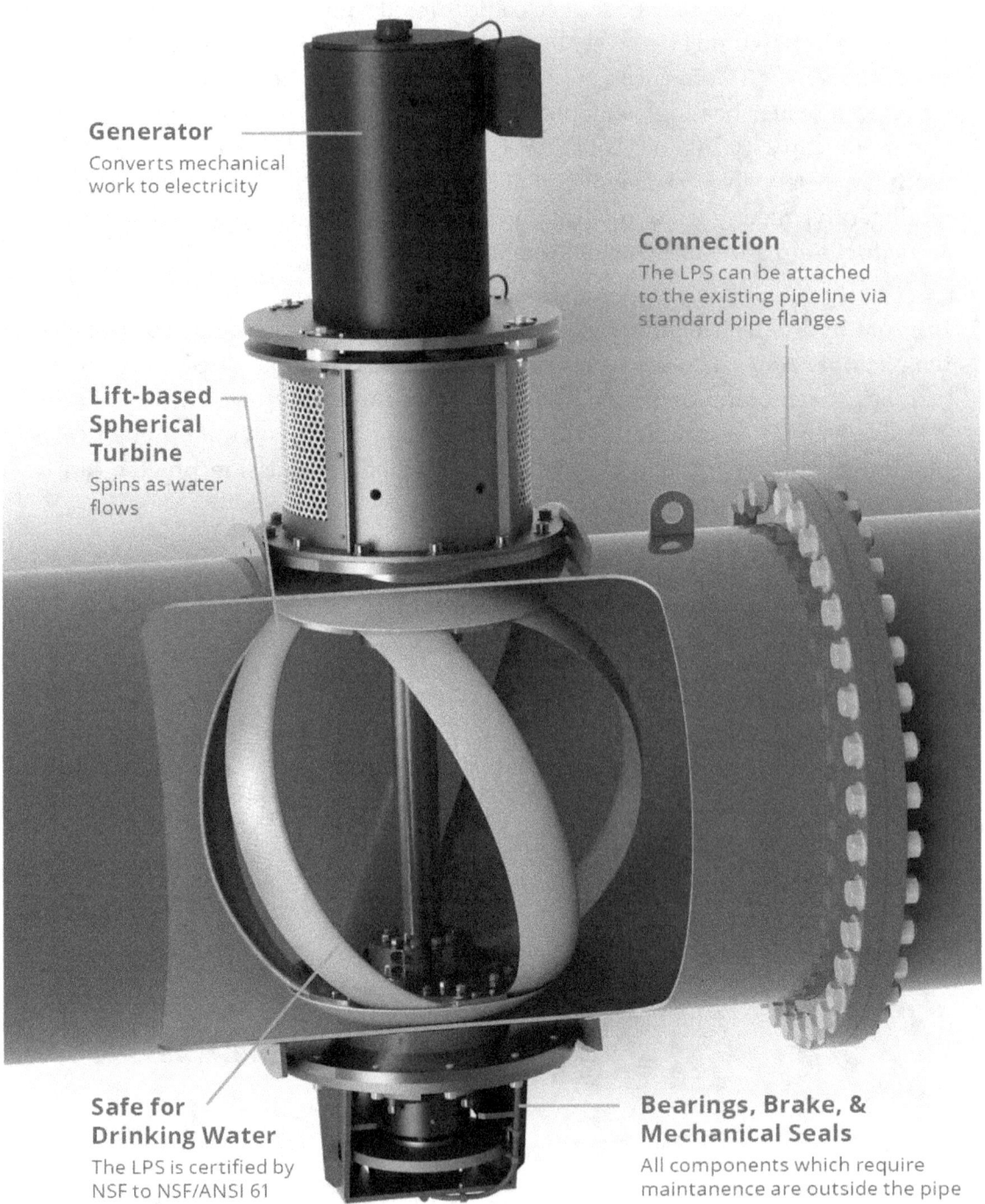

Image source: Lucid Energy

Using a similar approach, the National Aqueduct would generate electricity 24 hours a day, 7 days a week by virtue of its primary function: transporting water over thousands of miles. Yet unlike other hydroelectric power stations, like the Hoover Dam on the rapidly-depleting Lake Meade,[417] this method is environmentally friendly and highly reliable.

Hydrothermal power. As water has one of the highest specific heats in nature, once it gets hot, it stays hot for a long time.[418] At the scale of billions of gallons, water stays hot for *an extremely long time* – days, even weeks – all the more so if the storage mediums are well-insulated.

One of the key functions of a CHP Plant is to keep water hot when it comes from a Multistage Flash Distillation Facility, and in turn pump it into the National Aqueduct hot. But as it travels over distance, the water will eventually cool.

To address this problem, we'll need to rely on the National Aqueduct's energy-generating features. By using the excess energy generated by pipeline-mounted solar arrays, wind turbines, and internal pipe turbines, we'll have the energy necessary to keep water hot throughout the entire Aqueduct. This is beneficial for three important reasons:

First, water will reach its destination hot, sparing the energy needed to heat it within residential and commercial hot water heaters. By virtue of the Aqueduct's control component, not all water would necessarily be delivered to residences at high temperature, but it *can* arrive hot at any given time. This gives municipal managers more flexibility in how they route water, as well as save significant sums of money and energy on heating costs.

Second, keeping the water hot prevents it from freezing in pipelines during winter months. Although the pipes would be insulated and resilient, a residual amount of heat could be constantly emitted to melt snow covering solar panels, allowing for electricity generation year-round.

Third, if the entire Aqueduct were heated, it would store a tremendous amount of potential energy that could be converted into electricity,[419] allowing it to functionally act as a battery – the world's largest by far.

Several companies – such as Marlow Engineering's models in the images on the next page – make thermoelectric generators designed specifically to be placed over hot pipelines,[420] which can be placed in arrays up to three-wide. If installed throughout the entire system, they could generate immense power from such a large volume of hot water.[421]

Image source: Marlow Engineering

Let's say, for example, that we'll store 500 billion gallons throughout the National Aqueduct, and let's say we heat that water to 200° F to maximize potential energy storage. Using the worksheets provided by the helpful folks over at Engineering Toolbox,[422] we'll conclude that 1 gallon of water at 200 °F contains 1,660 BTU of energy. Across 500 billion gallons, that comes to 830 trillion BTU, or 875.7 billion megajoules.

Converted into electricity, that's equivalent to 243 billion kilowatt-hours of potential energy *from hot water alone.*

Combined with the hydroelectric and solar functions of water pipeline arrays, it's feasible for the National Aqueduct to store enough potential energy to generate trillions of kilowatt-hours over time. To put that in the proper scale, take another look at the nationwide road map we saw earlier:

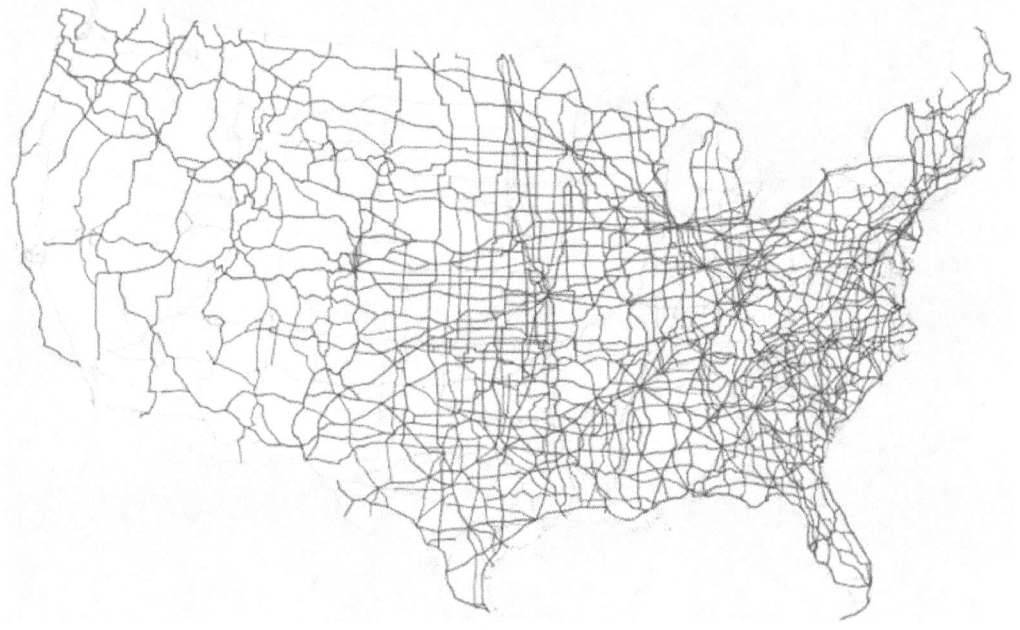

If each one of these highway lines represented arrays of water transmission pipelines with solar functionality, hydroelectric functionality, and hydrothermal functionality, we'd be generating a level of energy that words don't give justice to – particularly when combined with Liquid Fluoride Thorium Reactors, CHP Plants, and municipally integrated renewables. As adoption of Universal Energy's technologies spreads, the multiplier effect accelerates until we're well beyond the energy targets needed to support our way of life, with each step beyond that concordantly reaching a subsequent tier.

With the National Aqueduct implemented alongside integrated municipal renewables and CHP Plants, we have electricity, water, and fuel at our fingertips in indefinite abundance. That, once achieved, empowers us to evolve beyond the concepts of drought, water scarcity, and water-borne disease. And from there, we can do the same to famine.

Because as we can use effectively unlimited electricity to produce unlimited water and unlimited fuel, we can then use all three to produce unlimited food.

Clean water is a great example of something that depends on energy. And if you solve the water problem, you solve the food problem.

- Richard Smalley

Chapter Nine: Everybody Eats

At its most basic form, life requires energy, water and food. Historically, we've depended exclusively on nature to provide those first two resources so that we could grow the third. With Universal Energy, this would no longer need to be the case.

An effectively unlimited supply of electricity, water and fuel gives us opportunities to revolutionize our industrial and agricultural systems, allowing us to grow produce locally in indoor farms. This produce can be healthier and of better quality than much of what we cultivate today – and more of it can be grown at greater efficiencies with shorter delivery times (and reduced environmental impact) on less land.[423]

These systems can be built close to areas where food is consumed, which reduces obstacles to transportation and delivery, especially within cities. Indoor farms can also be climate controlled and operate 24 hours a day, 365 days a year – dramatically increasing output and efficiency compared to traditional farming. Past obstacles to their deployment stemmed primarily from material and resource costs,[424] problems that Universal Energy substantially reduces.

With indoor farms, as long as there is water, light and heat, the location and outside environment doesn't matter. This allows food to be grown anywhere on the planet at any time of year, with increased yield, higher efficiency, longer growing seasons, and greater food security.

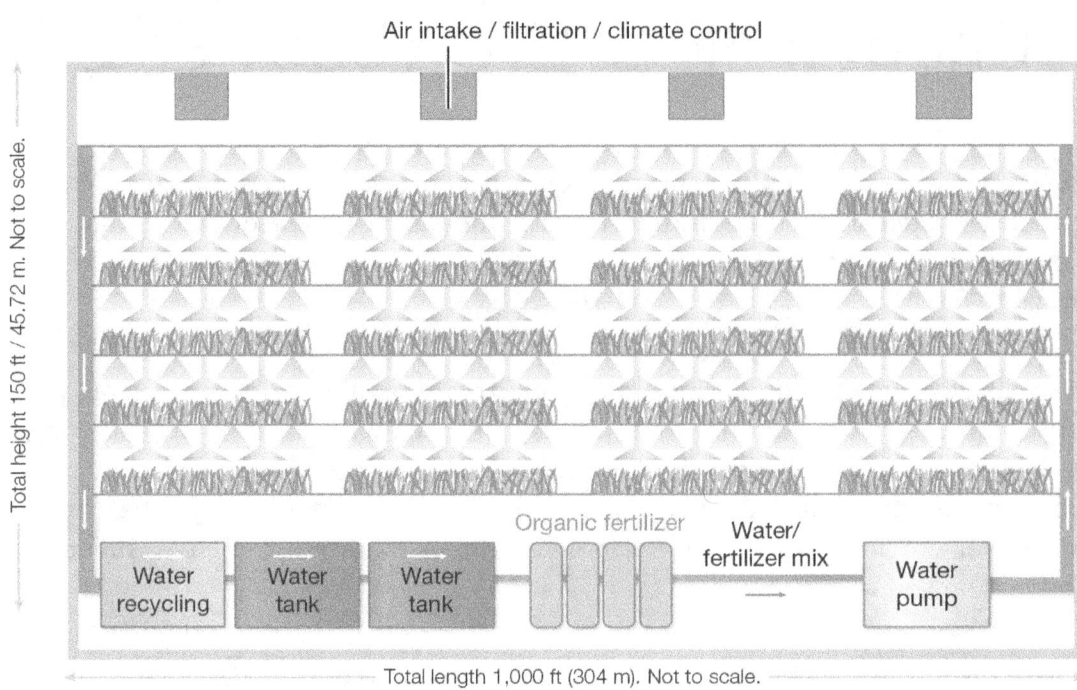

In this concept image of an indoor farming warehouse, water from storage tanks is mixed with organic fertilizer made from excess plant matter grown within the warehouse itself. This water is then pumped through the facility and dispersed over plots of crops that grow under high-intensity lights. These crops can grow on modular wheeled platforms that are easily moved, and the crops growing on them can be manually pollinated as necessary.

What water isn't absorbed by crops drains into a collection mechanism in the floor, which sends the water to the bottom of the warehouse where it is filtered and placed back into circulation. As we see today, the construction of large warehouses at acceptable cost isn't uncommon – take any Walmart, Target, Home Depot or other big-box retailer, for example. These buildings are huge, often encompassing a surface area into the hundreds of thousands of square feet. Similar structures present promising opportunities for indoor farming.

If an indoor farming warehouse had dimensions of 600' per side, that comes to a total surface area of 360,000 square feet (roughly 1/15th of the surface area of Tesla Headquarters).[425] Yet if the growing platforms were stacked, each subsequent layer adds that same surface area to the aggregate total to become a force multiplier. At five stacks, that warehouse now offers 1.8 million square feet of growing space. At ten stacks, 3.6 million. At twenty stacks, 7.2 million. Building two dozen of such warehouses, at twenty stacks each, would boast a total growing surface of 172.8 million square feet. That's 3,967 acres - some 6.2 square miles.

As these warehouses would operate 24 hours a day, 365 days a year, their aggregate output could grow large enough to provide food for large metropolitan areas – the capabilities of which would be limited only by the number of indoor farm clusters. This setup is becoming more common, and is in use throughout many of the indoor farms that have sprung up around the world today.

Combined, these systems create a cultivation mechanism that allows produce of effectively any kind to be grown locally. And this produce would be grown under controlled conditions – that is to say, each type of crop would be grown under ideal conditions for that type of crop. To elaborate further, here are some of the more remarkable benefits of indoor farming:

Total control of environment and constant operation. Since humans discovered how to farm, we've been limited to a growing season as determined by the local environment. Indoor farming completely bypasses this limitation, allowing us to emulate any growing conditions we wish. Moreover, indoor farms can be customized to the point where we'd have total control over the temperature, humidity, light spectrum, and soil composition in any given section of warehouse. And as indoor farms operate 24/7/365, they can reflect the ideal light cycle for any plant grown. This would dramatically increase overall efficiency, as there would be no seasonal slowdowns or environmental complications.

Technology-driven pest/contaminant prevention. Pests and weeds are problems in any open environment – problems we've tried to solve with chemicals of varying degrees of toxicity. As they offer total control of environmental setting, indoor farms allow us to manage the presence of weeds and pests without dependence on more toxic methods Examples include:

- **Positive pressure.** The indoor farm can be pressurized higher than the outside area, so that when a door opens, indoor air blows out of the building instead of outdoor air blowing in. Alongside worker sterilization and air filtration mechanisms, this would limit the presence of contaminants inside the farm.

- **Isolated sections.** In what would also benefit food security as a whole, isolating areas of the indoor farm would hinder the ability of a pest or contaminant to spread from one area to another. This would further assist in any necessary cleanup operations should an infestation occur. It would also make it functionally impossible for large-scale pests (such as locust swarms) to decimate crops.

- **Active anti-pest measures.** In the event that a pest did get in, we could respond more surgically or with more benign pesticides. We wouldn't need to rely as strongly on more toxic treatments, because the pests they combat would be less present in the first place.

Waste management. Indoor farms can be designed to minimize the use of artificial fertilizers through composting. Whenever a plant dies, sheds material, or leaves behind waste after harvest, that material can be collected into a composting mechanism (shown in the concept image at the start of this chapter) that can be mixed with other organic fertilizers and pumped directly into the water supply used to irrigate crops. As much of the world's soil is already facing varying degrees of contamination (such as arsenic in rice and steroids in runoff water from feed lots),[426] this method translates to healthier food.

Diversity of crops. As their components would provide ideal growing environments that are naturally pest-resistant, indoor farms can encourage greater use of heirloom crops that might not fare as well as genetically modified variants outdoors. This allows us to cultivate a greater variety of produce and shift our focus towards growing food with higher nutritional properties, expanding organic and farm-to-table markets.

Local operation. Indoor farms can be built in close proximity to metropolitan areas, so that food is grown close to the people who consume it. Food production in New York would be consumed by New Yorkers, food production in California would be consumed by Californians. This simplifies the delivery of food from farm to market, saving resources and allowing for fresher produce.

It can also present major improvements to how we provide food aid, as global anti-famine initiatives usually involve shipping food that's already grown. With indoor farms, the system itself can comprise the aid, allowing stressed regions to grow their own produce by themselves. This concept is discussed further in the "Collective Capitalism" section of the Appendix (page 295).

Food efficiency. As touched on earlier, the benefits of indoor farming present notable benefits for efficiency. Plenty corporation, an indoor farming startup, claims that indoor farms can produce 350 times as much produce-per-acre of land compared to traditional methods – with only 1% of the associated water usage.[427] AeroFarms, a New Jersey indoor farming company, claims they can produce upwards of 130 times the amount of produce grown per acre, with similar reductions in water usage.[428] Even if these numbers are inflated – which seems unlikely, considering the ideal growing conditions and 24/7/365 growing season that indoor farms provide – at even half the claimed efficiency improvements, the benefits would still be enormous compared to traditional farming.

In addition to growing efficiency, the efficiency of delivery, particularly as it relates to food quality, must also be considered. Approximately 35% of the fruits and vegetables we eat as Americans are imported from abroad, with leafy greens traveling an average of 2,000 miles from farm to plate.[429] Some produce travels for weeks before being served, losing, at some estimates, 45% of its nutritional value in the process.[430] Further, Americans rarely buy unsightly produce. Locally grown food would provide significant benefits in these areas.[431]

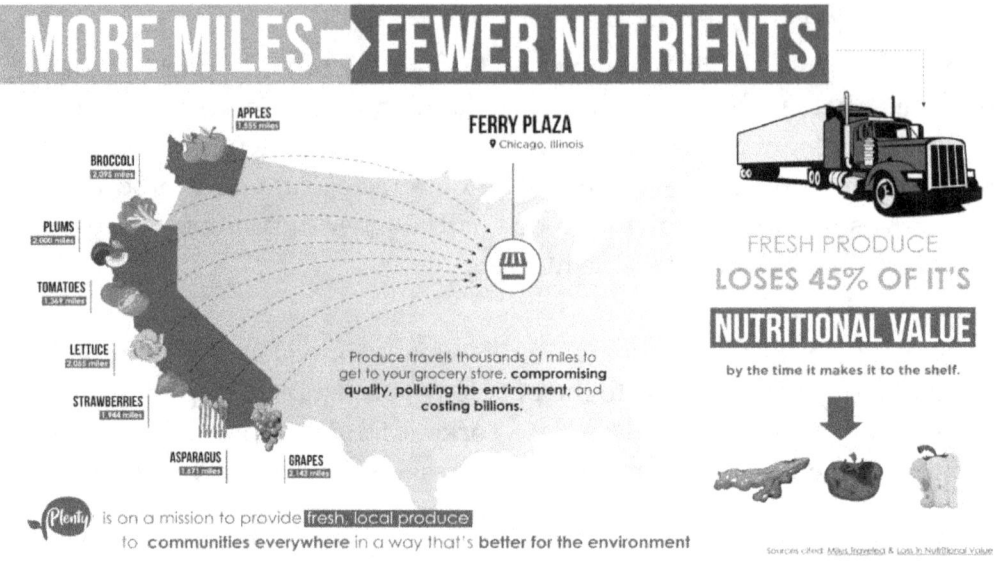

Food security. Local food production in isolated environments also allows us to reduce security risks to our food supply. In late 2011, a Listeria outbreak in cantaloupe killed more than 20 people in the United States,[432] and food contamination (and recalls) have proven relatively frequent as of late.[433] Supermarkets across the nation are stocked with produce that comes from different states, different countries, even different continents, and it's hard to keep track of where everything is coming from in real time. So, when an outbreak is detected, our food networks are thrown into chaos until investigators can pinpoint the source of the contamination and isolate it. With locally grown produce, security issues, however rarer, are automatically isolated since production environments are sealed.

This also protects our food supply from pathogens. Genetically modified crops, often with identical genomes, make up large swaths of our food supply – comprising approximately 90% of soy and corn.[434] Of the crops with identical genetics, any self-replicating pathogen that could infect one plant could potentially infect all of them.[435] By design, indoor farms exercise self-quarantine, which is likely the best defense they could have.

Indoor farming can provide local and sustainable food production while systematically avoiding most of the obstacles that exist in agriculture today. But beyond indoor growth in warehouses, we can integrate indoor farms directly within urban environments – not only to supplement food production, but also to serve as centerpieces for more advanced cities with next-generation infrastructure.

Urban Vertical Farming

Since the end of World War II, American agriculture has increasingly represented a centralized model where the majority of food is grown in one large region (the "American Breadbasket") and shipped elsewhere for processing and distribution. This has led to logistical challenges that indoor farms are designed to address, but providing a framework for climate-controlled indoor farming is only the first step. The next is urban vertical farming.

An urban vertical farm doesn't use a flat plot of land or a large warehouse, but instead floors of city buildings that have been either built or modified for farming. This approach has existed in varied forms for millennia (the hanging gardens of Babylon perhaps most famous of them) and has been used in Europe since the 1850s.[436] Today, urban vertical farms are being constructed in London, Chicago, Milan and Newark, with future farms planned in several other cities.[437]

In the past, the feasibility of vertical farming has been limited by constraints inherent to resource scarcity, namely the cost of energy and materials.[438] With the implementation of Universal Energy, these constraints would no longer be present in force, reducing the limitations to building large-scale urban vertical farms. The following concept images illustrate further.

At first glance, these systems might appear somewhat futuristic, but from an architectural or engineering standpoint, these designs present few challenges that have not been solved already in other industries with today's technology. They are well within the realm of feasibility, especially if energy costs and material limitations are removed from the equation. Even so, the following images show urban vertical farms that are already operating around the world today:

Urban vertical farm in South Korea:

Urban vertical farm in Chicago, Illinois:

Urban vertical farm in Singapore:

Urban vertical farm in Newark, New Jersey:

Further, urban vertical farms have additional benefits that cannot be provided by warehouse-style indoor farms, making them especially attractive for cities with higher population densities. These include:

Smarter food production. While indoor farms may grow more crops at the size of a warehouse, urban vertical farms can still provide a major boost to food production. As urban vertical farms would exist directly within city centers, it effectively zeroes out the distance between production and consumption. In doing this, vertical farms redefine the notion of "farm to table," as they would be within short distance of the millions of people they could provide food for.

Municipal water recycling. In addition to integrated water management, vertical farms could also solve problems that are not necessarily present outside of an urban environment. Today, it is standard for water treatment systems to filter water based on what its purpose was. Grey water (water that's been used for washing clothes or bathing) is treated differently than water that's been used for cleaning dishes, and both are treated differently than water that's used to process bodily waste. Using currently existing water filtration systems, semi-treated waste water could be used in place of fresh water to grow crops in vertical farms. This would aid in recycling municipal water and also reduce the stress vertical farms could place on a given area's water resources – unlimited water supply through the National Aqueduct notwithstanding.

This said, it's noteworthy that food production isn't the only application for vertical farming, as plants provide more value than just consumption. For example:

Soil is a great insulator. The idea of placing greenhouses on rooftops has been around for some time, and one of the most attractive supplementary benefits of doing so is their insulation potential. As hot air rises through a building, it escapes through the roof, which increases energy costs. Acting like a blanket, rooftop greenhouses work to keep as much energy as possible within a given building. This concept has already been applied in Italy with the Bosco Verticale towers in Milan, which were completed in 2015.[439]

Several cities have applied this concept further into law. New York City, for example, recently passed legislation requiring green roofs on new buildings within its jurisdiction.[440] Denver, Portland, San Francisco and Chicago have enacted similar measures to establish green rooftops for urban air purification and temperature control of buildings.[441]

Plants are air purifiers. Plants are highly effective in scrubbing air of impurities and contaminants, which are often concentrated in cities. Large-scale urban agriculture through vertical farms can act as a massive, constant air filtration system. As allergies and respiratory ailments have been on the rise, this can provide a boost to public health both physically and mentally.[442]

China has taken this to heart and has begun construction of the world's first "forest city" that will integrate plants and trees directly within city buildings.[443] Planned for a 2020 completion date, the city is expected to be covered in one million plants and forty thousand trees.[444]

Indoor parks and greener cities are socially beneficial. Regardless of how far we have come in terms of technological advancement, connection to nature is important for human happiness, lower stress levels, higher productivity, and positive life outlook. For those of us who live in cities, disconnection from nature can sow the seeds of depression, and depression is a social toxin. But could you imagine how much less stressful life would be if you could just take an elevator to the roof of your office building and hang out in a tropical park for 30 minutes on your lunch break? Or walk off a city street into a large indoor park where it was bright, colorful and warm?

Daily exposure to nature, however short, can make a major difference, which not only leads to a happier and healthier society, but also improves collective hope, and thus the collective drive to seek, build, and accomplish greater things. And as real estate for public parks faces high costs in cities, devoting floors of a building to a welcoming public hangout can significantly lower expenses for the same deliverable – especially if the park is on the roof of a building.

Farming More Than Food

Indoor farming would significantly reduce stress on the American breadbasket, requiring outdoor farmers to grow significantly less food than they do today to meet demand. At first glance, this might seem like a trouble spot for farmers, because growing less food means making less money. So, should indoor farms give them cause to worry? No – because growing less produce gives them an opportunity to grow other crops – crops that have both a higher density and commercial value than the kind that generally makes their way to supermarkets.

For example, instead of growing corn or soy, farmers could instead grow:

Algae for biofuel or plastics. The rising price of petroleum over the past 15 years – at least until the discovery of easier shale oil extraction – has led to a surge of investment in biofuels.[445] Biofuels are hydrocarbons that come from living plants, as opposed to oil that comes from fossilized plant remains. Even though Universal Energy would promote hydrogen as the nationwide fuel standard, biofuels and existing petroleum reserves can be devoted to a more appropriate purpose: advanced materials. Today, we use corn ethanol to make plastics,[446] but corn is not the most effective crop we have at our disposal.

That honor goes to algae.

Several forms of algae have properties that allow for hydrocarbon production. The biofuel company Algenol claims that with today's technology it can produce thousands of gallons of ethanol per acre of growing space, and we could increase that output with Universal Energy.[447] Further, since the energy and materials industries tend to be more profitable than the food industry, what money might be lost from a move to indoor food production could then be replaced – with profits gained – from the shift to growing hydrocarbon-producing algae.

For comparison, the Department of Energy estimates that if we were to use algae to replace petroleum in all respects in the United States, we would need an area of about 15,000 square miles (roughly the size of Massachusetts and Connecticut combined).[448] That's less than 1/7th of the space we use for corn, meaning if we were to move much of our food production indoors, we would have ample space to grow algae – and plenty of economic dividends for farmers to go along with it.

Concept image of an outdoor algae farm.

Algae for supplemental nutrition. Beyond uses in plastics and materials, algae can also be grown for added nutritional value in food products. The species chlorella, in particular, is among the most promising candidates.

By all definitions a "superfood," dried chlorella is comprised of 45% protein, 20% lipids (fats), 20% carbohydrates, 10% vitamins/minerals and 5% fiber.[449] This composition, combined with a high photosynthetic efficiency (how well something grows in sunlight), gives chlorella one of the highest protein yields of any crop.[450] That's why, after World War II, chlorella was considered as a solution to the then-global food crisis.[451] At the time, chlorella was difficult to grow outside of laboratories, but with advances in technology post-1950 – and the added benefits Universal Energy provides – we have the ability to grow a potent nutritional supplement that can be used to enhance any segment of the food supply. One that, just as importantly, can also provide supplemental nutrition in remote or isolated locations.

Alternative use of genetically modified organisms. One benefit to indoor farming is its ability to grow heirloom crops with yields that are similar to genetically modified crops in outdoor environments. However, this is not to say that genetic modification of plants is a negative thing in and of itself, but rather that its benefits can also be realized in other applications, something that bears special mention in this context.

Much of the anti-GMO movement[452] has focused on opposing *any* manipulation of crops at the genetic level, rather than genetic engineering that allows a plant to survive otherwise lethal pesticides and herbicides.[453] While this writing does not maintain a skeptical position on the current state of genetic engineering in our food supply, it suggests a moment of pause as to the wisdom of disavowing an entire scientific discipline because a corporation engineered plants with a genetic immunity to a relative of organophosphate nerve agents.[454]

People have been genetically engineering plants for millennia through splicing, cross breeding, and human (as opposed to natural) selection. We were modifying genomes then, we just weren't doing it under a microscope. It's what we do to a plant, and our guiding ethical standards when we do it, that ultimately matters. So, what if we instead extended the genetic modification of plants to a different focus? For example:

Efficiency of hydrocarbon production. Potential yields of hydrocarbon-producing algae for plastics and chemical stabilizers are already high.[455] However, we could configure the plant at a genetic level to produce even greater volumes of hydrocarbons that can induce a higher-percent yield when synthesizing plastics,[456] that are geared for more advanced polymerization[457] and that recycle more effectively (or biodegrade faster) than most plastics today. We'll go through this in more detail next chapter.

Inclusion of bacteria. Algae isn't the only organism that can produce hydrocarbons. Scientists in several countries have successfully modified the genetics of E. coli bacteria to produce diesel fuel that is nearly identical to the diesel derived from petroleum[458] – and, in theory, genetic modification could help us produce other hydrocarbons synthetically,[459] as well as accelerate the disposal of their plastic derivatives.[460]

Maximum growth. Beyond genetic engineering for industrial applications, food crops can be genetically modified in ways that do not raise as many concerns as the genetically modified crops of today. This might include engineering plants to maximize growth within indoor farms, produce larger and/or more nutritious products, or be able to optimally operate under a longer daylight-to-night ratio.

What's Next?

As a system, indoor farming delivers indefinitely abundant supplies of food. Backed by the auspices of Universal Energy and an effectively unlimited supply of both water, electricity and fuel, it can grow enough food to feed the entire planet. This can eradicate the concept of famine as we know it and greatly improve global stability and economic growth – saying nothing of the cascading humanitarian benefits. That, by itself, is a transformational goal to reach.

Yet the tools that make it possible have a secondary, vital function through the provision of the building blocks of next-generation synthetic materials. And that, once delivered, is the final piece we need to evolve beyond a zero-sum resource paradigm, and the final piece we need to build our civilization upward to ever-greater heights.

The next episode of 3D printing will involve printing entirely new kinds of materials. Eventually we will print complete products – circuits, motors and batteries all included. At that point, all bets are off.

- Hod Lipson

Chapter Ten: Materials and Recycling

Electricity. Fuel. Water. Food. Universal Energy and the systems it powers can sustainably deliver all of these resources indefinitely. Yet even with these four resources provided, we still need materials with which to power our ever-evolving economy and build our ever-advancing society. Materials are themselves a resource, and as such they remain subject to the same laws of cost and scarcity as energy resources. Forests get cut down, quarries run dry, and both metal and composites are subject to market forces driven by scarcity.

To address this problem, Universal Energy's next function is to power systems that can synthesize and recycle materials. With these materials, we can build things better, stronger, and less expensively than we can today. Just as importantly, these materials enable us to remove major limitations to our manufacturing capabilities and revolutionize the properties and performance of what we can build. Universal Energy's approach in this context begins with four concepts: advanced synthetics, sophisticated waste management and intelligent recycling, arriving finally at the bleeding-edge of next-generation manufacturing.

Advanced Synthetics

Previous chapters have alluded to technological breakthroughs that have allowed us to build sophisticated systems on scales and at prices that were previously thought impossible. Of these breakthroughs, several have involved the invention of high-performance synthetic polymers – commonly referred to as "plastics." When we think of plastics, we often imagine substances we'd see around our homes or workplaces: grocery bags, containers, appliance sidings, and the like. These kinds of plastics are commonly used in manufacturing because they are easy to produce at relatively low cost.

Yet plastics have problems. For one, they don't biodegrade well and they can't be easily recycled. Consequently – once we're done with them, they're stuffed in landfills, or worse, end up in our oceans at severe environmental impact.[461]

Additionally, plastics that can be recycled at all have a low "percent yield," which is the amount of final material produced from the original supply material.[462] That means it might take 10 lbs. of source material to make 1 lb. of plastic, which is too inefficient to be viable on a large scale – especially if the 9 lbs. of material lost as waste is environmentally toxic.[463] We've made progress towards solving this problem, but making plastics that have both a high percent yield *and* are easily recyclable has proven challenging. Recent advances, however, have made reaching this goal more feasible.

The first of these advances is the one that's made Universal Energy possible to begin with: sophisticated computing. Since computers became prevalent in industrial and research settings, their performance has increased at a truly exponential rate. Today, they're millions of times more powerful than they were in their first iterations.[464] And faster computers – especially boosted by artificial intelligence (A.I.)[465] and machine learning[466] – can help chemical engineers create models to manufacture ever-better synthetic materials.

This becomes all the more possible through quantum computing, a potentially revolutionary method that uses quantum physics to dramatically increase processing speed.[467] Computers today can only process one calculation at a time, whereas quantum computers could, in theory, process millions of calculations simultaneously.[468] While still in its infancy, computer scientists estimate quantum computing could surpass traditional computing as early as 2021,[469] presenting even stronger implications for the future performance of A.I.

Considering that modern computers – while only processing one calculation at a time – can still process quadrillions of calculations per second,[470] speeding this up by a factor of millions could allow for potentially instantaneous data processing of presently daunting calculations. What that allows us to do is model material synthesis with more insight and predict how to create recyclable materials with a higher percent yield at increased efficiency and reduced cost.[471]

The second advancement getting us closer to superior plastics is the genetic modification of algae and bacteria to produce specialized hydrocarbons.[472] Accessing such hydrocarbons can give us an increased capability to tailor synthetic materials to fit our needs. With customizable hydrocarbons, ever-better A.I. and quantum computing at our fingertips, we can have more granular control over the composition of source chemicals as well as detailed models that we can use to manipulate those chemicals into plastics and other synthetics that meet demanding performance requirements.[473]

Add in Universal Energy's ability to decrease energy costs, and we have promising new tools that can unite to revolutionize the materials we use to build and improve our world.

We're already on our way to realizing that future, even with current technology and material limitations. Consider a list of some standouts on the market today:

Nanocomposite plastics. Researchers have discovered that by layering ceramic nanosheets (very thin sheets made from clay) over each other and combining them with a polymer that works similar to elementary-school-style white glue, the ceramic nanosheets will interlace with one another like bricks at the molecular level, creating a structure as strong as hardened steel.[474] This suggests that nanocomposite plastics might have a wide array of potential uses, including high-performance applications within aerospace, transportation, defense, and civil engineering.

High-strength polyurethane (Line-X). Polyurethane is a type of polymer that has an extremely high impact tolerance, which makes it both durable and useful for shock absorption. While commonly used to finish wood products, more robust varieties perform herculean feats of durability. A practical example is Line-X, a nigh-indestructible spray-on coating used to line the beds of pickup trucks and other industrial machinery.[475] To demonstrate its performance, eggs are coated and dropped off buildings onto concrete, watermelons are coated and driven over with trucks, plastic cups are coated and stepped on by Sumo wrestlers – all without effect, as seen at the links in this citation.[476] While these are admittedly hyperbolic demonstrations of polyurethane's capabilities, they nonetheless allude to the potential for more practical applications in any instance where flexibility under stress and impact resistance are requirements.

FR-4. The common name for a high-strength, flame-resistant composite made from glass-reinforced epoxy, FR-4 is one of the strongest synthetic materials in the world.[477] Not only is it highly resistant to chemicals (including acids), ultraviolet radiation, and electricity, it is also lightweight and extremely strong. For comparison, the tensile strength of structural steel and aerospace-grade aluminum are ~40,000 PSI and ~43,500 PSI, respectively.[478] The tensile strength of FR-4 is ~45,000 PSI.[479] This allows components made with FR-4 to retain fine detail and be built to tight tolerances, enabling their use in high-precision manufacturing, aerospace and space exploration.

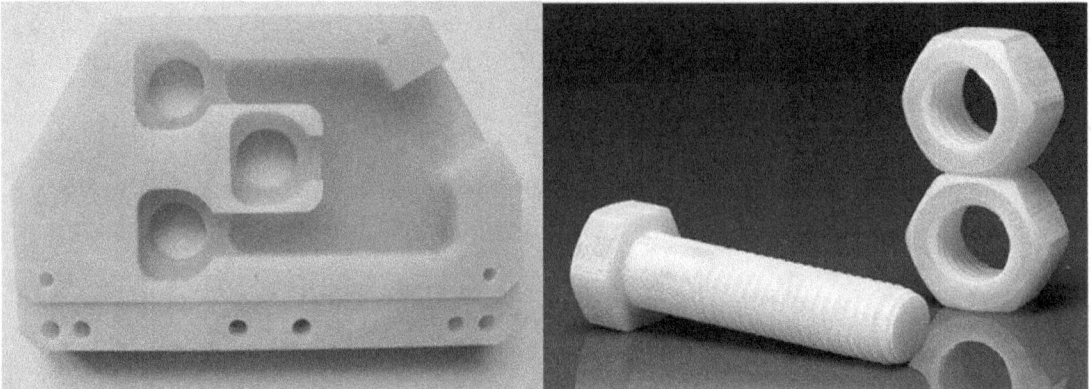

Synthetic wood. Improvements in material science have brought several brands of artificial wood to market. Made from "downcycled" plastics, recycled organic

wood and other composites, synthetic wood is used for decks, framing, siding and supports for millions of structures worldwide. It is already a $3.4 billion industry, and it's still growing.[480] Synthetic wood is fire-resistant and lasts longer than traditional wood. It's also strong enough to support the weight of a locomotive, as the above image of Ecotrax™ synthetic railroad ties demonstrates.

This strength allows synthetic wood to potentially replace traditional wood in the construction of houses, buildings, bridges – anything, really. Cost presently remains a limiting factor, but in a world powered by effectively unlimited cheap energy – and with abundant resources – these costs would likely fall substantially, allowing for an indefinite supply of yet another building material.

Superior metallurgy. Metallurgy has been an evolving science for thousands of years, allowing humanity to build ever-greater alloys from combinations of metallic elements. To create these substances, we've generally needed three components: base materials, a forge to contain heat, and a source of energy to provide that heat. Building forges that operate at specific heats and pressures has not been prohibitively difficult for some time, so delivering the second component is relatively straightforward.

However, the other two components – source materials and heat energy – have been more elusive, or, at least, more expensive. But we've accomplished much with the technology we have today. Researchers have developed new superalloys that push the boundaries of strength-to-weight ratios,[481] new steels as strong as titanium[482] and specialized alloys that can increase the safety and performance of nuclear reactors.[483] A world powered by nigh-unlimited cheap energy and an abundant supply of source materials only stands to accelerate this even further, with improved quality control and more-precise manufacturing following suit.

Although they are samples from a far longer list, these materials are all commercially available as you read this. Each stands to further benefit from the future performance improvements provided by subsequent technological innovation. Yet as impressive as they are, they each ultimately bow in honor of another synthetic material. One that, by itself alone, carries unrivaled potential to revolutionize our way of life. That material is graphene.

The Graphene Key

In Chapter Five, we briefly discussed graphene's potential to make hydrogen storage tanks. But that's only the start of its capabilities. As a base concept, graphene is a single atom-thick sheet of carbon that takes the shape of a hexagonal lattice. Extraordinarily thin in form, it has three distinct traits:

1. **Extreme strength.** With an overall strength roughly 200 times that of hardened steel, graphene is the strongest-known material on Earth – as well

as one of the lightest.[484] It's also highly resistant to both corrosion and heat, with a melting point in excess of 8,132°F (4,500°C). During a 2008 interview with graphene's inventor, Professor of Engineering John Home at Columbia University, he remarked:

> *"Our research establishes graphene as the strongest material ever measured, some 200 times stronger than structural steel. It would take an elephant, balanced on a pencil, to break through graphene the thickness of Saran Wrap."*

2. **Flexibility.** As graphene is formed by sheets of carbon as potentially thin as one atom it can retain a rigid structure or be as flexible as a sheet of paper. It can further function at full strength when assuming a range of shapes, even when designed to bend or stretch.

3. **Conductivity.** Graphene is a lattice of carbon atoms, and in such a form is an excellent conductor of electricity.[485] This enables graphene to lend its strength and flexibility to anything electrical – as a function of either transmission, structure, or storage.

The combination of these three traits make graphene useful for a wide array of applications that serve critical roles in our society.

They include:

Consumer electronics. As an ultra-strong and ultra-conductive material, graphene can be used to create sophisticated electronics that are highly durable. The following images show prototype mobile phones with graphene screens that are both flexible and thousands of times stronger than today's phone screens.

Electronics made with graphene can also store large amounts of data in small physical spaces.[486] The following images show a concept for a new type of jump drive that works like sticky post-it notes,[487] which while not currently in production are well within the realm of technical feasibility.

Graphene's conductivity also gives it high capacities for data transfer, some 7,000% faster than today's commercial methods.[488] Researchers have conducted experiments that show graphene antennas can transmit data at speeds of up to 100 terabytes a second.[489] To contextualize, a high-definition feature-length film generally ranges from 3-9 gigabytes in size. There are 1,000 gigabytes in a terabyte. Assuming an average size of 6 gigabytes, this translates to a transfer capacity of approximately 166 high-definition films per-second.

Medicine. High strength and conductivity with low weight and reactivity gives graphene excellent potential for medical applications.[490] Uses include stents to prevent arterial restriction, high strength and lightweight casts for injuries and providing the physical scaffolding to help paralyzed people walk again.

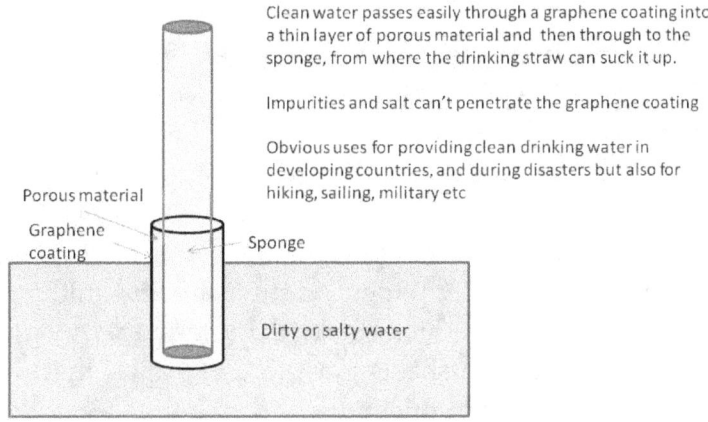

High-performance energy storage. Graphene is uniquely well-suited to hold energy in capacitors, which work through direct energy storage as opposed to batteries that technically require a chemical reaction to generate a charge.[491] Most portable electricity in our day-to-day lives involve batteries, and they're generally preferred to capacitors because they normally have a higher "energy density" – the amount of energy stored in a system per unit volume.

But graphene's different.

Because it's so thin and conductive, graphene can form "supercapacitors" that can function at equal to even superior levels than high-performance lithium-ion batteries.[492] These supercapacitors are made by layering sheets of graphene between an electrolyte that's then encased in non-conductive plastic. The following two images show a theoretical cross-section of a graphene capacitor and a real-world prototype of such a capacitor, respectively.

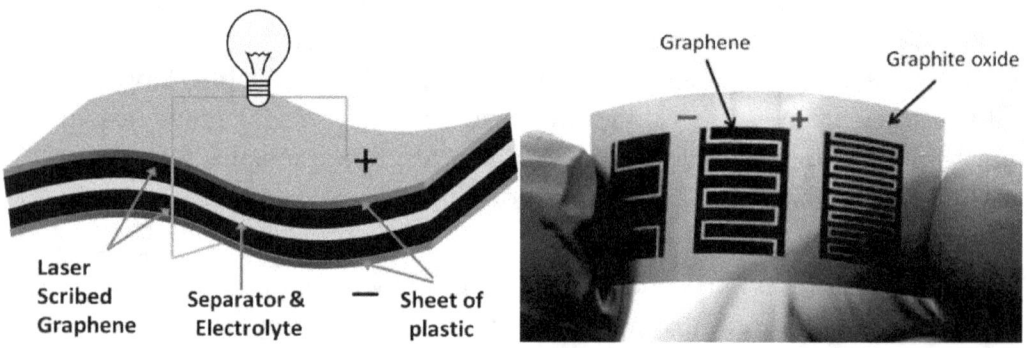

But these prototypes reflect the relatively new emergence of graphene research. As graphene can be made microscopically thin (theoretically to the scale of an atom), advances in manufacturing can eventually enable us to drastically increase the potential surface area of graphene used for direct energy storage.

To conceptualize: let's imagine a sheet of printer paper, which is 0.1mm thick.[493] Pretty thin, right? For human eyes, certainly, but not when we're thinking of a substance that could potentially be *microscopically* thin. Let's say that we become able to reliably manufacture insulated graphene capacitance sheets to the thickness of a micron (0.001mm). As a sheet of 8.5x11" printer paper has a thickness of 0.1mm,[494] it would take 100 micron-thick sheets of graphene, then, to comprise one sheet of 8.5x11" printer paper. As there are 25.4 millimeters to the inch, we can derive that a one-inch-thick stack of printer paper would have 254 individual pieces. At 100 graphene sheets per piece of paper, that translates to 25,400 sheets of graphene per inch of thickness.

It's not uncommon for a standard car battery to have a width and depth similar to a sheet of printer paper. If we were to assume a height of ten inches, we could layer as many as 250,000 graphene sheets in that same volume. At 8.5x11", a sheet of printer paper has a surface area of about 0.65 square feet. At 250,000 sheets strong, that would amount to a total of 162,500 square feet. A football field, for comparison (including end zones) has a surface area of 57,600 square feet.[495] That means this graphene box would contain the equivalent surface area of 2.5 football fields, yet compacted to the size of a car battery – and made with flexible, non-toxic materials that are 600 times lighter than structural steel and 200 times stronger.

A 2017 paper from *The Nanotechnology Journal* claimed that graphene supercapacitors today can deliver levels of performance that begin to eclipse industry-grade lithium-ion batteries.[496] Lithium-ion batteries are also much heavier, more expensive and have increased environmental and humanitarian costs.[497] That graphene supercapacitors in their infancy can already compete with high-performance lithium-ion batteries – as well as directly interface with hydrogen fuel cells – further supports their candidacy for future deployment as a solid-state energy solution.

Structural material. Extremely strong. Completely flexible – or reliably rigid. Capable of storing tremendous energy in a tiny footprint. Graphene isn't revolutionary because it's impressive in each of these categories, it's revolutionary because it's impressive in *all* of these categories.

Consequently, it can transform the way we build things. Take the electric car for instance.

Instead of an electric car storing packs of heavy batteries, the car's chassis can be interwoven layers of graphene and **can itself be the energy storage medium.** When you think of the 162,500 square feet of surface area that could be potentially stored within the size of a car battery, an entire vehicle chassis is a subsequently higher order. We could potentially reach points where we could interweave millions of square feet of energy storage medium within a single electric vehicle – increasing range, reliability and structural integrity at far less weight.

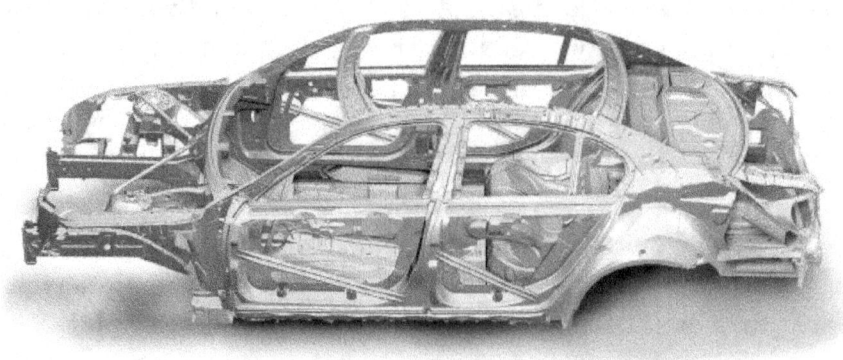

Graphene enables us to wield strength, flexibility in form, and energy storage all at once, each potentially at a higher degree of performance than we can manage with leading commercial solutions today. Tomorrow's skyscrapers, bridges, public infrastructure, transportation systems – each can also serve as electricity storage. And through Universal Energy's core offerings of electricity, fuel and heat, we have the means to synthesize graphene and other revolutionary materials easier and at less expense than we can today.

While this is key to a clean energy future, making materials is only one part of the equation. Whatever we construct must in turn be recycled or disposed of in an end-of-life cycle. The framework accounts for this through several approaches on both land – and sea – that can present major environmental improvements.

Next-Generation Waste Management

Technology has brought us the litany of conveniences that define our modern era, yet these conveniences have come with a massive waste footprint – a problem that has proven deeply challenging to manage. We've taken several approaches to process our immense volumes of trash: burying it in landfills, attempting to recycle old materials into new materials, "downcycling" waste into usable products (such as making park benches from shredded plastic bottles), or simply burning it. But none have truly succeeded in managing this issue on a global scale.

Universal Energy intends to change that. While it's true that a goal of the framework is improving our ability to recycle things like plastics from the design stage, recycling complex polymers is inherently difficult. And when we consider the extent of waste across the globe, the problem seems daunting, no matter how sophisticated our technology becomes.

Worse, even while advances in recycling certain materials show promise, and getting rid of single-use plastic products is a laudable goal, each ultimately does little to turn the tide of our global waste accumulation.[498] The solution to this problem must be massive and immediate, effective and achievable – and all on a planetary scale. There presently aren't many technologies that can meet these requirements, but one called plasma gasification can.[499]

In concept, the "gasification" of waste isn't much different than incineration – simply burning it as humanity has done for millennia. But gasification uses a high-intensity electric current to create a plasma – an ultra-hot state of ionized gas – to separate trash into two separate states: the first, a material known as "syngas" that can be used as a fuel in myriad applications, as shown on an accompanying overview diagram. The second is an inert slag that can be used to make other useful materials like concrete, roads and insulation.[500]

Plasma gasification itself can also be harnessed as a potent source of energy, which can be used to both sustain its electric current and power other processes.[501] Normal incineration requires fuel and works at much lower temperatures, leaving behind harmful byproducts that present dangers to both public health and the environment. Gasification avoids these issues completely.

Overview of Plasma Gasification:

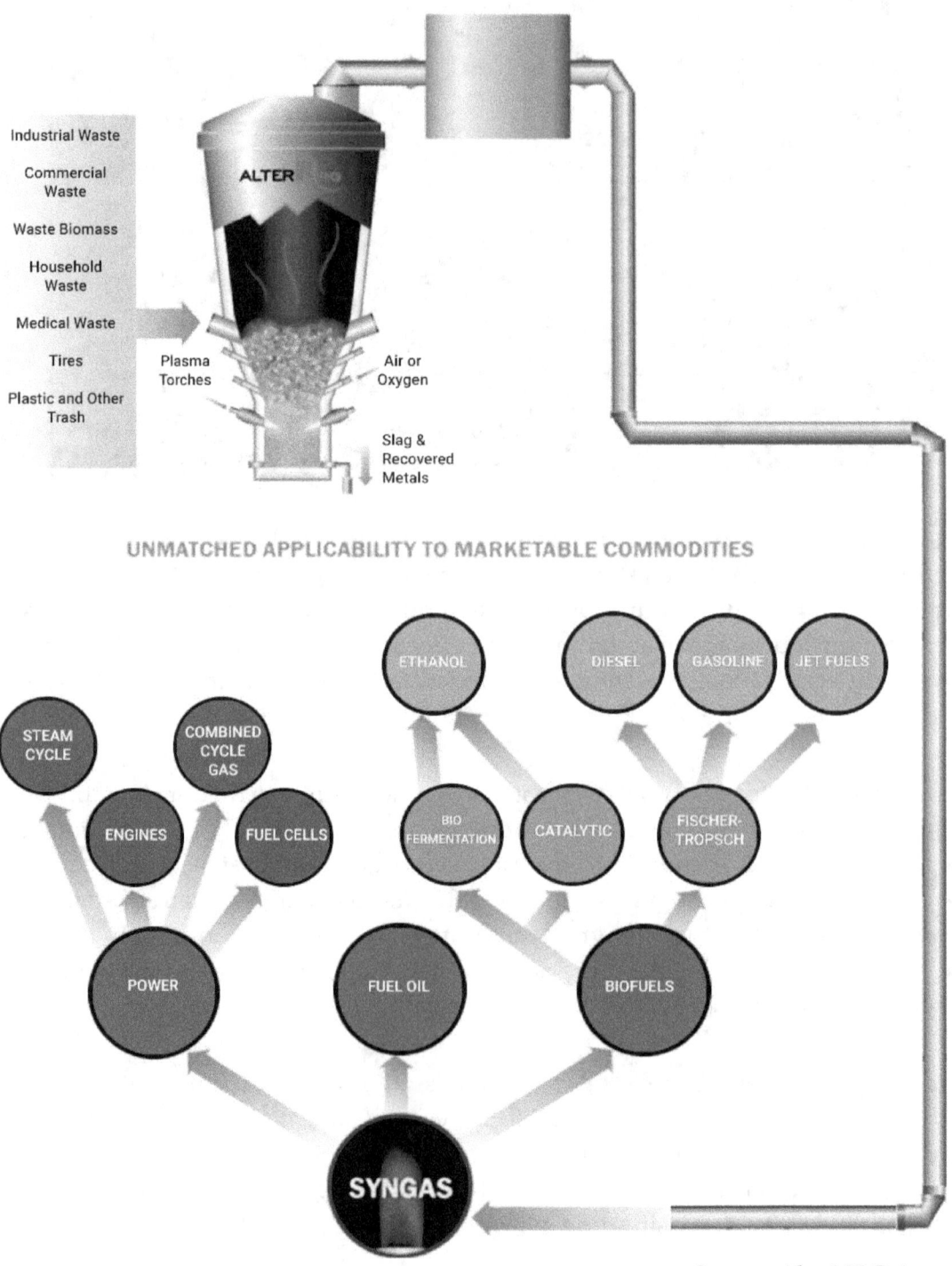

Source: AlterNRG, inc.

Gasification vs. Incineration

	Plasma Gasification	**Incineration (burning)**
Temperature	High temperature, 1,400-1,600 °C	Medium heat (870-1200 °C)
Emissions	Does not create air pollutants.	Allows toxic atmospheric pollutants to form.
Byproducts	~99% of waste material becomes usable material.	~30% of the waste material remains as toxic ash.
Useful Products	Syngas and an inert slag which is used in building materials and roads.	None. Incineration is among the most expensive and polluting methods of waste management.
Power	Generates power. [502]	Requires fuel.

Table source.[503]

Plasma gasification is an integral part of Universal Energy, that when partnered with other technologies in the framework, brings sophisticated waste management to any energy-generating ensemble. This application is vital to a clean energy future that presents minimal impact to Earth's ecology. But it can be applied further to an even more critical task: cleaning our oceans.

Over decades of mass extraction, international consumerism and irresponsible waste disposal, ocean trash has become a major problem on a global scale. There's so much garbage floating in Earth's seas today that it's now accumulated into patches that are thousands of miles across.

Plastic all at sea
79,000 tonnes of plastic is floating in one patch of the Pacific Ocean

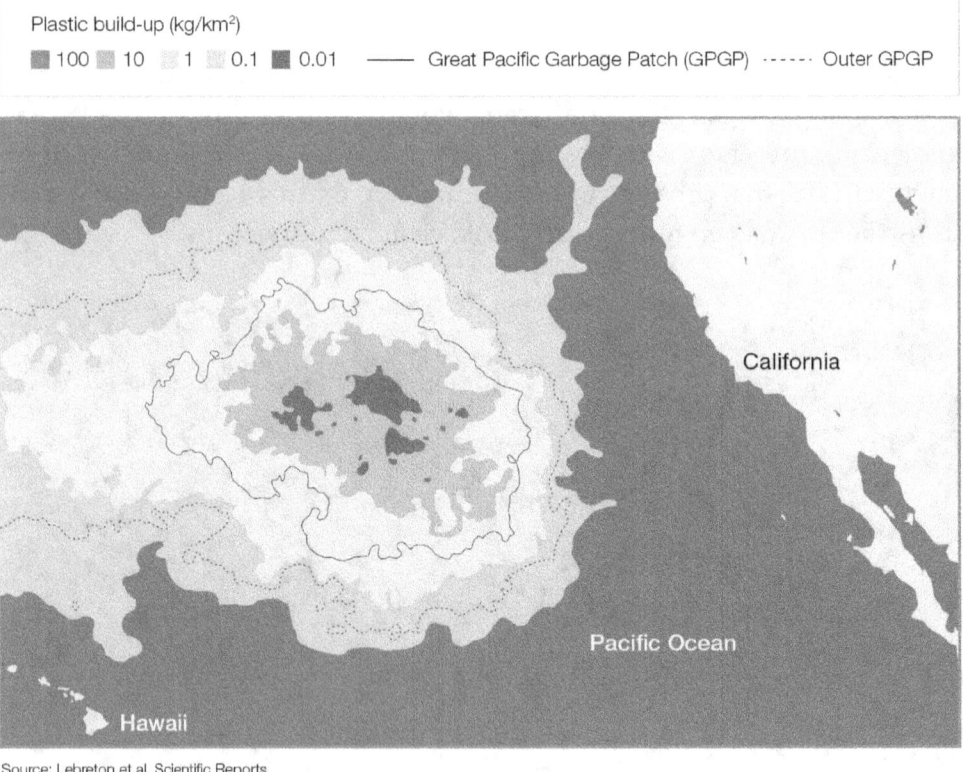

Source: Lebreton et al, Scientific Reports

Each major ocean has a large accumulation of trash within primary ocean currents (referred to as "gyres.")[504] The picture above outlines the "great pacific garbage patch," one of five worldwide.

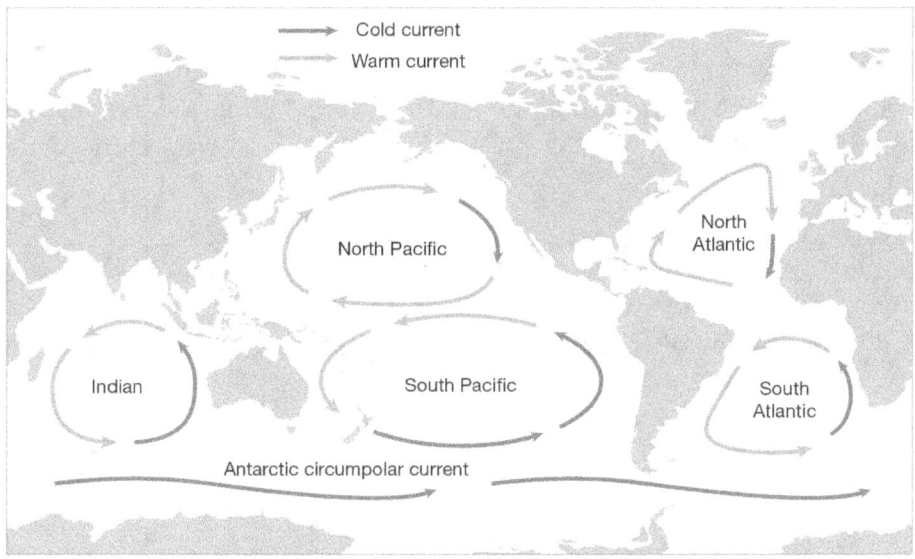

Worse, waste from these massive piles is continually decomposing into smaller pieces – and these smaller pieces are being consumed by marine life. This poisons the creatures it doesn't kill, which eventually makes its way into humans when we eat seafood. Currently, some 3.2 billion people rely on seafood for almost 20 percent of their animal protein intake.[505]

Science and industry are aware of this problem and have invented promising tools to help reverse course. The Ocean Cleanup Project, for instance, recently launched a gigantic sieve consisting of floating pipes and netting that corrals trash into a U-shape for future processing.[506] In concept, several of these sieves would float with ocean currents to slowly accumulate trash over time, with the hope of eventually cleaning up marine environments completely.

But that step of "accumulation" presents another important question. Once the trash is corralled, what do we do with it? And how does cleaning up ocean trash in one place prevent it from being reintroduced to marine environments elsewhere? Universal Energy's answer to this question is something called a "Trident Facility."

A Trident Facility is built like an offshore oil rig – a large floating facility that can navigate any ocean in the world. Yet instead of drilling for oil, it would both synthetically produce resources and dispose of ocean trash cleanly and safely.

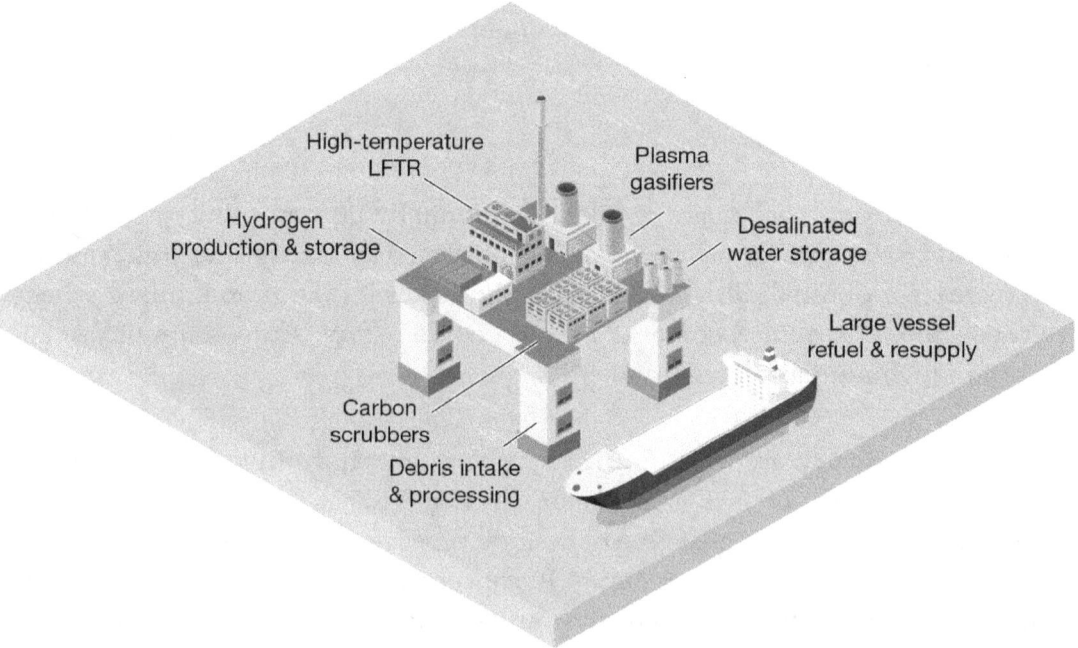

Here's how they would work:

1. The primary power source of a Trident Facility is a small LFTR, which provides the core power for the plasma gasification of ocean trash and the residual energy needed to power auxiliary, resource-producing systems.

   A cooperating series of Trident Facilities would scour the oceans and suck in trash from one of their four floating legs or a central lift. Ideally, these would work in conjunction with floating trash collectors like those from the Ocean Cleanup Project. From there, trash would be processed via a plasma gasifier to become syngas or the slag that can be turned into useful materials.

2. The excess energy generated from the LFTR and the plasma gasifier would be further used to extract fresh water and hydrogen fuel from seawater, just as with CHP Plants. This could be used to resupply large ships on-demand, enabling Trident Facilities to serve as minature ports or emergency safe havens.

3. Beyond trash gasification, hydrogen production, and seawater desalination, any excess energy produced by Trident Facilities would be used to power atmospheric scrubbers to reduce air pollution and greenhouse gasses in the atmosphere.

Trident facilities earn their namesake because they perform three unique roles: ocean cleanup, resource production and atmospheric scrubbing: three points of Poseidon's trident. They enable us to work to eradicate ocean trash, while using the byproducts to make useful materials. This helps us come ever closer to the material revolution Universal Energy seeks to make the new normal.

The Circular Economy

We build thousands of products on an industrial scale today, but once those products become obsolete, the materials used in them have often not been recycled or repurposed outside of scrap yards. Engineering and financial limitations, difficulties with transporting materials to recycling locations, extant systems with which to recycle them, and the energy necessary to power those systems are all limiting factors. With Universal Energy, it becomes easier to take a sophisticated machine, strip out non-recyclable materials and send them off for plasma gasification, and extract recyclable substances that remain – retaining them for use elsewhere in manufacturing and fabrication.

The underlying concept behind this approach is commonly referred to as a "Circular Economy" – an economic system aimed at reducing waste and maximizing efficiency through the continual re-use of resources and materials.[507] In most instances, implementations of circular economies seek to minimize the use of new resources and the creation of waste, pollution and greenhouse gas emissions, and instead employ concepts of refurbishment, remanufacturing and intelligent recycling at scale to keep already-present resources and materials in play.[508] The underlying idea is that "waste" is contextual, and opposed to a traditional "linear economy" that functions on a production model of "extract, manufacture, sell and dispose" a circular economy leverages as many technological methods as feasible to enable regenerative resources and materials for the "circular" reproduction – and upgrade – of systems in a model that's as closed-loop as possible.[509]

Linear vs. Circular Economy

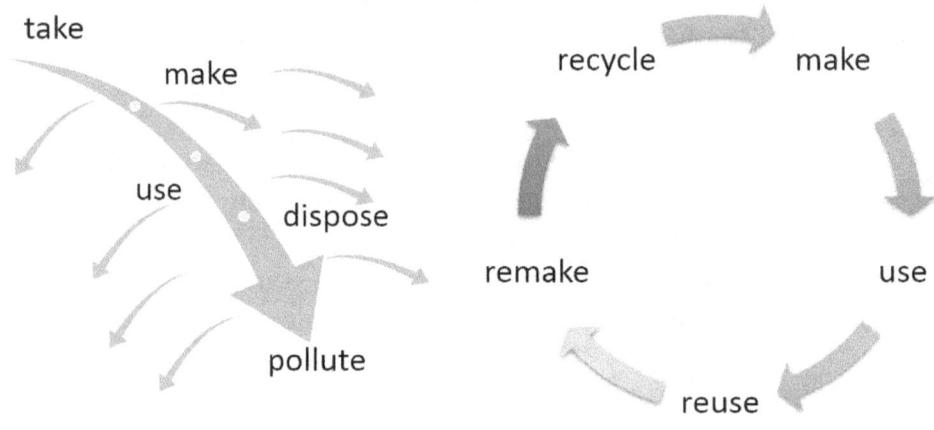

CC 3.0 Catherine Weetman 2016

Implementations of circular economies are not necessarily novel concepts, and the systems that can perform such processes already exist today. But Universal Energy significantly increases the scale at which this can be accomplished while lowering energy and resource costs. As a result, we would be able to supplement next-generation manufacturing with recycled materials more effectively than we can today, both reducing the need to acquire new materials and enabling the expansion of circular economic models on larger scales – nationwide, or even global. Not only does this improve how we can build things from the design stage, it fundamentally transforms our capabilities to manufacture, recycle and upgrade advanced systems. Examples include:

Intelligent deconstruction. Many products today are simply too complex to be recycled cost-effectively. Consequently, once discarded, they are often thrown in landfills. Reduced energy and material expenses, modern computer systems, and novel engineering methods have lowered the costs of recycling consumer electronics, vehicles, ships, buildings, and airplanes. And as we improve how those things are recycled – or more specifically, disassembled and recycled – this gives us the opportunity to tweak our manufacturing methods to build products that can be disassembled, recycled, and/or upgraded in a modular capacity.

Reduced costs and material requirements. Beyond removing energy as a significant expense, the ability to repurpose older models into newer ones means we don't have to build new products from scratch. For example: instead of a vehicle going to a scrap yard once it gets old, we could instead engineer that vehicle to be stripped down to its basic components by a factory and refurbished into a newer model – a concept that could be applied to essentially any product.

This form of recycling benefits both consumers and manufacturers. It gives consumers the ability to exchange obsolete models for credit toward a newer model and reduces the expenses manufacturers pay to acquire materials and create products. These benefits contribute to cost savings for both parties, and also increase the agility of manufacturing by allowing us to do more with what we have currently. Plenty of companies strive to reach this goal, but increasing the number of companies who do so alongside advances in energy generation and material procurement increases the efficacy of this recycling method – promoting its adoption within a greater share of market sectors.

Reduced waste footprint. Intelligent waste management, an abundant supply of cheap energy, indefinite supplies of synthetic materials, and deconstruction-focused engineering reduces the waste footprint of manufacturing processes. With Universal Energy, we wouldn't need to build as many products at the expense of the environment, nor would we need to build products that are destined for landfills. By upgrading and reconnecting our energy production, material procurement, manufacturing methods and recycling infrastructure, the lifecycle of a product is contained from start to finish – reducing nature's presence in the equation.

From Here, We Have Now Reached Three Important Goals:

1. With Universal Energy, it becomes much easier to make next-generation synthetics that can outperform even the most advanced materials we have today – synthetics that we can use to build and improve practically anything.

2. With Universal Energy, we can dispose of waste and recycle products easier and at far greater scale than we can at present.

3. With goals #1 and #2 met, we can now design systems and products to eventually be disassembled, repurposed, and/or upgraded – turning products into long-term investments instead of one-use discardables.

Combined with computer-aided manufacturing methods like additive/3D printing, this can transform the way our society manufactures products.[510] Traditionally, such manufacturing approaches have allowed companies to rapidly prototype new designs or hobbyists to make improved or replacement parts for various projects.

However, advances in commercial applications of this technology have revolutionized the capabilities present at our fingertips, enabling us to manufacture complex shapes and even the preliminary foundations for replacement organs.[511]

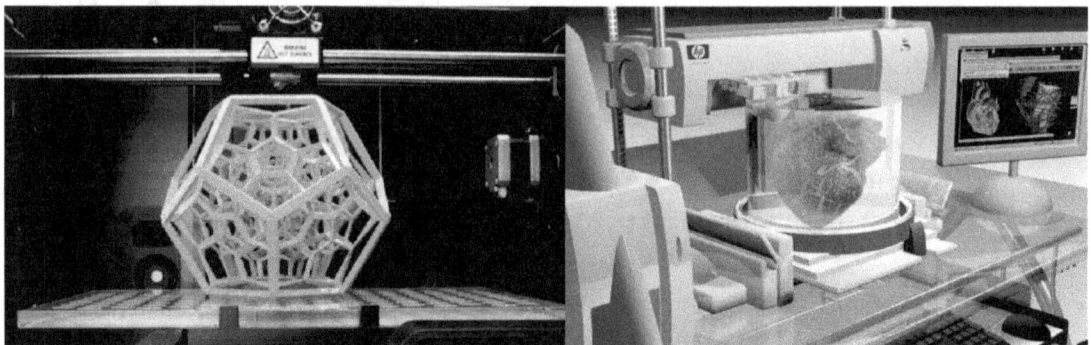

Left: 3D-printed multidimensional object. Right: concept image of 3D-printed heart.

One example with strong industrial potential is present in modern metal 3D printing, such as models from the MarkForged company – a leader in 3D metal printing innovation.

MarkForged's Metal-X models are capable of building complex shapes to the micron or better tolerances out of high-strength thermoplastics, fiberglass, Kevlar, carbon fiber, titanium and both stainless and high-strength tool steel.[512] According to their already-at-market commercial data sheets, their printing methods can build shapes up to 100 times less expensively than traditional casting or machining.[513]

The following three images respectively show a pump impeller, camshaft sprocket and aircraft bracket printed with their technology.

Another example is "selective laser melting" – a method of applying a high-intensity laser to a bed of fine metallic powder (steel, aluminum, titanium, metallic alloy or graphene composite) to manufacture highly detailed shapes, also at micron-or-better tolerances, that meet the highest material strengths commercially available.[514]

Like 3D metal printing, these shapes can reflect practically any attribute that can be envisioned on any axis, and be either solid, hollow, or solid with embedded pathways for isolated fluids.

Image source: DMG Mori

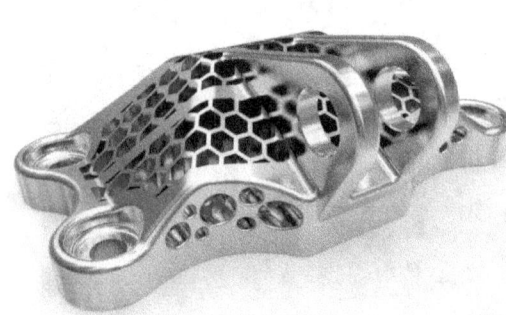

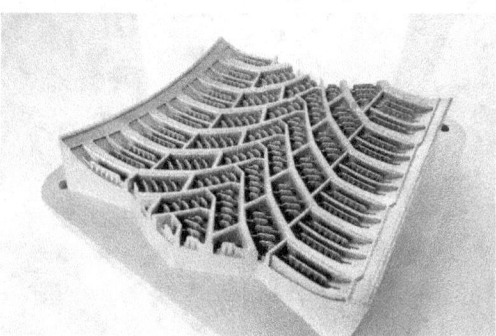

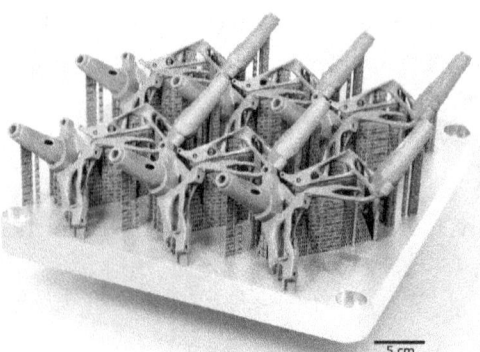

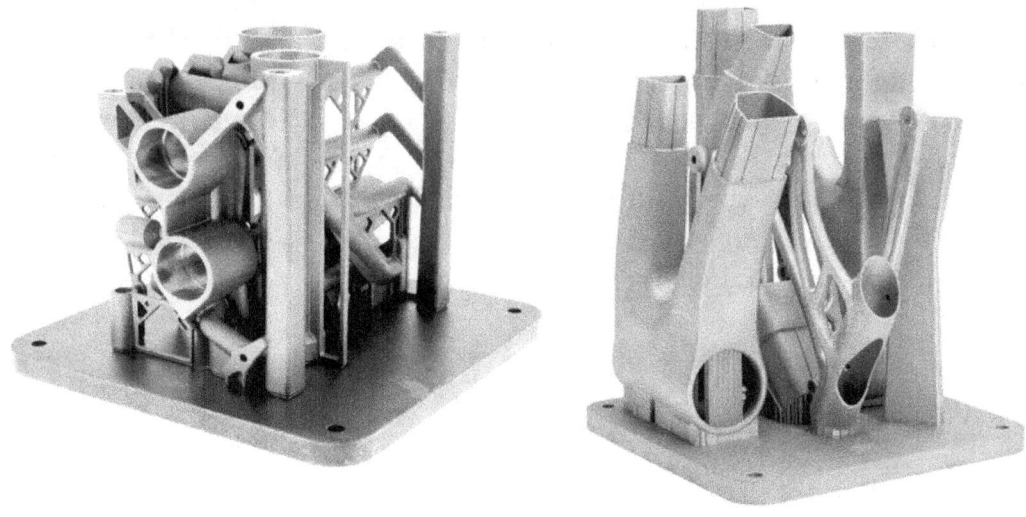

Image source: DMG Mori

Without question, the sophistication and precision of these capabilities raises the bar of our manufacturing prowess to uncharted heights. Yet because their deliverables can now also be manufactured as iterative models at the push of a button – and can also further be built using the strongest and most resilient materials presently available – these advances *can be integrated into any aspect of a modular, standardized and scalable manufacturing chain.*

Combined, this changes the game completely – all the more so since these technologies are in their infancy and stand to advance over time with proportionally higher sophistication and proportionally reduced costs. It's the last piece of the puzzle that we require to build whatever we could need or want. When backed by a technical framework that provides effectively unlimited energy and critical resources by design, this enables us to extend that capability to any social sector we could imagine. And that, in sum, allows us to land the final blow to defeat resource scarcity outright.

When I think about creating abundance, it's not about creating a life of luxury for everybody on this planet; it's about creating a life of possibility. It is about taking that which was scarce and making it abundant.

- Peter Diamandis

Chapter Eleven: The End of Resource Scarcity

We've frequently touched on how Universal Energy is based on a mindset of standardization and modularity. If you recall, modularity is the idea of designing a system to be flexible yet *standardized* in deployment. A great example of this are USB accessories. Devices of all kinds – from webcams to smartphones to external hard drives – can connect to your computer via the same *standardized* USB port, yet can be *modularly* added, removed or swapped with ease.

Another familiar example is an AC power cable. Every electronic device in your home connects to power via the same type of standardized plug. Each device doesn't have its own unique connection – that would be a crazy mess! That's why they're all built to a universal standard. Audio/visual ports (HDMI), and Bluetooth devices are all extensions of this idea.

Standardizing a function to be modular reduces complications for building things and lowers the bar (and research + development costs) for manufacturing. For these reasons, standardization and modularity are driving principles when building sophisticated products. But these principles have only been taken so far.

We saw earlier how most power plants are built as unique entities – they might standardize a doorway, railing or stairwell, **but the system as a whole is essentially *made to order.*** The same is true with most larger-scale things in our society. With few exceptions, every bridge built, tunnel dug, railway laid or building erected was done so as a custom entity – made to order, each and every time. This is because we are presently living in a world with technical limitations that would make it difficult to build something like a skyscraper or bridge on a factory assembly line. Removing this limitation is one of the final functions Universal Energy is intended to perform.

With a nigh-unlimited supply of all critical resources – especially energy, fuel and materials – we have the building blocks to build as much as we want, however we want. As we further have sophisticated computing and modeling, next-generation manufacturing with advanced synthetics and precise tolerances, we can automate the construction of sophisticated systems on a larger scale.

The application of this idea is commonly known as "prefabrication" – building something in a factory and assembling it at a final location instead of constructing it from scratch with basic building materials. It's an approach we've been improving for years, but recent advances in manufacturing have enabled us to increase its scale, sophistication and potential applications.[515]

For example, this is a prefabricated house:

This house was not constructed at this location, *it was <u>assembled</u> here*. There were no workers on-site cutting wood for framing or nailing in subfloors. Pieces of this house were built on a factory assembly line, just like we build vehicles. They were delivered by a truck to a construction site, and this house was assembled in a matter of days.

This house came with a finished interior, with all electrical, plumbing and heating elements pre-installed. Should the homeowners decide one day that they want to expand the size of their home, it would be a matter of bringing in a new piece, removing modular components from the original house, and fastening the new piece into the whole. If they wanted to move, they could disassemble their house, put it on trucks, and assemble it again somewhere else. Essentially, we can now build houses with life-sized Legos.

Prefabricated houses have been growing in popularity,[516] especially since they offer high energy efficiency and durability. However, the price of these homes is still comparatively steep. Today, the cost to deliver a fully finished prefabricated house ranges between $140-$200 per square foot or more, considerably higher than the $125 per square foot national average for a traditionally constructed home.[517]

But this price range includes expenses inherent to any fledgling industry – initial research and development, prototyping, and marketing among them – and those costs have to be recouped through fewer sales in a smaller-if-growing market. The energy and materials needed to both construct and transport these prefabricated homes are also a considerable expense. Universal Energy helps to mitigate these cost factors, and with future advances in synthetic materials, the total cost of these houses could drop significantly.

Houses are only one example of the potential benefits of prefabrication, as practically anything can be built this way: LFTRs, renewable energy, Multi-Stage Flash Distillation Facilities, hydrogen production systems, CHP Plants, National Aqueduct and urban vertical farm components, Trident Facilities, and even larger buildings and civil infrastructure.

To understand the implications of prefabrication, take a look at the following two images. The image below and to the left shows the Boeing Corporation's Everett, Washington facility that can mass-manufacture a complete 737 jet aircraft *every nine days*.[518] Their flagship 787 Dreamliner aircraft can be manufactured to completion in as few as *seventeen days*.[519] (Don't let the indictable corner-cutting[520] of the 737 MAX scandal fool you, either – both Boeing and Airbus have been mass-manufacturing commercial aircraft with 100% safety reliability for years). The image below and to the right shows a 30-story prefabricated structure built by the

Broad Sustainable Buildings corporation in Changsha, China that was assembled on-site in 15 days.[521] That's two stories of a building, per day.

For comparison: *both of these feats were accomplished faster than the time it takes Budweiser to brew a bottle of beer.*[522]

The start-to finish timeline of the four pictures below is *15 days*.

These achievements were accomplished with today's technology. With the energy cost reductions and improvements to both manufacturing and materials Universal Energy brings, the possibilities expand. We can sustainably prefabricate advanced systems on massive scales, and we can build things better and less expensively than we can today. This enables us to dramatically advance our economy, society and infrastructure. But we can also ensure shelter as a resource – which brings us back to housing. At this scale of manufacturing prowess, building small residences on assembly lines becomes trivial.

For example, the images below show houses made from shipping containers – the same kind used to transport goods on trucks and cargo ships. Shipping containers are so inexpensive to make that in some cases, it's actually cheaper to use new containers than it is to ship the empty ones back to their origin.[523] Thousands of containers nationwide are routinely left near shipping yards, prompting innovative architects to use them as housing:

Shipping containers are plentiful and, naturally, easy to transport. Each container is made from steel, which is extremely resilient and boasts a high load strength. With today's energy, shipping, and manufacturing costs, a single-container home can be delivered for between $20,000-$40,000.[524] The last home on the previous page, for example, was fully constructed for less than $40,000.[525] If energy and material expense reductions are applied by way of Universal Energy, these costs would likely drop substantially.

Shipping containers are far from our only prefabricated housing option. U.S. tech startup ICON, for example, can prefabricate a 650-square-foot house in less than 24 hours at a cost of $10,000 or less.[526] They use a large 3D printer to pour a concrete mix layer by layer, creating a solid structure that's significantly stronger than traditional stick-framed construction. The company has already built more than 800 homes in partnership with local communities in Bolivia, Mexico, Haiti, and El Salvador.[527] In developing nations, the company estimates homes like the ones below could be manufactured for less than $4,000.[528] ICON's market sector is shared by companies in Russia,[529] Dubai[530] and Amsterdam that manufacture comparable models.[531]

The MADi corporation in Italy has taken a different approach, using folding joints to create small residences that can be set up in hours.

As a modular, standardized structure with a base cost below $40,000, MADi folding house shapes can be integrated together and extended to feature a wide array of configurations:

These advances in prefabricated residential construction have given our society a cost-effective method of manufacturing and transporting houses practically anywhere. This especially includes homes built small enough to be deployed on smaller plots of land that are either publicly owned, extended via land grant or purchased using charitable funds – further increasing social utility and philanthropic value.

But what are the greater implications of the expansion of materials and construction methods for smaller-scale residential dwelling? Most importantly, for a modest investment we can now provide quality living spaces for anyone who needs a home, such as:

Victims of natural disasters. As events like Hurricanes Katrina, Harvey, Maria and Dorian have demonstrated – alongside tornados, flooding and ever-worsening wildfires around the world – millions of people can be displaced from their homes after natural disasters. Displacement traditionally leads to depression, social unrest, higher crime, and reduced economic activity, among other social problems – all of which are often cyclical in nature.[532]

While temporary FEMA trailers have granted some relief in the U.S., these shelters are only free to use for a limited time, and at $70,000 per unit, each costs several times as much to produce than a prefabricated living space of similar size.[533] Rather than using more expensive temporary FEMA trailers, we can now deliver inexpensive prefabricated homes that can have integrated heat and hot water, providing a comfortable, warm, and private space to displaced people in both the U.S. and abroad.

For example, during the 2010 Haiti Earthquake and its aftermath, roughly 105,000 homes were destroyed and another 208,000 badly damaged.[534] International governments devoted millions of dollars to assist, with some $93 million going to build some 2,600 homes – a cost of roughly $36,000 a house. Though approximately $13 billion in total international aid was donated so that Haiti could rebuild, much of the country today looks little different from how it did in the immediate aftermath of the earthquake.[535] Had we been able to purchase 2,600 prefabricated homes at $30,000 each, it would have cost $9.3 billion – meaning that we'd have provided living spaces to replace every destroyed home with another $3.7 billion to spare.

Low-income/fiscally reserved individuals. The average price for a single-family home in the United States is nearly $300,000 – an obstacle for even the median wage earner in this country.[536] The millions of families who are forced to rent are in an increasingly precarious financial situation, as rent prices have largely increased in inflation-adjusted dollars over the past 30 years while median income has not.[537]

Perhaps a family can't afford to buy a house and are forced to rent at the expense of their ability to save money or invest in something they own. Conversely, perhaps a family wishes to purchase a modest home on a larger plot of land with more cash on hand as opposed to a more expensive house with a heavier mortgage. Prefabricated homes make either possible, allowing people to take advantage of the value of home ownership at lower prices than are possible today.

People experiencing homelessness. There are currently an estimated 565,000 homeless people in the United States,[538] and every year the Federal Government spends approximately $4.5 billion on efforts to reduce that number.[539] Assuming a cost of $30,000 for a small prefabricated home, we could provide a comfortable and private living space for every homeless person in this country for $19 billion. Assuming $10,000 for a 650 square foot 3D-printed home, we could provide the same for just $6.5 billion. That's what the Federal government spends on preventing homelessness every 2-5 years. Also of note: that's between 1-3% of the annual defense budget.[540]

It's worth mentioning that providing a private living space to get people off the streets isn't necessarily going to fix any underlying reasons for their homelessness, as afflictions like drug addiction and mental illness are often factors.[541] But reaching the ability to extend the most vulnerable among us a place to live and rebuild their lives is key to solving major social problems. The lowest a person in the United States (and perhaps abroad) would thus be able to fall is a private living space with heat, food and hot water – a historical first.

Being able to accomplish these goals in aggregate is a milestone of major significance. It represents a massive leap in our societal advancement, and more critically, it's the final nail in the coffin of resource scarcity.

By combining the systems described in this and previous chapters, we would have the means to synthetically produce everything we need to exist: electricity, fuel, water, food, advanced building materials, and now shelter, and we would have the means to produce them far less expensively than we can today.

Indefinitely sustainable production of the crucial resources and amenities our civilization needs to function would be revolutionary, completely changing how we relate to people within our neighborhoods, our nation and our planet.

Critically, this would allow us to reset our relationship to nature.

Since we evolved from hunter-gatherer tribes and started building societies, the environment around us has paid the price. We have razed forests, destroyed ecosystems and altered our planet's climate. The rise of human civilization, in and of itself, has been an extinction-level event. Universal Energy allows us to chart a different course because it can provide every resource that we need to exist and advance without relying on perpetually invasive extractive technologies. This alone greatly reduces the damage we inflict on nature by decreasing our reliance on what are essentially finite resources.

It's true that technological advances might increase the extraction of certain materials. But with superior recycling and manufacturing methods, this can be minimized – and would ultimately pale in comparison to the other environmental benefits we would see with Universal Energy. We would no longer need to cut down forests for building materials, extract finite sources of oil and gas for energy, or devote swaths of land for farming. We would no longer need to deplete natural water sources for drinking, industry, or agriculture. We would no longer need to pollute our atmosphere or destroy waterways with toxic chemicals. Over time, the Universal Energy framework would allow nature to return to its natural state, and heal to a point before our hands scarred it.

And, this would remove the primary cause of human conflict.

For thousands of years, for thousands, we have butchered each other. Whether by the sword, the arrow, the bullet or the bomb, we have exterminated our brethren in every horrific manner we could think of. In this doing we have told ourselves *lies,* and allowed ourselves to believe that we were justified in killing and dying by the millions for causes that boiled down to nothing more than resource scarcity and the pursuit of the money, power and economic might it bestows on the winner of its zero-sum games. We have believed these lies and lived with these horrors because we thought we had no other choice. And whether near or far from the results of their manifestation, we have been powerless to prevent it all from repeating for time eternal because we had no means to truly change how the world worked.

Now, we do.

Technology can finally empower us to evolve beyond the zero-sum game of resources. No matter how much energy, water, food or materials are consumed by society, we can always generate more. With that, we can not only build transformational things, but further transform the very means and tools with which we build them – and change the world from the old model to the new.

Paradigm Shift

Universal Energy is first and foremost a framework, and its ultimate purpose is to make a new model for our society. It doesn't seek to use money to pay for social programs that mitigate social problems. It instead seeks to use money to build systems that make those problems irrelevant.

10,000 years ago, making fire was a problem. Today, you use a lighter. 300 years ago, transportation over distance was a problem. Today, you hop in a car, bus, train, or plane. 100 years ago, communication was a problem. Today, a phone call can reach any corner of the globe. Right now, today, energy and resources are a problem. **Through technology, they don't have to be a problem tomorrow.**

For millennia, resource scarcity has been central to human interaction, chaining and binding us to its restrictions. With its bounds removed, we can focus the entirety of our social strength towards transforming our civilization into something unprecedented.

By dramatically lowering the costs of energy, resources, and materials while improving the quality of life for everyone, costs fall, as does the amount of resources that have to be devoted to addressing social afflictions. This frees up collective funds that could be devoted to social advancement, and the same goes for industry, which would have greatly increased capabilities to build ever-greater accomplishments.

In a scarcity-free world, we would have unlimited potential to discover, create, construct and achieve. That world, and the economy it would power, is a future that we can begin building today.

And that, above all else, is a future worth having.

A future worth having. That is what we strove for once, and it's something we can strive for again. Yet in order to do so we must recapture the vision we once lost, a vision we once cherished: the drive to build great and amazing things. Minus the shiny new weapons systems that consume trillions of dollars decade in, decade out, that collective drive has been forsaken.

The past six decades saw us build the interstate highway system, put a man on the moon and invent GPS and the internet. We didn't care about difficulty or political opposition – we achieved those goals because we could and because they proved to ourselves that we were worthy of rising to the occasion as both a people and a species.

Today, amidst the backdrop of our crumbling roads, collapsing bridges and aging skyscrapers, we are living within a decaying testament to the greatness we once sought and collectively built. And we've lowered ourselves to bickering, bitterly, over ideological squabbles and petty partisanship about how we're going to build anything of actual societal value, for ourselves, for those who will come after, and for those who once looked to us as models to follow.

That is not who we are, and that is not where we came from. We deserve a better future than Ozymandias, and Universal Energy is how we may see it realized. We can see it realized because the framework has a tertiary function that becomes possible when we move past a finite-resource dynamic. Once we stop chasing our tails and wasting untold blood and treasure in the name of resource dominance, it allows us to devote our full strength as a civilization to advance ourselves even further to a new stage of technological and humanitarian evolution – one that ultimately sees us build a world we thought possible only in dreams.

Whatever good things we build end up building us.

- Jim Rohn

Chapter Twelve: Advanced Infrastructure

Up until now, this writing has primarily focused on the technologies Universal Energy proposes to solve resource scarcity. The reason for this is unambiguous: resource scarcity is the core human malady. It's the primary cause of large-scale conflict, the major driver of environmental destruction and climate change, and both the facilitator and accelerant of economic decline, poverty and state-level enmity. Scarcity has been a determining factor in our existence since civilization became civilization. It's solution, therefore, is the central factor in determining our evolution beyond the dynamic that's limited us from that time-onward.

The question thus becomes: what happens once we reach that zenith?

Concretely, we'll have a game-changing abundance of resources. Resources sourced not only from next-generation and highly scalable technology, but also from freeing up immense resources that were previously devoted to mitigating the consequences of scarcity, both at home and abroad. Mainstays of state spending will experience a paradigm shift: defense, security, finance, agriculture, manufacturing, energy, healthcare, construction, communications, education and beyond will all be forever changed by such abundance.

This cyclical effect has the potential to reshape our future and improve our quality of life on a scale unrivaled. This goes beyond thinking bigger and building larger. The very foundations of our civilization will be on a trajectory of collective ascent to heights that were never before possible until we reached this threshold.

To put that statement in perspective, recall that humanity has existed for about 200,000 years, although some estimates say it's as long as 300,000.[542] For 95% of that timeline, we were basically cavemen. If we characterize "actual" civilization as the start of the Bronze Age, that period started only 5,000 years ago. From the year 200,000 B.C.E. until the mid-1800s, the fastest a human could travel was on horseback. Yet by the start of the 20th century we had the train, automobile and aircraft, and we landed on the moon less than 70 years later. The light bulb, internet, cellphone, computer, skyscraper, satellite and spacecraft were all invented in roughly the past 1/2,000th of our history.

We achieved each of these advances through technological ascension – breakthroughs that empowered our species to rise higher in both capability and knowledge – fundamentally expanding the extent of the latter through a greater command of the former.

Universal Energy accelerates our rate of ascension by providing effectively unlimited energy and resources in which to build practically anything to a superior civilizational scale. In doing so, we are presented with unique potential to advance our social infrastructure – especially within areas of civil engineering, transportation and aerospace.

Civil Engineering

Social infrastructure – what we can build, how we build it and how long it lasts – makes our society possible, and beyond that, makes it enduring and inspiring. Universal Energy can revolutionize our recently-neglected social infrastructure to great social benefit by allowing public works projects to complete faster, less expensively and on larger scales.

You'll recall that repairing the decaying infrastructure across our nation is expected to cost several trillions of dollars with today's methods – even in the most conservative estimates.[543] These are repairs that *need* to be made, yet we have neither the earmarked funds nor the political willpower to pay for them. However, as with energy and resources, technology provides an opportunity to solve the problem for us by leapfrogging limitations cost-effectively.

How exactly? First, let's cover some givens:

Most heavy machinery today is powered by diesel, which makes fuel a considerable expense of any construction project.[544] And while diesel engines have legendary reliability, the pumps, belts, and hydraulics in heavy machinery aren't generally as dependable, which leads to delays and additional costs when they eventually fail. Electric construction equipment on the other hand is mechanically simpler, and as such can avoid many of these complications while delivering the same standard of performance as their diesel counterparts.[545] As prices for electricity drop with the implementation of Universal Energy, fuel becomes less of a construction expense (all the more so once hydrogen becomes more industrially viable).

Building materials also command significant percentages of construction budgets.[546] With Universal Energy, construction projects could source better materials for lower prices. This would enable us to build lighter and stronger structures with less expense than we can today; and as a structure's maximum size is limited largely by strength-to-weight ratios, these materials would also increase the scale of what we are capable of building.

Once computer modelling, 3D printing, and factory prefabrication are added in, however, is when we truly start building the future. These technologies have been around for only the past decade, meaning that the majority of structures in our society were built without the aid of computers, and anything built before the late 1970s didn't even involve a calculator. Today, architectural software allows us to design structures virtually on computers. This provides engineers with 3D representations of what they're constructing along with highly accurate predictions of material requirements and limits of load and scale. The following images, for example, respectively show a bridge being designed on a computer and another being 3D printed in real time:

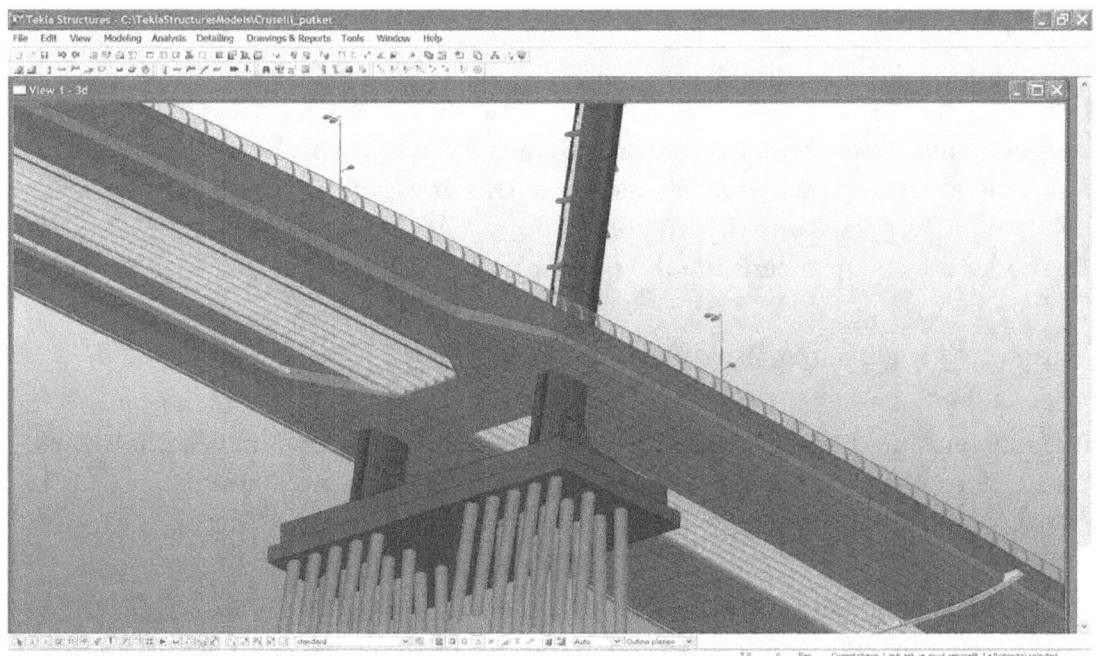

Considering that many of these technologies are in their infancy with regards to civil infrastructure, there is ample room for them to grow in the future. Integrating improvements in material science, nigh-unlimited energy and resources, and design-first principles of standardization and modularity only stand to serve as accelerants. We previously saw how we can embrace these advances to build houses, buildings and prefabricated systems, but this concept applies to the manufacture of practically anything on a large scale.

If we can prefabricate jetliners and LFTRs, why not bridges, tunnels, apartment buildings, and skyscrapers? Aerospace-grade engineering carries the highest requirements for reliability in the world, and today we already can completely assemble a flagship jetliner every nine business days – or 3D print sophisticated components for the same in a matter of hours. A world powered by Universal Energy grants us the means to raise the bar higher still.

Within civil engineering, examples include:

Next-generation roads, bridges and tunnels. Railroads and paved highways rank among the greatest marvels of human engineering, revolutionizing travel, transportation and commerce on global scales.

Yet in many ways, such infrastructure is only as useful as its ability to overcome obstacles in the landscape, something made possible through bridges and tunnels

– engineering accomplishments that we don't often think about when impressive structures come to mind. And yet:

- While not the longest bridge in the United States, at 4.8 miles the Chesapeake Bay Bridge is one of the most important, as it connects Delaware and Maryland's Eastern Shore with the Baltimore-Washington Metropolitan Area. Roughly 25.6 million vehicles travel on it every year, each one saving time and fuel that would be devoted to longer routes should the bridge not exist.[547]

  How much time and fuel? Assuming each of these 25.6 million vehicles traveled between Washington, DC and Dover, Delaware, they need drive only 93 miles for 1.8 hours if they use the bridge. If not, they would need to drive 134 miles for 2.75 hours via I-95.[548] This means that over the past 10 years, assuming consistent traffic and 21 miles per gallon fuel economy, the Chesapeake Bay Bridge has collectively saved motorists a total of one billion miles of driving distance, 224.3 million hours (2,776 years) of driving time, and roughly 500 million gallons of fuel.

- The Colorado I-70 corridor splits the Rocky Mountains with a highway, allowing motorists to avoid slow and often precarious mountain passes. The corridor is made possible through the 1.7-mile-long Eisenhower-Johnson tunnel, which was completed in 1979. It takes approximately four hours on I-70 to travel the 235 miles from Denver to Grand Junction on opposite sides of the Continental Divide. Without the corridor, it would take approximately 8.6 hours[549] to travel the 432 miles via U.S. Route 40.[550]

  To put those numbers in perspective, the highway and tunnel has saved each vehicle 4.6 hours of driving time and a driving distance of 197 miles. As roughly 13 million vehicles travel through the tunnel annually,[551] we'll conservatively assume that from 1979 to 2018 a total of 400 million vehicles have traveled this route to Grand Junction. At an assumed average fuel economy of 21 miles per gallon, this tunnel system has collectively saved drivers 79 billion miles of driving distance, 1.8 billion hours (210,000 years) of driving time, and 3.7 billion gallons of fuel.

Under those assumptions, these two public works projects, alone, have collectively saved motorists a total **of 80 billion miles of driving distance, 213,000 years of driving time, and 4.2 billion gallons of fuel.**

Both the Chesapeake Bay Bridge and the Colorado I-70 corridor were built with technology from the 1950s-1980s – a far cry from what we have available today, which in itself is a far cry from the capabilities we would have with Universal Energy. Under the framework, we would be able to increase the scale of our bridges and tunnels – connecting places in ways that were never before possible.

Megabridges: as the name suggests, a megabridge is a bridge of large size and scale. Built with the strongest materials available, a megabridge spans longer distances and supports more lanes and thus heavier loads. They can also enable travel of both road and rail, increasing diversity of use and overall social utility.

Megabridges have already made their debut on the world stage. The above concept images respectively show the proposed Fehmarn Belt Fixed Link,[552] connecting Germany and Denmark, and the Sheikh Rashid bin Saeed Crossing megabridge in Dubai.[553] The image below shows the recently completed Zhuhai-Macau Megabridge connecting mainland China to the island of Macau:

As impressive as they are, neither of these projects can yet utilize large-scale factory prefabrication with next-generation synthetics, meaning they are ultimately constructed ad-hoc with less advanced material options than would be possible with Universal Energy. Future megabridges can avoid these constraints. Imagine if construction crews didn't need to pour concrete, lay cable, steamroll asphalt or spot-weld junctions by hand, each and every time? What if they instead could take prefabricated pylons, platforms, support arches and integrated renewables, and assemble the entire bridge like a hobby kit, just on a larger scale?

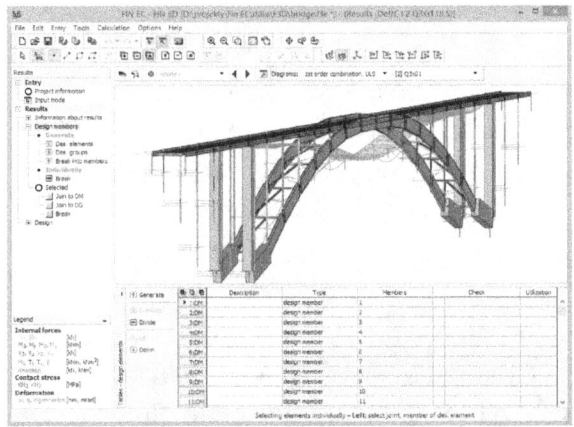

This approach is not only possible, it can be a hallmark of tomorrow's manufacturing capabilities. Such an approach further allows architects to expand their vision, as it reduces several of the problems with modern construction and material sciences. As civil engineers can hash out the technical details of a bridge with ever-more sophisticated software at the design stage, it could be rapidly determined what it would take to enlarge the bridge to greater scales of size, should the materials and manufacturing methods be present. Such efforts could lead to a day where bridges eight to twelve lanes wide with lengths upwards of 100 miles or more enter the realm of possibility. That is the future made possible by effectively unlimited energy and resources.

Megatunnels: the megatunnel is the evolution of subterranean / underwater transportation structures. Of the megatunnels in existence or in planning stages today, perhaps the best examples are the 30-mile Channel rail tunnel connecting England to France, the 33-mile Seikan tunnel connecting the Japanese islands of Honshu and Hokkaido, and the 35-mile Gotthard Base Tunnel under the Alps.[554]

Gotthard Base Tunnel

These tunnels are rightfully considered among mankind's most impressive accomplishments. But challenges remain to increasing their scale, especially in the context of submerged tunnels. There are unique complications to building submerged tunnels that are not present with bridges, namely the presence of extreme water pressure.

Building a submerged tunnel between England and France, for instance, is possible today, as the depth of the English Channel doesn't exceed 150 feet.[555] Yet building a submerged tunnel from, say, Tokyo to Beijing, or London to New York, is far more difficult. When water depths reach thousands of feet, pressures are so great that hardened steel structures can crumple like paper bags. The material advances made possible by Universal Energy allow us to significantly extend our capabilities to build structures that can withstand such pressures. They also allow us to reach an even more achievable goal of floating megatunnels that would serve the same effect as deep water variants in function.

Through a combination of buoyancy control mechanisms and cables/weights tethering tunnels to the ocean floor, these tunnels would stay close to above-water atmospheric pressures to avoid complications with structural integrity or surface breaches in the case of emergencies: [556] Two concepts shown below:

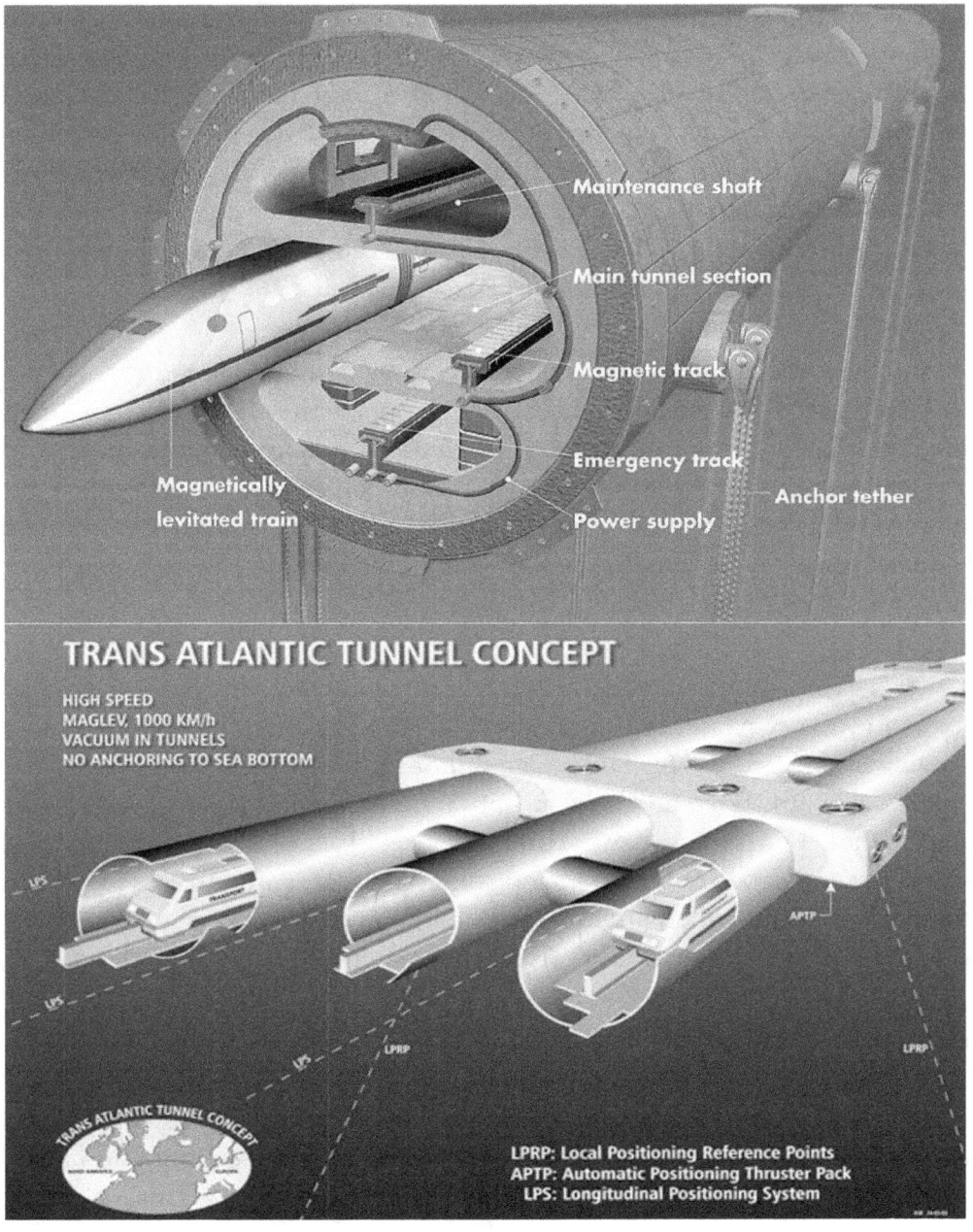

As a structure's weight displacement is different when submerged as opposed to on land, the buoyancy of these tunnels can be calibrated to maintain high degrees of stability – strong enough to support vehicle and even high-speed rail travel. Norway is already considering building submerged tunnels to cross fjords, a model that could be applied to more ambitious projects over larger bodies of water as our technology improved.[557] Universal Energy-underwritten energy cost and material advancements bring this possibility closer to reality.

Luminal communication networks. While perhaps not to grandiose scales in physical terms, the information networks we have built over the past three decades rank among the most advanced infrastructure in history.[558] In the United States, however, these networks are becoming ever-more dated. While corporate monopolies and broken politics certainly don't help,[559] the primary problem is one of distance. The vastness of the United States presents challenges to providing high-speed internet nationwide at low cost – costs that have to be paid over and over again once outdated technology needs to be updated.

As we saw throughout much of this writing, municipally integrated renewables and the National Aqueduct provide ample opportunity to run utility lines, including those for communication. This gives us a natural platform to run internet cables over any distance effectively, as their generated electricity could power amplification systems to prevent signal loss. And instead of traditional cables that transfer data through copper wires, we can now install fiber-optic cables that are 30-100 times faster.[560]

If internet service were embedded throughout renewable-integrated highway networks and the National Aqueduct, we'd effectively turn the country into a giant antenna. This process further becomes more cost-effective, because running fiber cables through above-ground conduits is far easier and less expensive than today's method of running cables underground. If municipal internet were provided through road networks and the National Aqueduct, we would effectively have nationwide wireless internet, low in cost an expansive in scale – cementing a next-generation information backbone for everyone in society.

Prefabricated buildings. We've thus far talked at length about the concepts of prefabrication and 3D printing, with attention to how we can use them to build advanced systems quicker, better, and with less expense. Megatunnels and megabridges are good examples of how we can apply these concepts to larger-scale infrastructure, but there are also other promising applications – such as extending residential prefabrication beyond what we saw last chapter. The following three images show the "One9," a nine-story prefabricated apartment building in Melbourne, Australia that was installed in just *five days*.[561]

This apartment complex in Kansas City features 80 modular units that was finish-assembled on-site within four months:[562]

For comparison, the average time to construct a single-family home in the United States is between 6-11 months, once a building permit is issued.[563] Prefabricated structures such as these are attractive options for reducing housing shortages, a problem that is expected to increase as large numbers of people continue to migrate to cities.[564]

Prefabrication also works on even larger scales. We saw in Chapter Eleven how China's Broad Sustainable Buildings corporation assembled a 30-story tower in 15 days. But that's only a pioneering example of the potential of prefabricated structures. The company has since outdone themselves by building a 57-story skyscraper in nineteen days, which, at three stories per day, is 33% faster than their previous performance. Named "J57 Mini Sky City," the structure is one of the tallest modular buildings in the world.

For a time-lapse video, YouTube "How to build a 57 floor building in 19 days."

Back in the United States, Skanska, a Swedish construction company, has recently completed "461 Dean," a 363-unit apartment building in downtown Brooklyn. The 32-story building, completed in approximately 18 months, saved 20% on construction costs when compared to traditional building methods.[565]

These cost savings are then passed on to prospective tenants, as the studio units in this building start at $560/month in a neighborhood where the median rent is roughly $2,700.[566]

When we consider the reduced costs and shorter construction timelines on these buildings, it's important to be mindful that they represent the first variants of this emerging technology. Further, these feats of engineering are also performed with today's technology and material limitations. If we think about how far we've come in other areas over just the last 20-30 years, there's no telling how much more advanced this type of construction can become in the future – especially if the advances of Universal Energy were incorporated.

Supercities

Prefabricated, standardized, and modular construction has incredible potential to revolutionize how we build things, allowing us to raise structures far larger and far faster than we can today. When taken with the other advancements of Universal Energy, this stands to transform humanity's approach to cities, and how we live in them.

As cities have evolved, they have expanded in population and sprawl, drawing in people by the billions to their economies, amenities and culture. Modern technological advancements in infrastructure, however, didn't exist until the early-to mid-1900s, even in affluent areas. Today, even the most modest city dwellings have amenities such as running water, plumbing and electricity, as well as means of transportation and communication that would have been unthinkable for the past 99.99% of human existence. Further advances in technology can take us to an even higher stage of city living: "supercities."

Conceptually, a supercity is an urban center that provides residents with an unprecedented quality of life by leveraging next-generation technical capability. As it has no fixed definition within our lexicon, we'll define a supercity as having the following six criteria:

1. **Population:** a supercity has a total population of 10 million or greater, or the ability to readily scale to support that population. This is the only requirement a supercity shares with a "megacity," presently defined only by having a population of greater than 10 million people.[567]

2. **Energy and resources:** although integrated with external power grids and resource production systems, a supercity is able to produce the majority of its energy and resources through municipally integrated renewable infrastructure, supplemented by external LFTRs or CHP Plants.

3. **Utilities as public provisions:** taking advantage of inexpensive energy and simplified installation of utilities through renewable-integrated infrastructure, a supercity provides electricity, water, heat, and high-speed internet as publicly funded municipal services.

4. **Advanced construction:** building new structures and upgrading existing ones are top priorities for supercities. Buildings, bridges, and tunnels are rapidly constructed using prefabricated, modular methods with high energy efficiencies, and are further integrated with renewable technologies. Supercities, therefore, have a high percentage of new, modern buildings and infrastructure.

5. **Transportation:** a supercity features advanced transportation technologies, such as maglev rail and autonomous vehicles. These are discussed in the next section of this chapter.

6. **High quality of life:** a supercity provides excellent education, healthcare, employment and recreation at low cost with a high quality of life index.[568]

Life in a supercity would stand in stark contrast with today's urban environments, where most areas range in quality from fantastic to poor, usually with good to mediocre mixed somewhere in between. But polarized distribution of wealth becomes less of a social malady if all necessities of life can be inexpensively provided through technology.[569] Higher-quality amenities would become more affordable, enabling everyone to increase their quality of life without necessarily having to spend more money. Essentially, Universal Energy would allow things to be built to the quality of the fantastic at the cost of the modest. This capability supports businesses, venues, attractions, and thus jobs – allowing any given area of a city to thrive, and in turn, advance.

As a result, everyone is afforded a greater sense of community, which translates to reduced crime and a collectively greater life experience. Utilizing this approach in all areas of a given city *raises the floor* and in turn enables a city to devote greater amounts of resources to continually advance, improve and evolve.

Supercities are closer to reality than one might think. Cities have been rapidly developing over the past 100 years, and tomorrow's technology is only going to accelerate that pace. For example, take a look at the New York City skyline over the past century, starting from 1914:

1914

1948

2014

In this 100-year period – a blink of an eye in historical terms – we see that the New York City skyline has grown immensely in both scale and sophistication. With the technical breakthroughs Universal Energy provides, we can advance urban construction at proportionally reduced costs. As Universal Energy would enable us to prefabricate and rapidly construct effectively any type of urban infrastructure, we can grow cities to scales that are not yet possible today.

The question now becomes: what does the skyline of New York City, or any, look in a world powered by a dynamic of effectively unlimited energy and resources, 20, 50, or even 100 years from now? Futuristic concepts notwithstanding, they nonetheless represent the potential futures made possible by the advanced technology on our near-term horizon.

This is especially important because as humanity's population is rapidly expanding, billions of people are expected to flock to cities within the next few decades.[570] For reference, roughly half of the planet lives in cities today. By 2050, that number is expected to exceed 70%.[571] Such environments must be able to scale in size in order to accommodate this shift, and supercities can do so while also supporting internal energy and resource production, advanced systems of transportation and rapid construction of social infrastructure.

Next-Generation Transportation

Until the invention of steam power, the only options for moving people or goods over distance were either horses or sailboats. Today, we have cars, trains, and planes that can carry us thousands of miles across the planet, some in a matter of hours – advances that are fewer than 100 years old. Future improvements in technology will make the world even more accessible, saying nothing of what lies beyond our terrestrial home.

One of the first of such improvements is already here: the production of vehicles and mass-transit systems that run on sustainable fuels, which Universal Energy helps extend through electricity, hydrogen and graphene. But in a world with nigh-unlimited energy, sustainable resources, advanced manufacturing methods, and synthetic materials that are both lightweight and ultra-strong, the possibilities multiply to the limit of imagination. Some notable standouts include:

Ultra-efficient, self-driving vehicles. The past decade has made substantial headway with autonomous vehicles that are able to drive themselves without any human interaction. They work via arrays of sensors and short-range laser-enabled radar (LIDAR) that instantly relays data – like road direction, location of other vehicles, obstructions, and weather conditions – to the vehicle's computer, which handles the actual driving and steering.

As this data is processed instantaneously, the vehicle reacts instantaneously as well – much faster than human reaction times. Autonomous vehicles are then exceptionally safe, especially since they are programmed to follow speed limits and obey the rules of the road.

One of the most extensive autonomous vehicle programs in the nation is run by Google ("Waymo"), although Tesla, Audi, Uber and several other car manufacturers have made significant investments in driverless technology.[572] Google's program has completed over 700,000 autonomous-driving miles with 12 separate vehicles, with only one safety incident that was caused by human error.[573] Uber's model has been slightly less successful, with one fatality deemed unavoidable due to a lone pedestrian jaywalking.[574] To compare this safety record with human drivers in the United States: 268 million vehicles annually drive 3.17 trillion miles per year[575] and are involved in roughly 10 million accidents, averaging to one accident per every 26.8 vehicles or every 317,000 miles driven.[576]

In comparison, the safety record of autonomous vehicles is a dramatic improvement. The implications of this achievement are especially important because the 10+ million auto accidents occurring annually claim the lives of roughly 33,000 people and injure some two million others.

Fatalities and Fatality Rate per 100 Million VMT, by Year, 1975–2016

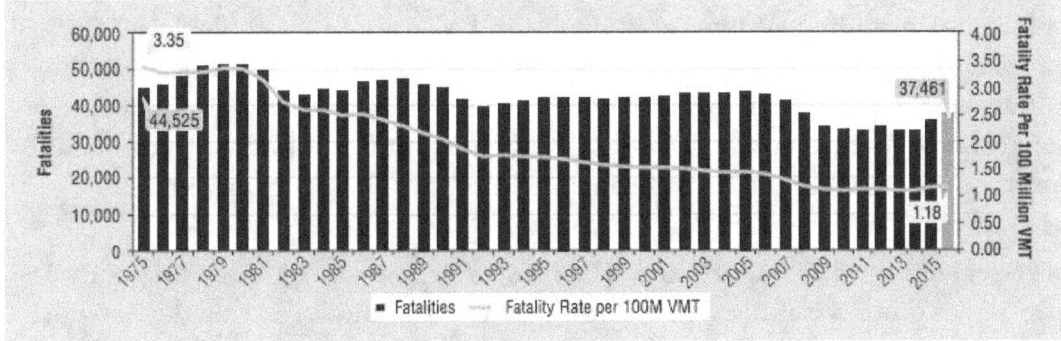

Sources: FARS 1975–2015 Final File, 2016 ARF; Vehicle Miles Traveled (VMT): FHWA.

People Injured and Injury Rate per 100 Million Vehicle Miles Traveled by Year

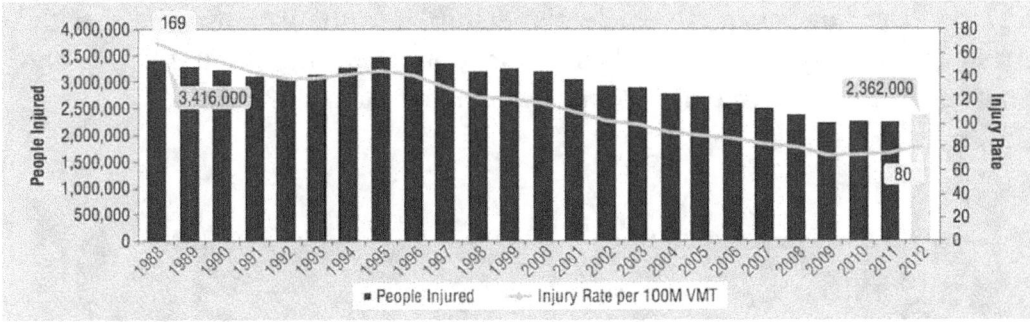

Source: NASS GES 1988–2012; Vehicle Miles Traveled (VMT): Federal Highway Administration.

While safety is the most important factor when considering the benefits of self-driving cars, it's not the exclusive selling point:

- As all speed limits and road rules are obeyed by vehicle software, self-driving cars are highly efficient, as they can maintain a uniform speed without having to constantly accelerate or decelerate in reaction to other vehicles (assuming all others on the road are also autonomous). This ultimately reduces traffic congestion.

- Of the 33,000 road fatalities every year, nearly a third of them come from drunk drivers; presumably the same is true of the 2.6 million injuries that occur annually as well.[577] Self-driving cars make this problem go away effectively overnight.

This is merely the state of current technology. If Universal Energy's advancements are applied, our capabilities increase accordingly. For example, instead of just having self-driving vehicles track road surfaces through internal sensors, they can also use the built-in Wi-Fi of renewable-integrated road surfaces to navigate, providing system redundancy and security. As such vehicles would increasingly be electric, they could also be charged on roads through wireless emitters embedded within municipal infrastructure. In the increasingly unlikely event of an accident, emergency crews could be instantly notified with automated reports of the extent of the damage and the number of passengers injured.

Large-scale ground transportation. The most effective way to transport goods or people over ground is rail, and the latest versions are known as "maglev" – short for electromagnetic levitation.[578] While much of the developed world has already employed this technology, America is far behind. There are several reasons for this state of affairs: the petroleum lobby, the geographical size of the United States, the extent of personal vehicle ownership, and the expansiveness of the interstate highway system.[579] But it's time for America to embrace the future, and when it comes to large-scale transportation, maglev rail is the future.

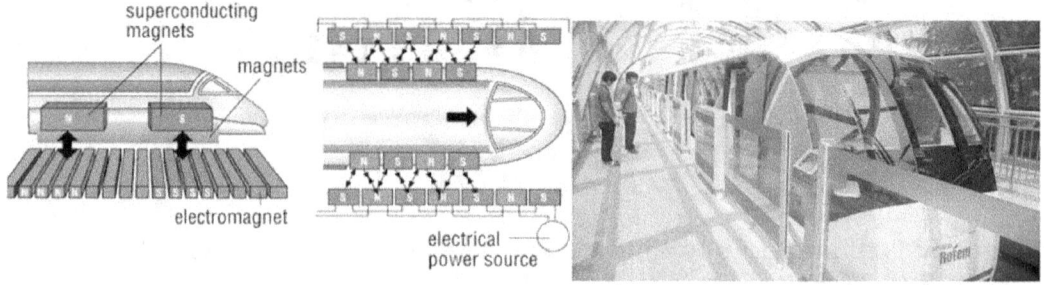

As maglev propulsion is nearly frictionless, maglev trains can travel at speeds exceeding 300 miles per hour (482 kilometers per hour). With Universal Energy, building prefabricated, modular train cars and track systems becomes more straightforward. This can enable us to build trains near renewable-integrated roads or the National Aqueduct – creating an insulated mass transit system with constant connectivity to power sources. And if these trains were built on prefabricated pylons next to highways, it would also remove the need to purchase additional land for their construction, further reducing costs.

Providing improved nationwide rail networks that can transport people or goods at 300 miles per hour is a four-fold improvement over most domestic rail technology. The implications this presents for trade, transport and tourism stand to return extensive economic benefits, all the more so if considering the new job opportunities opened by America gleaning expertise in this market sector.

Yet even though such speeds are impressive for a train, maglev technology can theoretically propel trains much faster. The biggest obstacle to doing so is air resistance, which at high speed becomes especially significant for safe operation. A new system called the Hyperloop can enable us to change that.

Hyperloop. Originally envisioned by PayPal, Space X, and Tesla founder Elon Musk, the Hyperloop is a theoretical extension of the tubular transport systems used in banks, where a capsule rides a wave of air inside a tube from one location to another. With the Hyperloop, instead of transporting a capsule it would instead transport a train, integrating maglev technology for high-speed travel.[580]

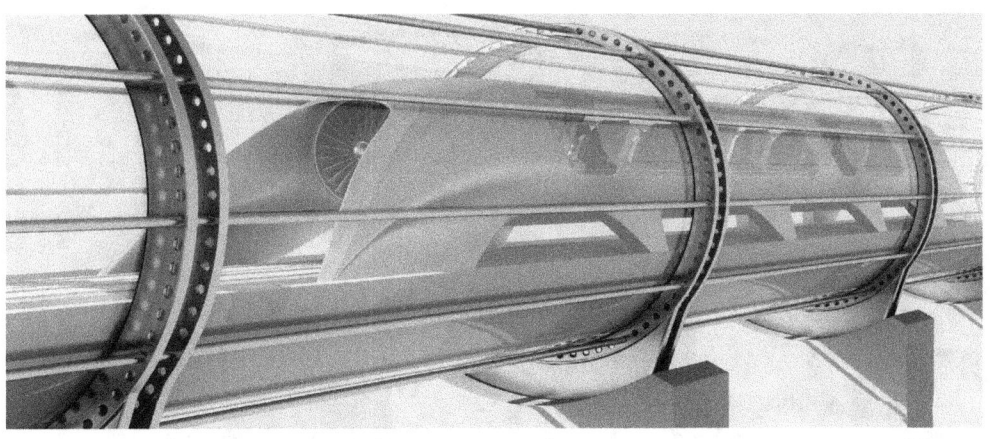

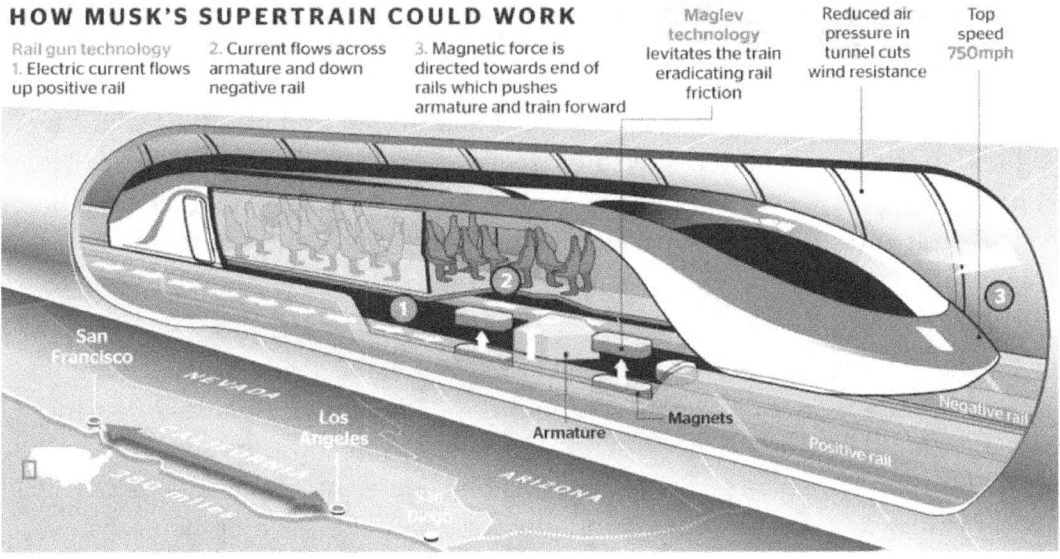

As with submerged megatunnels, the Hyperloop would operate in a partially depressurized environment. Mounting a high-strength air compressor at the front of the train would remove forward-facing air resistance and, at the same time, provide a frictionless air cushion around the train body.[581] Reduced air pressure translates to reduced air resistance, permitting the Hyperloop to travel at far higher speeds than currently possible, with reduced effects of breaching the sound barrier due to a lack of air density.

Conceptual image of a hyperloop station:

The Hyperloop concept is envisioned to be prefabricated and built on pylons by design. Connecting Hyperloop technology with integrated renewables or the National Aqueduct would also provide constant power the system as a whole, which, if built with advanced synthetics, would be stronger and lighter than most commercially available materials today.

The Hyperloop has already started construction and has demonstrated initial successes in early tests and national competitions.[582] Several companies have since emerged to build functional models in Dubai, California and Europe.[583] A world with nigh-unlimited energy and resources alongside easy, sustainable access to high-performance materials only stands to accelerate the development of next-generation technologies and their arrival to market – presenting yet another transformational addition to capabilities of human movement.

Personal Flight. The price of passenger vehicles has steadily dropped over time, making them affordable to most Americans. Airplanes, however, remain unaffordable to a majority of people, even though they've been around nearly as long. For a few thousand dollars, you can get a working, used car. The cheapest used aircraft starts at ten times that, given that the mechanical tolerances for aerospace are more stringent than for land-based cars, and the demand for small aircraft is substantially less.

Further, private aircraft have not enjoyed most of the safety advances seen in motor vehicles. Life-saving features such as airbags and crumple zones are rare, and only a select few planes feature roll cages and emergency parachutes.[584] As most private planes have forward-mounted engines, the lightweight fuselage (skeletal structure) of the aircraft often lacks the structural integrity to prevent the engine from crushing the occupants upon forward impact. The survivability of a crash, therefore, becomes a dubious prospect in many cases – the safety measure, by and large, is to not crash in the first place.

There are debatable causes to this slower progress of innovation in aviation, but what's certain is that circumstances are changing. Some of these changes involve the greater inclusion of common-sense safety features into light aircraft.[585] Yet other companies have embraced advances in technology that allow flying craft to be redesigned from the ground-up. Some of these are already seen today in the form of quadcopter drones.

 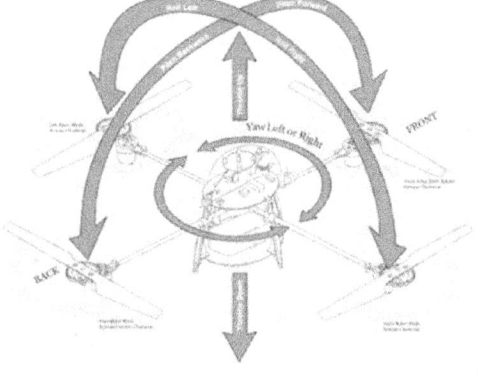

As opposed to helicopters that use two or four blades on a single rotor, quadcopters have four rotors on opposite and balanced points, thus removing the need for a stabilizing tail rotor. This makes quadcopters extremely well balanced, which in turn makes them both more maneuverable and easier to fly than traditional aircraft.

The concept of a quadcopter large enough to carry people (if only for short distances) has already been proven to work with today's materials. The SureFly personal flying vehicle is one example of quadcopter innovation reaching commercial viability.

The Scorpion-3 hoverbike is another recently developed prototype:

These images below show the Ehang 184 personal quadcopter:

While such quadcopters are possible with today's technology limitations, their capabilities can be extended via graphene's capacity for structural strength and energy storage. As discussed throughout this writing, graphene is an ultra-strong, ultra-light, and ultra-conductive material that Universal Energy can cost-effectively synthesize to effectively unlimited scales.

With it, not only can we store the requisite energy to power a large quadcopter with minimal added weight, we can also integrate the storage medium into the fuselage while ensuring uniformly high strength. This would make quadcopters light and large enough to transport both people and materials.

After tackling energy storage and weight capacity, the next obstacle to building a large quadcopter is a design that makes the quadcopter cost effective enough to work for transportation – ideally in a way that can both fly and drive.

While we don't have a human-sized solution yet, smaller-scale models have been recently released that demonstrably prove the concept as workable.

Approximately two feet long and made with the same polycarbonate materials comprising bullet-resistant glass, this vehicle can drive over obstacles at speeds exceeding 20 miles per hour *and* engage flight rotors at a push of a button.[586]

While there are of course obstacles to scaling a smaller prototype to the size of a vehicle large enough to transport people or cargo, they are absolutely solvable as they center on areas that Universal Energy gives us the means to address:

1. Ultra-strong, lightweight materials at acceptable cost – made possible through graphene and next-generation synthetics.

2. Lightweight energy storage mediums – made possible through a graphene-interwoven polymer/polycarbonate fuselage.

3. Manufacturing capability – made possible through next-generation additive / 3D-printing, virtual modeling and precision automation.

The actual flight of any such vehicle would be as simple as driving a car. Large quadcopters of 50 pounds or more are already being flown long distances using two joysticks and a first-person video screen – far less than the array of functions necessary to fly a helicopter or airplane. Any serious production of flying vehicles  would of course carry more options, safety features and regulatory controls, but those are not obstructions to their delivery. It's also true that certain automobile drivers might not demonstrate the requisite proficiency to pilot flying vehicles. Besides the application of automated flight controls, these concerns can be addressed through more lengthy training and more stringent licensing requirements. Fundamentally, these aren't details of potential capability. They're details of process, that and only.

It's difficult to overstate how much this could improve how we live and move. The overland distance between destinations "as the crow flies" is always shorter than the meandering roads we have to take today, and that shorter distance can significantly decrease the response time of emergency crews, help deliver rapid aid to areas without road access, and, in the case of drone-sized quadcopters, even deliver consumer goods on-demand.[587]

We are only one technological step away from having personal flying vehicles that are safe, strong and easily flown. Vehicles that can use autopilot programs that already fly and land commercial aircraft today.[588] Vehicles that, like the autonomous cars before them, can be charged via wireless power over municipally integrated renewables and through designated relays – potentially enabling effectively indefinite flight.[589]

Many of us spent our childhoods dreaming of the day when we would see flying cars. The technical capabilities that can be at our fingertips tomorrow make this dream possible, along with all others that have allowed this writing to portray a vision for a future brighter than the one we face today.

The areas it places focus, like many of the possibilities opened up by Universal Energy, are simply the beginning. We can extend these advancements to anything we can imagine, from our day-to-day lives, to cutting-edge aerospace that can revolutionize not just travel within our world, but well and far beyond it.[590]

This is the nature of advanced infrastructure as this writing refers to it, and the realization of the future it brings is now at our command. We now have the tools to make this – all of this – real. To manifest a reality where we can provide the abundance, advancement, achievement and ascension we told ourselves was the path we were destined to walk. We now have the means to build such as a world as a testament to the choices we made in its furtherance, and the promises we made to those who came before who brought us to this singular threshold, where it could be transcended, at long last, by our own hand.

A future worth having. The start of something new. And the next giant leap.

"There are no passengers on Spaceship Earth. Only crew."

- Marshall McLuhan

The Next Giant Leap

Universal Energy is a framework designed to solve core human problems. Its intent is to provide blueprints for a system that can permanently address the maladies that have plagued our species since the dawn of time; software to support the foundations of the human condition. Further, it is intended to allow – as I believe it to be true – the best parts of our nature to flourish in a world where their absence has caused no shortage of rue and woe. But as a writing, *The Next Giant Leap* is intended to provoke thought and the consideration of ideas. It's meant to be a conduit for a conversation among ourselves about how we might evolve the hindrances of our existence, and build, truly, a better world.

To build a better world.

I have rewritten this chapter many times. Yet each instance I read those words I must admit I become pained in a way that is profound. That such a conversation needs to happen in the first place is, on its face, a travesty of potential and an abdication of promise. It was fifty-one years ago, nearly to the day these words were written, where mankind took its first giant leap and set foot upon our lunar surface. It was a moment that reflected the culmination of unquantifiable sacrifice, and immeasurable investment, into our ability to accomplish the impossible. 200,000 years of human evolution and the lives of billions converged at a single moment in time – and we leapt forward.

That was the hope our future was meant to be built on. That was the light by which we were meant to find our way.

Now in the decades hence, where the problems of our time command headlines of newspapers instead of chapters of history books, I am haunted by a deeply lonely and desolate sense of shame to think this future might be forsaken. That for all of the nameless sacrifices people made throughout history, the future they died for would nonetheless yield to a world where billions of others wallow in needless, purposeless suffering. Where our ecological home is dying. Where the possibility of nuclear extinction is an everyday fact of life. Where we are dominated, again, and again, and again, by the petty conflicts that for millennia have devoured the brightest elements of human potential.

Yet even in the face of these circumstances, I refuse – to my core biological basis – to grant them surrender. Because our future is not yet forsaken. Because that light has not yet left us. I believe it is still a guiding beacon, waiting to be once again embraced as torches in our hearts, small as they may be against the shadows of our time – yet together bright enough to beat the darkness. And with the tools that can today be in our hands, I believe sincerely through every fiber of my soul and being that we can accomplish exactly that, once and for all.

Because "the way things are" and "the way the world works" are not reflections of incontrovertible destiny. They're forces of circumstance, reflections of an old model that, for all its faults, got us this far – yet is now too broken to carry us further. We now require a new model, one that upgrades our existential framework on a civilizational scale. A task that, at the pinnacle of our technological prowess, is now at last possible.

It only takes our choice, like those who came before, to make that leap.

Thus, I am now speaking to you – as one person to another – in perhaps the only opportunity I can in such a context to tell you *that there is still a way to fix this.* We can still build a future where we can strive for higher aspirations and retire each evening feeling legitimately hopeful for the days ahead. To hold belief that we can embrace our full potential as a people and reach a harmonious plane of existence with our environment, with our planet, and, most of all, with each other.

These aspirations are not "lofty," nor are such appeals evocative of flowery rhetoric or emotional cliché. **They are core perspectives.** If there is meaning to this life, if life is precious and worth cherishing, worth empowering and worth saving, then there is no greater goal we should have for ourselves. There should be nothing more important that we would see achieved. *This is the foundation of existence.* And by engaging these newfound capabilities at this critical time, we can evolve the fundamental structures on which that foundation stands.

This mindset encapsulates an expansion of perspective that our nationalist, tribalist tendencies might consider unrealistic – cynically ignoring that the blood spilled and resources wasted by their tenets may, in fact, have been better invested in causes other than our own destruction, or annihilation. This mindset looks beyond that, into something greater and something deeper.

In 1964, a Soviet astronomer named Nikolai Kardashev postulated the idea of civilizational "tiers" – quantifiable metrics of how objectively advanced a civilization has become or could become in the future – based on the perspective of a sentient, carbon-based biological lifeform.[591] Known as the "Kardashev" scale, his model had three tiers:

Type I: A civilization that sources its energy and resources from its planet.

Type II: A civilization that sources its energy and resource from its star.

Type III: A civilization that sources its energy and resources from its galaxy.

Other scientific philosophers, Carl Sagan, Michio Kaku, John Barrow – and several others[592] – have made their own models using their own insight and expertise.[593] Even if I had the intellect or standing to disagree with any of them, I don't. Yet deep within my mind is another tiered model, one that has influenced my worldview and perspective starkly throughout my life. It's not much different than others like it, but it is a reflection of who I believe we are, what I believe we are capable of, and what I believe we can become – should we so choose.

The Ten Tiers of Civilization

Tier 1: Fire and Stone: control of fire and the ability to craft stone tools, subsisting exclusively on a hunter-gatherer diet.
This tier represents approximately 95% of human history.

Tier 2: Agricultural: the ability to grow crops and raise livestock, accelerating population growth. Social hierarchies and customs form, and the possibility of organized conflict becomes a fixture of life. Humanity reached this tier during the Neolithic Revolution, around 10,000 B.C.E

Tier 3: Pre-industrial: command of simple metallurgy with a basic understanding of math, science and astrology. Written language and laws are established, as are formal relations between governing regions. This tier was reached at the founding of Sumer in ancient Mesopotamia, roughly 4,000 B.C.E

Tier 4: Industrial: complex self-powered machines are invented, including mechanized assembly and transportation systems. Economic trade becomes globalized and conflict carries consequences of increased severity. We reached this tier during the Industrial Revolution, approximately 1760.

Tier 5: Atomic: civilization discovers atomic energy and has the ability to build large-scale infrastructure. Population grows exponentially. Potential for resource conflict increases, as does the potential for mass destruction. We reached this tier on 16 July, 1945 when the first atomic bomb was detonated.

Tier 6: Orbital: civilization can defy gravity and even orbit. Electronics and globalized communications emerge. Transportation over terrestrial distances becomes trivial. Population continues to grow exponentially. Potential for resource conflict is extreme, which for the first time can potentially be an extinction-level event due to nuclear arsenals and global delivery mechanisms. We reached this tier on 4 October, 1957 at the launch of the first satellite. This is the tier we are in now.

Tier 7: Ascendant: civilization has developed technology capable of synthesizing unlimited energy, resources and materials, thus ending resource scarcity and resource conflict. Maslow's needs are met,[594] addressing most social problems and stabilizing population growth. In turn, civilization is able to devote the entirety of its resources to collective social advancement with ever-more sophisticated infrastructure. This is the tier Universal Energy bring us to.

Tier 8: Transcendent: civilization has crossed the biological threshold and is able to store and transport consciousness outside of a physical body (sophisticated brain to computer interface). Complex artificial intelligence exists and both biomass and bionic structures can be synthesized effectively, leading to the possibility of synergy between organic and synthetic life.

Tier 9: Interstellar: civilization has reached the mastery of planetary existence and becomes capable of inhabiting other planets. Intersolar and interstellar transportation is invented, as is greater command of nanoengineering.

Tier 10: Intergalactic: A hypothetical Tier 10 civilization is capable of intergalactic space travel and can artificially create habitable worlds. It would furthermore command a comprehensive knowledge of universal physics, both on micro and macro scales.

In this model, we first and foremost see that humankind has ascended at an accelerating rate. It took us ~190,000 years to go from Tier 1 to Tier 2, yet only 12,000 years to go from Tier 2 to Tier 6 – the tier we remain in presently. And while this is an impressive reflection of our capabilities, we've only come far enough to be forced to take another leap, for a critical attribute of our tier is that it's inherently precarious.

Due to exponential population growth and the environmental changes and resource scarcities that come with it, a civilization can only stay in this tier for a limited time. It either ascends, or it falls to resource conflict and/or ecological collapse, which in the nuclear age carries extinction-level consequences.

It's a reality that pays homage to "The Great Filter," which is a derivative consideration of "The Fermi Paradox" – an essential philosophical question when discussing our seemingly isolated existence within the vast cosmos our planet calls home.

Postulated by the great physicist Enrico Fermi in the 1950's, the question can be succinctly paraphrased as the following:

How can our universe, in all its unimaginable vastness, present such an immense likelihood for sentient life, yet at the same time we can't seem to find it?

I'll frame this another way to help clarify:

Our planet, Earth, orbits our Sun along with seven other planets, comprising our solar system. It is only one out of roughly 100 billion solar systems in our galaxy, the Milky Way.[595] Our galaxy, itself, is only one out of some *two trillion* galaxies in our observable universe by the last known estimate.[596]

In another way of saying, if we were to take every person alive today and send them each to a unique galaxy, we'd only be able to visit about 0.37% of them. As each galaxy has hundreds of billions of stars and there are trillions of galaxies, it makes the odds of Earth being the only planet to support life in the cosmos to be nigh impossibly low.

Presently, astronomers estimate that our universe contains at least a septillion stars (that's 24 zeros).[597] At this scale, if even only one out of one million stars had orbiting planets that sustained life, it would still leave ten quintillion planets that did (10,000,000,000,000,000,000). *That's ten million individual groups of one trillion planets.* Think about that for a second.

It's so impossibly unlikely that we're alone, yet at the same time, we haven't heard from another lifeform beyond our planet that we know of.

"The Great Filter" is a proposed answer to this paradox, theorizing that all intelligent life faces a threshold it must cross for it to ascend beyond its planet and survive for the long term. That in order to truly advance as a species, it must overcome a series of obstacles which would otherwise stop its ascent – or be the harbinger of its destruction.

Consider it another way, if you will:

If all of Earth's history were reduced to the scale of one year, humanity did not emerge until 11:55PM on New Year's Eve. We only reached the modern era at about 10 seconds to midnight. By 7 seconds to midnight, we had invented the means to cause our own extinction. By 5 seconds to midnight, we will have run out of the resources that sustains our rapidly expanding population. And if our dynamic remains unchanged, it will destroy us before the clock strikes twelve.

That is "The Great Filter." It is something that we are facing right now, today. And it is our generation, our time, that is tasked with passing this gauntlet of unforgiving truth.

Yet it's a reality that makes us brethren against the forces, natural as they may well be, which would otherwise erase what we have built and accomplished. That would extinguish the stories of who we are and who we could become, casting the ashes of the aggregate into a void where the culmination of our memory is to be forgotten. For this reason, this writing has conveyed a sense of touch in language that appeals to the power you bring to bear as a person, in hopes that it would inspire your choice of action towards building a better world and a brighter future.

I believe there is an ether in this world that embraces our sense of soul and what it means to be human. An ether that connects us on a core wavelength, where we all wish the best for ourselves and for others, and ascribe value to our sense of collective meaning and shared purpose. We find it, in moments, at high points in our lives and seek it still maybe through churches and community organizations, as acolytes of sports teams, countercultures or ideological tribes; striving for warmth and affirmation through what strings of connection we can grasp and hold close. Yet another torch in our hearts, perhaps, one that keeps the cold and loneliness at bay.

This ether is not esoteric. It's not something that can be made, bought or bartered for – it's been within us all along, an aspect of the human condition that we can engage in ourselves and others should we be willing; that like love, friendship, respect and honor is a conduit for connection, for inspiration, for belonging. It does not come from acquisition. **It comes from choice.**

The best parts of ourselves are made possible because we choose them. We choose to be better, to give, build, create and forgive, just as we choose to hate, steal, forsake and destroy. While perhaps inclined towards one aspect or another by our natural dispositions, we are ultimately manifestations of our choices – defined one way or another by the actions we take, the ethers we embrace, and the natures we feed.

The world that we live in, therefore, is a collective reflection of those choices – even if we didn't realize that we made them, or that those who came before us made them, or those who came before them still. The world we wake up in tomorrow, accordingly, will reflect the same. But uniquely in our context is an unprecedented capability to expand the impact of our choices and the scope of options we have available. And of them, the most vital is the neutralization of the concept of need.

Human nature is commonly described in dualities: binary opposites that interact together as dichotomies of circumstance, character or choice. Rich versus poor. Success versus failure. Strong versus week. Good versus evil. *Us versus them.*

Such dualities define both ancient and modern frameworks of theology, law, nationalism, decorum, ideology – even technology, itself made possible by the application of binary numbers, 0 and 1, processed through transistors on a massive scale. When applied to the realities of our nature, each of these dualities have abstract merit in our world past, present and future. Yet none of them truly identify the core dichotomy of the human condition. Our world is not bound by good versus evil, or right versus wrong, or strong versus weak. It's bound by **supply versus need.**

On any scale, (especially national), need and scarcity – in either perception or reality, can be attributable to the cause of most any conflict, social fracturing, environmental pillaging, or lust for greed, power or oppression. I studied war crimes in university. To this day, it still haunts my dreams. What we have done to one another, what horrors we can justify, what choices we can accept and what abominations we can call weapons make any other logical reason incapable of defining the forces directing and binding collective human action. The horrors of such actions aren't reflective of how far we will go, they're reflective of how far we *won't* go – limits that evaporate and emerge expanded, pushed time again by the merciless realities of scarcity-driven need.

While its presence is undoubtable and at times unyielding, we as a people too frequently make the mistake of attributing to malice that which can be attributed to need because it's easier – a simple designation of "evil" that avoids us having to look inward to what drives our adversaries in any given context. Outside of the subjective application of moral relativism or revisionist history, it's much harder to see "them" as human and their needs as logical, to them – even if we are the target of their antagonism as we perceive it. It's harder still to realize that most people, even in times of strife, are not acting out of wickedness or cruelty but are rather doing what they think they need to do in the context of what they perceive their needs to be – even if their actions manifest respectively as such. It's simply our uncompromising reality as pawns of a zero-sum game.

Need is the driving adversary of the human condition; the core malady that has continually held us back as a species since the dawn of time, and has kept us fighting amongst each other instead of enabling us to realize that we can reach our true potential, and a new plateau, should we enact the means to make the concept of need irrelevant. Today, we now possess those means, and we can choose to wield them for that end – to defeat our ancient adversary with finality, and build a better world upon its ruin.

And to make that choice is what would I ask of you now.

It may not be easy, yet it's incumbent on us to shake off the apathy and despondence our time has given us cause to adopt, and to embrace the better parts of ourselves the world's cruelties have taught us to suppress. It's incumbent on us to consider reinvesting in ourselves and our future with tools that can even the odds in our favor. To invest in the potential of each other, to understand their perspectives and forgive their prejudices, and seek to find common ground strong enough for us to once again start building. We may not be able to solidify all ground to find commonality, but we can solidify enough ground to build platforms on which we can extend our hands and, maybe one day, a bridge.

At the end of the day, deep within ourselves, who we are, what we care for, what we value, is that not the choice of life? Is that not what we want for ourselves, for our children, and for theirs? For thousands of years, people just like us gave *everything* they had for our future. Our time is the culmination of a billion sacrifices – every soldier on every battlefield, every martyr, every king, tyrant, slave, warrior, artisan, philosopher, lord or peasant. The sum of all their toils, all the sacrifices of their hopes and lost dreams are boiled down to this moment, here, and now, and the choices we make with the time we have been given.

Simply stated, I can't think of anything worse than failing them. To not carry the torch they have lit and carried for us to the victory they never could reach, the victory that we uniquely can. To me there is nothing more important, and I'm tired of being encouraged to ignore that. I'm tired of glorified ignorance to the reality of our world and our potential to change it. I refuse to continue granting meaning to society's dog and pony shows: celebrity news, celebrated complacency and fleeting materialism – the choreographed wrestling matches of today's bribed political dynamic – all washed down with diet cola and light beer.

We have one life to live, one life to interact with the framework of existence, and we find ourselves at the zenith of our capability to evolve the foundations of our biological constraints. To choose to take this leap, to reach a higher tier, and to live knowing that this was when our species made it. Where we passed the test life gave us and earned the right to continue our evolution not just within our world, but far beyond it.

That is the choice of our time – one that faces every one of us. And as one of us, I made a promise to devote my very best efforts to propose an actually effective way to choose for that end. Something that could be given away to anyone who wished to adopt – and evolve – these ideas, and begin discussing how we can work together to see them made real.

Nobody asked me, paid me or qualified me to make this promise. I did this on my own, and I made this promise, to myself, and to you, to see this task fulfilled because I chose to. I made this promise because I don't answer to the cynics and the apologists of the status quo – *I answer to you*.

I answer to you because we are all in this together, and I sincerely and honestly believe in our shared capabilities, in the potential that we can have if we set aside our contrived differences and work together to build what can become of the best of ourselves.

It's the only thing that can save us. It's the only thing that should save us.

And should we choose to make that leap, then our feet will land in uncharted terrain on a brilliant frontier. As technology expands and satisfies ever-more needs through indefinite resource production, conflicts will reduce, economies will grow, as will relationships and trade agreements. Development and modernization will begin in regions that were once war-torn, and the echoes of resource conflict will begin to fade into memory, just like all other plights of our nature that technical means have allowed us to banish into the past. From there, as technology greater connects us and brings us closer together, exploration beyond Earth will become ever-more sophisticated and we will find what there is to discover in the vastness beyond our planet.

We will reach not just the next tier of civilization, but also an essential realization: that we are not just members of individual countries, as this isn't the label that should define us. We are all human beings; we are all people – *that* is the label that should define us. That is because we all share this rock in space together. And whether we live on it together, or die on it together, one way or the other, ultimately, it will be so together.

It is my greatest hope that we can be able to realize that one day. We place boundless faith in gods we cannot see to form our fate and future. Perhaps we could strive instead to see the day where we might place faith in each other.

So I will start by placing my faith in you.

"The Earth is the only world known so far to harbor life. There is nowhere else, at least in the near future, to which our species could migrate. Visit, yes. Settle, not yet. Like it or not, for the moment the Earth is where we make our stand.

It has been said that astronomy is a humbling and character-building experience. There is perhaps no better demonstration of the folly of human conceits than this distant image of our tiny world. To me, it underscores our responsibility to deal more kindly with one another, and to preserve and cherish the pale blue dot, the only home we've ever known."

- Carl Sagan

Appendix

A1: Universal Energy Implementation Strategy

While the purpose *The Next Giant Leap* is to outline blueprints for Universal Energy as a framework and describe the technologies and capabilities that make it possible, funding and implementing Universal Energy are wholly separate questions. In furtherance of their consideration, this writing proposes a boilerplate implementation strategy that doesn't wade into the oft-divisive and precarious subjects of partisan governance or partisan economics. Instead, this approach assumes a good faith analysis by mechanisms of economy and state – the world we should live in, and could live in, should we as people look past partisanship-driven self-dealing in favor of solutions that *actually* improve our way of life and our civilization as a whole.

This implementation strategy will hinge on several factors: how much Universal Energy is estimated to cost, the logistics and management of implementation, how it can be funded, and how it can overhaul our economy. This boilerplate strategy is intended to be a proposed path for actionable efforts to implement the framework and a starting point to encourage both commercial and state enterprises to begin investing in a next-generation clean energy future.

First, we will consider how much Universal Energy is estimated to cost at full, nationwide implementation – quantified by an energy generation potential of 300% of our present capacity (including current infrastructure). This figure is estimated to be **$6.63 trillion USD** paid over a period of 10 years ($663 billion annually), a detailed pricing breakdown of which can be found on page 315.

While this figure is substantially less than other energy overhauls that have been proposed by both public and private initiatives, it still carries a degree of "sticker shock" that's important to dispel in the context of nationwide infrastructural projects. This is all the more true since Universal Energy carries myriad social benefits that are not presented by most of the primary consumers of public funds (like endless wars).

In that mention, it's worth reviewing some of the larger expenditures our time has seen paid, whether we as a people focused on it directly or not:

- **$38 trillion:** the inflation-adjusted total the Federal Government (alone) has spent over the past 10 years.[598]

- **$6.8 trillion:** the aggregate sum of U.S. Defense Budgets for the past ten years. This figure **does not** include military spending paid outside of the defense budget (veterans affairs, pensions, homeland security, clandestine operations, interest on war debt, nuclear arms, etc.).[599]

- **$6.4 trillion**: the aggregate sum the United States has spent on wars in the Middle East and Asia since 2001.[600]

- **$3.5 trillion:** the total sum we have paid on the interest on the national debt over the past 10 years. Not principal – just interest.[601]

- **$1.5 trillion:** the estimated cost of the F-35 fighter jet program over its operational lifetime. That's for a *single* class of military aircraft.[602]

- **$6.5 trillion:** the total sum the U.S. military is unable to account for in a recent audit.[603]

In aggregate, these fantastical expenses have given our society very little in tangible value, and could have paid for Universal Energy several times over. In doing so, it could have instead built the foundations for a world that would consider scarcity – the prime reason for the accumulation and utilization of hard power to begin with – as a relic of a past, and have enacted a shift in circumstances that avoided the need for mass military spending *in the first place* – a true value, at comparatively modest cost.

Who Pays to Implement Universal Energy?

The short answer is a mixture of public and private entities backed by a tax and investment incentives. The longer answer is more nuanced and specific to the areas of implementation, which we'll go over in more detail throughout this section and Appendix. We'll begin by outlining the allocation of spending responsibility for the resource sector in question. In doing so, this estimate separates resource production into two categories: *public* resources and *commercial* resources.

Public resources. Critical resources that require implementation over a large scale, and/or operation with existing public infrastructure – making them ideal candidates for municipal management: water, electricity, and hydrogen fuel. These systems comprise the $6.63 trillion cost estimate for Universal Energy (page 315). In this model, the systems that produce these resources are developed by a mixture of public and private enterprise (explained below), yet purchased and implemented by public services, delivering their produced resources as managed-profit municipal functions.

Commercial resources. Commercial resources are not considered as appropriate for implementation as a public function and are intended to be delivered by private enterprise in a competitive market, with companies operating in resource production enjoying attractive tax benefits both in operation, employment and investment. These resources include food, building materials and materials for next-generation infrastructure.

The systems that provide these resources: indoor farms, prefabricated manufacturing facilities and advanced material synthesis should be funded primarily by the private sector as a commercial service that is aided by tax incentives that we'll discuss within the upcoming sections of this Appendix. The resources they produce would remain a function of the *private sector*, regulated by the *public sector*, and sold to the public in a regulated free market which would have increased purchasing power as a result of the social improvements discussed herein.

Waste management is assumed to function as either a public or a private function, depending on the locale, powered by systems developed by joint public-private enterprise under attractive tax incentives. The energy they generate could be sold on a regulated commercial market or used for supplementary resource production as deemed prudent by the operating entity in question, but is not focused on within this model.

To recap: the provision of **public resources** are the responsibility of the public sector. The provision of **commercial resources** are the responsibility of the **private sector**, even both will use technologies developed in large part by private enterprise.

A fundraising approach to reach $663 billion per year to pay for Universal Energy's initial deployment would come from several sources: government spending cuts (with intensive, honest and public investigations on wasteful programs and expenditures), mitigation of social problems requiring government expenditure and additional revenue generation from modest tax increases. A more detailed breakdown of the possibilities inherent to this approach can be found on page 307.

Who Manages Implementation and Operation?

In this model, Universal Energy's implementation and operational management is performed by "The Public Interest Company," (PIC) which is a new type of corporate entity that's a unique mixture of private enterprise, government agency, the non-profit sector and the American electorate.

To explain exactly what that is, I think it would be helpful to first illustrate why The Public Interest Company is different from other business entities – especially entities that resemble some form of marriage between the public sector and private business. To that end, let's consider some of these entities, and more importantly why they are not capable of managing Universal Energy's implementation or long-term operation:

Corporate entity. The primary goal of a corporation is to profit. Its purpose is to make money and grow to a point where it can maximize profits by any legal means necessary – with all other concerns as secondary. Corporations are excellent innovators that increase job growth, but in the absence of significant competition and effective regulations, they ultimately grow to a point where they monopolize their market sector and seek to increase profits at the expense of ever-diminishing services.

This is why, for example, telecommunication companies in the United States have a virtual oligarchy for mobile, television and internet service, which is why we pay significantly more than the rest of the world for those services at significantly reduced quality.[604] We cannot have this happen with something as important as Universal Energy.

Therefore, as corporations are *profit* motivated as opposed to *service* motivated, they are unsuitable for managing the operation of Universal Energy's systems over the long term. However, as corporations have superior engineering prowess due to the competitive nature of capitalism, they are the natural entity The Public Interest Company would contract to develop Universal Energy's systems, even though they wouldn't manage the services these systems would provide to the public.

Public entity (bureaucracy). A public entity is a function of government, generally the executive branch. It may provide abstract services that most of us never interact with (FDA, EPA), services we know about and hope to never interact with (FBI, Department of Justice), and services that we know all too well about and dread interactions with (Internal Revenue Service).

However, bureaucracies can tend to look dimly upon the notions of efficiency and accountability, even if what services they provide are ultimately of ostensible social value. This is rooted in the fact that there is no competition for a bureaucracy, and they are relatively insulated from outside audit to hold their performance accountable or change the regulations they write by themselves.

Because of this, bureaucracies have little motivation to improve or adapt, as even the admission that they might need to do so carries the implication that they are operating at substandard performance. These are problems that can be reformed in a *regulatory* capacity, but it is much harder to reform the ability of bureaucracy to undertake a large-scale project in an *operational* capacity. Universal Energy needs to be implemented with efficiency and effectiveness as paramount considerations, leaving bureaucratic agency as an unsuitable candidate.

Not-for-profit company. Many organizations of "nonprofit" operation have made marks on the world: The Red Cross, World Wildlife Fund, Salvation Army (and the NCAA, interestingly enough) being well-known examples. However, not-for-profit designation is largely a tax determination and non-profits do not necessarily funnel money back into themselves for internal growth – nor do they necessarily raise sufficient revenue to do so as a core competency.

For non-profits centered around charity work, much of what money they raise is simply given away or spent to mitigate social problems. For other "non-profits" that are set up for pass-through income (like the NFL used to be), proceeds go to various stakeholders or administrative functions as opposed to expanding the organization's scale.

But The Public Interest Company is different – its goal is to expand the reach of Universal Energy and ever-improve the quality and value of the service it provides. It needs to profit to some extent because systems will eventually need to be repaired, upgraded and replaced over time. Also, the scale of Universal Energy's implementation will need to increase as wide as possible to effectively end resource scarcity and climate change – both of which cost money.

To this end, whatever profits raised will need to be directly re-invested into The Public Interest Company, making the traditional non-profit model less than ideal.

Public/private hybrid. In the past, our government has jumped into bed with various private businesses to create conglomerated, state-owned entities, and in many cases, the results were monstrosities. Indeed, the marriage of "big business" with "big government" frequently ranks among the finest examples of government buffoonery, even among the other resident experts in this department (see: The Pentagon).[605] But while the execution is lacking, the idea is ostensibly well-intended.

Yet the reason this model fails on execution is because it is fatally flawed from the onset, as the entity is run in the same capacity as a bureaucracy: seeking not to rock the political boat, resisting accountability and efforts to self-reflect, self-reform and self-advance. Universal Energy's implementation needs to be performed by an entity that shares opposite traits and yet still retains all of the positive aspects of corporate business and not-for-profit companies – which brings us back to The Public Interest Company.

The Public Interest Company is intended to combine all the positive aspects of these business models while discarding their drawbacks.

To maximize efficiency, a Public Interest Company is structured like a corporation, with an executive leadership and board, offering services for a profit. However, unlike a corporation that seeks to profit as greatly as possible, a Public Interest Company profits only as much as it needs to in order to invest money back into itself to expand and improve the quality of offered services, except in the issuance of dividends.

Like a bureaucracy, a Public Interest Company provides a public service for the public interest and is (initially) funded by public funds, but unlike a bureaucracy or public-private hybrid, a Public Interest Company isn't owned by the government. Rather, *it's owned by the public*, collectively, and remains directly accountable therein. Each U.S. citizen, upon reaching voting age, is given a single share of the company and is paid dividends from the company in the event of financial surpluses. As opposed to a corporation where the majority shareholder has the most sway, all shareholders have equal voice.

In this model, The Public Interest Company would be run by a board of directors that serve eight-year terms and are elected every Presidential election cycle by the public via simple majority, ranked-choice vote. These board members, in turn, would appoint a CEO of the company to manage it in the same capacity as a corporation today. The CEO would remain accountable to the elected board, which could also vote to issue bonds to raise money for future initiatives, or in the case of revenue surpluses due to international sales of Universal Energy's systems (which are sold at greater profit margins abroad), issue dividends to shareholders. In this model, all votes and meeting minutes of The Public Interest Company are transparent by design and made public.

In operation, The Public Interest Company would be funded by congressional appropriation each fiscal year for 10 years (see page 307 for details) and would manage the implementation of Universal Energy's primary resource production systems by soliciting bids from private companies through a process that is public and transparent by law.

Upon selecting a bid, The Public Interest Company pays a private enterprise to develop, deliver and implement the systems that provide the Universal Energy framework. It's no different than how the government buys a fighter jet, except this expenditure now instead goes to systems of higher social value. And once the system was owned by The Public Interest Company, it would be implemented as determined by a nationwide implementation plan The Public Interest Company would publicly issue every year.

This process would work in one of three ways:

- In cases where The Public Interest Company purchases *already existing technologies*, the technology itself would become property of the PIC and the system developer would retain ownership of all relevant intellectual property and would have the right to sell additional models to whomever allowed by law.

- In cases where the PIC paid contractors to engineer *new systems*, the PIC would retain ownership of all intellectual property pertaining to the system, the agreement of which would be a prerequisite to awarding any contract.

- In cases where the PIC deems appropriate, it would have the authority to purchase intellectual property from private entities should the entity be willing to sell them.

- Companies who build technologies relating to Universal Energy as a primary business model would be subject to lower income tax burdens, as would their employees. Further, private investors in these companies could enjoy lower capital gains taxes than other private industries.

Depending on which case applies, The Public Interest Company would manage any delivered systems to provide energy and resources to society at low cost, quantified as no more than 2 cents per kilowatt-hour. In operation, The Public Interest Company would have four funding sources:

Congressional appropriation. In this model, this allocation is approximately $663 billion for 10 years. After that, The Public Interest Company would be self-funded through either direct energy sales or international sales of equipment.

Direct energy sales. The target price of electricity under Universal Energy is 2 cents/kilowatt-hour – 84% less than what it costs today. Assuming electricity consumption increases ~50% to 6.5 trillion kWh annually, this will generate an annual $130 billion to The Public Interest Company. This model does not include pricing models for hydrogen fuel, but expects it to sell at a comparatively lower rate.

Corporate bonds. To fund future initiatives, the Public Interest Company could issue bonds with a fixed rate of interest on an open market, subject to shareholder vote. These bonds would be sold similarly to any corporate / treasury bond today, with the exception that capital gains taxes on profits would be lower for The Public Interest Company.

International sales. The Public Interest Company, through coordination with the State Department, would sell energy technologies to foreign governments at an increased profit, generating significant revenue.

With direct energy sales, corporate bonds and international technology sales, The Public Interest Company would ideally be operationally self-sustaining after the 10-year initial funding period. As its scale expands, it would first pay into a surplus fund to cover any future cost overruns. With this fund in place, it would continue to re-invest profits over time into energy-producing infrastructure, either to expand the scale of implementation or maintain systems that have already been deployed. Once profits reach a point where there are continual budget surpluses, these surpluses would be evenly divided and issued as dividends to all company shareholders (every American citizen of voting age until death).

A2: Collective Capitalism

This writing has placed strong emphasis on how resources are linked to the health of a society and its economy, first with how resource scarcity inevitably causes economic hardship and thus conflict, and second, how we can use technology to fundamentally prevent that problem from occurring.

Yet while the scarcity of resources carries negative economic impact, an abundance of resources inversely causes economic growth – all the more so if that abundance is indefinite. To fully embrace that growth, this writing advocates spending significant sums of money to build systems on a nationwide scale to make nigh unlimited energy and resources a reality.

The model that this expenditure operates is neither capitalism, socialism nor even "social democracy" – an oft-referenced capitalist/social hybrid that's seen in many European democracies. Rather, this model takes a wholly distinct approach to a society's economic framework because it establishes parameters outside of a finite resource paradigm.

This model is called "Collective Capitalism" because it hinges on the belief that the systems of capitalism work best when all social sectors are operating from a place of maximum strength. This idea in and of itself is not controversial – few would argue that a collectively stronger, more educated, more prosperous and healthier society functions at greater performance than not. Yet using an ideological approach (liberal, conservative, technocratic) to achieve this status is far more elusive than a *technological* approach that can help build the foundations by means of indefinite provision. Indeed, it's much easier to provide water, energy, food or fuel to everyone when you can synthesize them inexpensively to effectively unlimited scales.

But the core resource provisions for our social operation are simply the first step. Collective Capitalism, as a mindset, seeks to extend those provisions to everyone at a low cost so everyone, collectively, can operate from a place of security and strength and engage the mechanisms of capitalism to invest, achieve, discover, invent and, in turn, continue to perpetually improve the collective by virtue of capitalism's inherent rewards. It's the system we should have had, and wish we had, but was missing the key component of unlimited energy and resources.

For example, here are some of the more noteworthy aspects of social and economic improvement that can be realized through Universal Energy:

More Disposable Income

According to the U.S. Census, the median household income in the United States is $60,336.[606] Not accounting for state and local sales taxes (as well as property taxes if a homeowner and/or extra payroll taxes if self-employed), the average wage earner has a tax burden of 29.6% of their 2018 pre-tax income.[607] That would mean that the median U.S. household takes home roughly $42,500 (rounding up for easier math). Note citation for rationale of average versus median figures in this context.[608]

That comes to **$816.30 per week**, or **$3,541.66 per month.**

With that broken down, we'll source some routine life costs:

- According to the USDA (as cited by USA Today), the weekly cost to feed a family of four ranges from $146-$289.00.[609] The median cost of that range is $218.00 / week. That's $942.50 per month, or **$11,310 per year.**

- According to the American Automobile Association (AAA) using data from the U.S. Energy Information Administration, fuel costs average approximately 11.6 cents per mile,[610] and the average American adult drives roughly 13,476 miles per year.[611] Assuming each household of four includes two working parents with two vehicles, that's 26,476 miles driven per year at a fuel cost of **$3,071 per year.**

- As of 2017, the average U.S. household consumes roughly 10,400 kilowatt-hours of electricity per year (as of 2018, it's 10,972).[612] While the average national price of electricity is 10.53 cents per kilowatt-hour for all sectors, it averages 12.87 cents for households[613] with a contiguous high of 20.60 cents for New England and a national high of 32.47 cents for Hawaii. That comes to an average of **$1,155.35** nationally, although New England and Hawaii are two to three times that.

- According to Rocket Mortgage, a subsidiary of Quicken Loans, the average natural gas bill is $661 per year ($55/month).[614] The company further estimates that the average water bill is roughly $845 per year ($70.39/month) for households consuming the national average of 88 gallons per day.[615] That comes to a total of **$1,506 per year.**

Adding up the costs of electricity, fuel, water and heat, that comes to $5,732 per year, arriving at $17,042 when the cost of food is added in.

If we were to assume a corresponding reduction in the cost of electricity, fuel, water and heat to match Universal Energy's target reduction of 83% to 2 cents per kilowatt-hour, that would present a cost savings of $4,757 per year. Let's assume further that these cost reductions translated to food in the form of reduced energy / resource costs for irrigation, cultivation and transportation, as well as the advent of vertical farming within urban infrastructure powered by inexpensive energy. If we were to suggest a **30% reduction** in food cost, that would come to a cost savings of $3,393 per year. Added to the savings from utilities, and that derives a sum of some **$8,150 per year** for family of four.

Assuming further that this cost saving could be extrapolated on a per-capita basis, that comes to a total of $2,037 per person of any age. Across a society of 330 million (according to the U.S. census), that comes to a total of **$672 billion** that can be injected into our economy, saved for retirement or education, invested into real estate, businesses or other financial strategies.

Across our society, that translates to **$6.72 trillion** per decade.

The scale of such a figure presents massive implications for all income classes, but especially the lowest – for each extra dollar in their pocket in this context is not only tax-free (effectively making its real value 29% higher), it functions as a force multiplier to their financial mobility. Indeed, an extra $500 per month to a family living paycheck to paycheck matters far more than it does to a family that's independently wealthy.

Yet in this application, the distribution isn't necessarily weighted to certain economic classes – everyone gets the same bonus based on reduced costs society-wide. It's not socialism that takes from the wealthy to give to the poor, it simply uses technology to raise the collective floor.

Reduced Business Costs

Energy is an inexorable aspect of the costs of doing business today in our global economy. This is quantified across several key areas: energy costs of material and/or resource procurement, manufacturing, transportation, processing of both material and data, heating, lighting, etc. These factors each manifest one way or another into nearly all business sectors, and consequently are absorbed into the price of the product or service that's offered at market. If reduced energy costs can provide a cost savings for residential households, the purchasing and consumption scale of incorporated business would see corresponding reductions on subsequently higher orders.

This is all the truer since the cost reductions would cascade across an array of business sectors and their interoperating supply and manufacturing chains. For example: if Universal Energy makes a raw material 20% less expensive to source at equal quality, reduces the energy cost to manufacture systems with it by 20%, and further enables a 20% cost reduction to transport the manufactured product to market – those cost savings combine across the supply chain. Across our national – or global – economy, the financial implications are enormous.

To determine just how impactful, we'll take a look at figures from the Energy Information Administration, specifically their Commercial Buildings Energy Consumption Survey (CBECS) for the year 2012 (the last year in which full data is available).[616]

According to the survey, American business consumed a total of 1.243 trillion kilowatt-hours of electricity.[617] At a national average cost of 10.67 cents per kilowatt-hour for commercial enterprise and 6.92 cents per kilowatt-hour for industrial applications (heavy manufacturing),[618] that respectively totals **$132.63 billion** and **$86 billion.**

The CBECS survey further assessed that American business consumed 2.193 trillion cubic feet of natural gas in 2012[619] at a cost of $7.78 per thousand cubic feet for commercial enterprise and $4.21 per thousand cubic feet for industrial applications. That respectively totals **$17 billion and $9.23 billion.**

Now, we'll take a quick look at fuel. According to the Department of Transportation, commercial vehicles (including trucks and busses) accounted for about ten percent of all vehicle miles driven.[620] Further data by the Energy Information

Administration, as curated by Statista, estimates that the U.S. consumed 8.98 million barrels per day of gasoline and 3.13 million barrels per day of distillate fuel oil (which includes diesel).[621] That translates to 3.277 billion barrels of gasoline and 1.142 billion barrels of distillate fuel oil. Extrapolated into gallons (at 42 gallons to the barrel), these figures respectively translate to 137.66 billion gallons of gasoline and 47.98 billion gallons of diesel.

Leveraging the Department of Transportation's estimate, we'll assume that 10% of that gasoline consumption was from commercial and industrial applications. Yet since diesel is the fuel of choice for trucking and heavy machinery, we'll assume 95% of diesel consumption came from commercial and industrial enterprise. This leaves a figure of 13.77 billion gallons of gasoline and 45.58 billion gallons of diesel that can be attributed to businesses.

At a national average price of $2.70 / gallon for gasoline and $3.06 for diesel,[622] that comes to a total aggregate cost of **$176.65 billion.**

With this established, let's add up our totals. Based on the calculations above, we've estimated that:

- American commercial enterprises annually spend **$132.63 billion** on electricity, falling to **$86 billion** for heavy industry.

- American commercial enterprises annually spend **$17 billion** on natural gas, falling to **$9.23 billion** for heavy industry.

Added up, that comes to **$149.63 billion** for commercial enterprises and **$95.23 billion** for heavy industry that's added on top of the estimated **$176.65 billion** shared by both for diesel and fuel costs.

This creates a total cost liability of between **$326.28 billion** and **$271.88 billion** depending commercial or industrial application. If we were to split the difference, that would come to a figure of roughly **$300 billion per year**.

If that figure, for sake of argument, were subjected to an 82% cost reduction, that would present a cost savings of **$246 billion** per year. As this translates to $2.46 trillion per decade, such cost reductions present a capital abundance that can further enable businesses of all sizes to invest in their own growth and future success.

This approach can be scaled further through targeted tax incentives. When discussing Universal Energy's management and implementation strategy in the previous section of the Appendix, mention was made of the possibility of dramatically lowering the tax liability for businesses operating in Universal Energy's sectors, along with corresponding incentives for the employees and investors of such companies.

This is a key component of a "Collective Capitalism" mindset, hinging on the notion that industries and personnel that provide critical – and extremely beneficial – services to society's long-term operation and improvement should face a correspondingly reduced tax liability to fund society's public functions. It defies reason that a company making cigarettes or hawking payday loans at predatory interest rates should operate under the same tax burden as companies developing energy technologies for The Public Interest Company, growing food in vertical farms or building next-generation infrastructure to provide an improved quality of life for society.

The same is true for the employees of such industries, as well as their investors across equities, bonds and other financial products. If an industry provides empirical social benefits on a transformational scale, why should an employee face the same tax burden as an employee of an industry that doesn't directly deliver the same degree of social improvements? Further, why should an investor seeking to inject capital into a socially beneficial enterprise pay the same capital gains taxes as someone seeking a quick profit by shorting the same stock, or by throwing their money into shadier organizations like private prisons or conglomerates with abysmal human rights records?

Here's what this could look like in practice. Let's say that we establish an empirical threshold (defined specifics, not abstract opinions) of social benefit within varied industrial sectors, and assign "classifications" to such industries using a transparent assessment criteria. Beyond participation in Universal Energy, this criteria could include a lower ratio of executive to average worker compensation, demonstrated ethical track record, external social outreach and investment, operational transparency, and/or quality of benefits offered to their workforce in aggregate.

Based on this corporate classification (not unlike a "B-Corp" designation), the company, its employees, and its investors could enjoy special tax incentives. An example might reflect the following table:

Corporate Classification	Corporate Income Tax Rate	Capital Gains Tax Rate
Class A Corporation	0-5% on a progressive scale based on income. Maximum tax rate is 5%. Employee income tax is capped at 15% up to $1M.	Short term: 5%. Long term: 0%. After $1M, gains are taxed as income.
Class B Corporation	0-10% on a progressive scale based on income. Maximum tax rate is 10%. Employee income tax is capped at 20%, up to $1M.	Short term: 15%. Long term: 5%. After $1M, gains are taxed as income.
Class C Corporation (Current tax rates as of 2019)	0-21% on a progressive scale based on income. Maximum tax rate is 21%. Employee income tax rate unchanged.	Short term: 25%. Long term: 15%. After $1M, gains are taxed as income.

Under this model, current companies do not pay any more in taxation than they do today, yet Class A and Class B corporations would receive significantly more attractive tax incentives to engage in business sectors earning such classifications – of which Universal Energy would be a primary qualifier. This reasoning can extend further to other industries that deliver an empirical social benefit: making bionic limbs for amputees, investing in next-generation medical research, building advanced transportation infrastructure, and so on.

In such reasoning, there is a clear distinction here between "picking winners and losers" and incentivizing investment to industries that make the world an objectively better place. Collective Capitalism doesn't seek methods that punish industries, companies, or personnel that do not choose to invest in empirical social progress, but it does seek to reward them through the establishment of frameworks that makes this task easier and less expensive. In turn, this would incentivize and encourage a social impetus to continually invest and be part of industries that deliver a strong social benefit – and, further, seek to expand that benefit at higher rates of return than otherwise.

Consequently, these incentives could see reprioritizations across our economy. Defense contractors, for instance, aren't really companies that specialize in building high-tech weaponry so much as they are expert engineering firms that specialize in building high-tech systems. There are few obstacles, in real terms, from shifting primary focus from military infrastructure to domestic infrastructure. If you can build a Generation-V fighter jet, you can build effectively anything. Providing both a public funding impetus and tax incentive to shift from weapons to, say, sophisticated energy technologies, next-generation rail travel, hyperloops or civilian aerospace can help facilitate this transition.

In conjunction with lower operating expenses from energy cost reductions and public funding allocations for Universal Energy and its underlying technologies, this can keep current flagship industries in play building critical American infrastructure, while also enabling opportunities for start-ups to gain a foothold and accordingly prosper. This would transform the "military industrial complex" into an "energy/resource industrial complex," delivering cascading social benefits at only moderate costs.

Increased Job Growth

Reducing the operating expenses of businesses, both through drastically lower energy costs and enhanced tax incentives, stands to provide several avenues for job creation beyond the obvious potential presented by large-scale investment in next-generation energy and resource technologies. Remaining competitive in this new frontier will require as much investment in personnel as technology, both in terms of manufacturing, maintenance, marketing, managing and deployment logistics. This means jobs.

It's true that in the past, certain historical instances of operating cost reductions have seen some companies choose to pay for increased executive bonuses and stock buybacks,[623] as opposed to workforce expansion and salary increases. But these have primarily occurred in instances where a company had entrenched dominance in their market sector and could comfortably afford to rest on their laurels to hold off emergent competitors who came to market with more nimble and disruptive approaches.

This possibility would be substantially harder with the market sectors opened up by Universal Energy, not only because the technologies (and thus industries) are in a degree of infancy that hinders monopolization, but also because The Public Interest company (in this model) would seek to prioritize contracts with companies who a) hire American, b) pay competitive wages and benefits, c) competes fairly and refrains from anticompetitive strategies, and d) takes strides to achieve a corporate classification that would only be granted after demonstrating a higher tier of operating ethics both to their society and workforce.

In such instances, the possibilities for job growth are enormous.

To see how, let's quickly circle back to the estimated $246 billion American businesses would save under Universal Energy. At an assumed 25% tax overhead to hire a salaried employee at $50,000 per year ($62,500), this reduction in *energy and fuel costs alone* would be sufficient to create 3.94 million new jobs. Adding on the potential savings due to increased corporate classification, investment in next-generation industries and technologies, and the advent and functions of The Public Interest Company, and the potential job growth increases substantially.

Of the $663 billion congressional appropriation that would be devoted to implementing Universal Energy over a 10-year period, it's essential to note that this money isn't just sent into a void – it's paid to enterprises who win contracts to develop and deploy the technologies inherent to Universal Energy under a transparent bidding process.

The first area to receive these funds would be engineering companies that develop the underlying technologies for Universal Energy. This will create job demand within a multitude of technical skills: physics and engineering, metallurgy, computer science, software development, graphic modelling, automated manufacturing, 3-D printing, quality control, human resources, project management, advertising and marketing (among others).

As these companies expand along with job demand, they in turn will need to expand their acquisition of materials and resources to develop the systems they were contracted to deliver. They will need to buy materials, tools, vehicles, fixtures, office space, uniforms, amenities and everything else that comes from the manufacturing world. All of this will create jobs.

It will also create job demand in industrial sectors these companies depend on to operate, which in turn will create job demand in all of the support and promotion positions that make their own business possible. This job demand will create additional demand for schools and the educators to staff them along with all of the support positions that make their jobs possible. The result is a cascading increase in job demand corresponding to a cascading reduction in operating costs – not only creating a next-generation economy, but also revitalizing the state of American manufacturing to a subsequently higher tier.

Revitalized American Manufacturing

With the exception of advents in information technologies and weapons development, the overall state of American manufacturing has been in decline since the height of post-WWII boom years.[624] While there are myriad causes of this state of affairs, an investment Universal Energy enables us to chart a fundamentally different course towards a future where the American economy can reclaim its status as a global leader in technology, infrastructure and innovation.

Sparking a social drive to build advanced energy technologies and corresponding infrastructure gives us the head start on research and development within this new economic frontier. Further, by implementing them first in our country we would become the foremost experts in this sector and those surrounding it. And as we have engineered the technologies therein and have perfected their ideal means of deployment, we position ourselves to be their best purveyors to other countries. The potential size of this market internationally is easily in the trillions of dollars over the long term, as our expertise with these technologies would translate to repeat business in contracts for maintenance, upgrades, etc., providing future revenue streams to American business and our economy.

Reduced Social Afflictions

The social investments and derivative results inherent to Universal Energy and Collective Capitalism fundamentally makes life easier and less expensive. An abundance of inexpensive energy and resources reduces the cost of living and increases economic, career and social mobility. It further mitigates the scarcity and desperation-driven impetuses to engage in criminal activity. Life's just "better" and people have more time and opportunities to engage in activities that they find value in – be it their family, hobbies, side projects or new vocations. In these circumstances, it is significantly easier for someone to start their own company, innovate a new idea or product, and/or invest either time or money in another venture they believe will ultimately have social value. It expands the purchasing power, investing power and financial influence of the middle class, and extends to them luxuries that were once available only to the social elite.

Effectively, Collective Capitalism leads to a world that is powered by resource abundance, a removal of need and a social safety net that is technology-driven, not by limiting the ceiling or redistributing from the top – but rather by using technology to simply raise the floor and the foundations on which it stands.

The chart below represents a concept known as "Maslow's hierarchy of needs." It breaks down the core needs of human beings, with the lower end of the pyramid represented by critical need that is resource-driven – the needs Universal Energy provides as a core function. This allows society to place increased focus on the remaining categories of need, encouraging us to extend ourselves for more meaningful things that result in greater social enrichment, culturally, intellectually and spiritually.

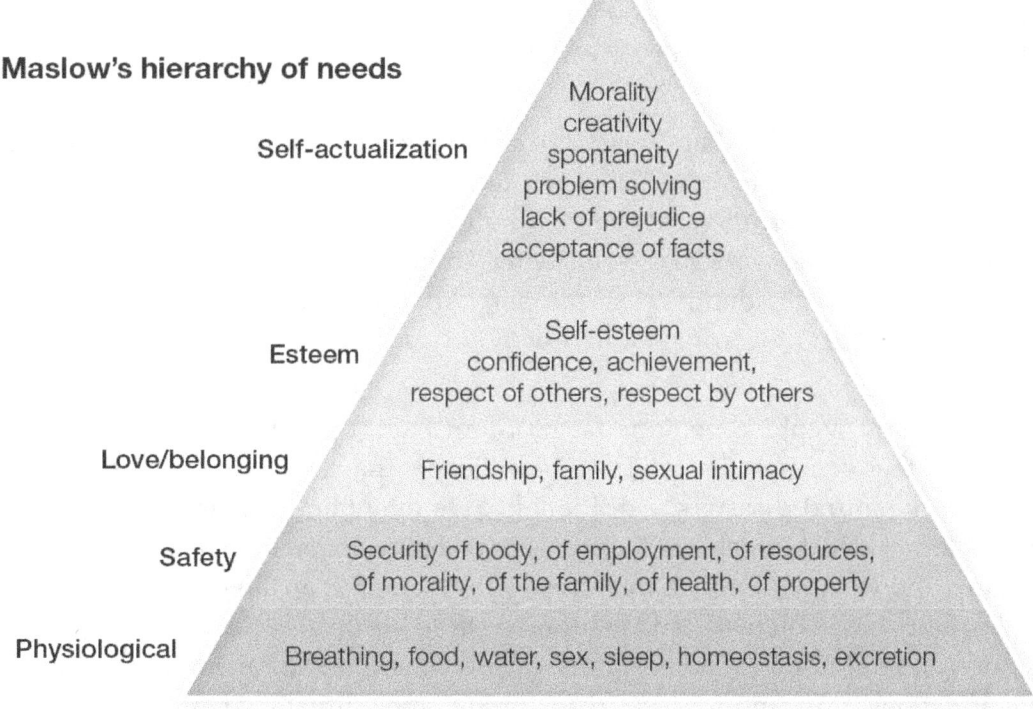

In this line of reasoning, it's worth mentioning that much of our culture today is provided by artists and writers who thrived in generations past. (Indeed, many of the comic book heroes who grace movie screens today debuted 50-80 years ago). There's little market for today's storytellers, artists, poets, artisans or philosophers because it's difficult for them to make a living from doing so. These people once gave society great value, but today their creative capabilities are sidelined by the demands of a cutthroat economy.

The financial benefits inherent to the Collective Capitalism afford the provision for abstract social concepts and enlighten the subjects we discuss, goals we set and behaviors we value. Because all core needs are met in this model through technology, the stresses and efforts that were previously required to meet the demands of life are no longer present. Concordantly, we no longer need to distract ourselves with fleeting content to make ourselves feel better in the face of these stresses and efforts, allowing us to finally concentrate on what we truly want as people, as we truly want it.

Over time, this will further mitigate existing social problems and increase the scale of shared economic prosperity on multiple fronts which allows our society to look inward and mend its wounds to become stronger and form a more cohesive cultural identity based on an improved quality of life.

Notably, this is an identity that can refresh our reputation abroad. When combined with the provision of Universal Energy's technologies internationally (allowing other countries to further increase their own quality of life) this affords us a degree of appreciation that can work to re-solidify the United States as the center for global economic and cultural identity. Additionally, it can extend our humanitarian outreach and also lead to stronger alliances.

Universal Energy would have done far more for Haiti than the largely ineffective relief efforts that consumed a total of $14 billion,[625] the same with any other area victimized by natural disasters. Additionally, we maintain allegiances with other countries today based largely on inexpensive resource acquisition and the security assurances and weapons exports that come with them. But an allegiance of security is an allegiance based on fear…and fear knows little loyalty. Rather, we could create allegiances based on social and economic improvement and a functional end to the social afflictions inherent to resource scarcity and climate change. These are allegiances based on far healthier terms amid a global climate of heightened economic prosperity and easier conditions for peace.

A3: Revenue Allocations

To fund Universal Energy's target cost of $6.63 trillion over a ten-year period, this model suggests several reallocations of federal spending, alongside certain revenue raising approaches, to obtain the necessary funds to implement the framework in the manner proposed. As with this model as a whole, these approaches will not be inherently partisan or politically ideological, and will rather reflect a good faith analysis under the assumption that our society, in turn, can get to a place where its leadership and social institutions can operate under such faith. Further, this model will primarily look to aspects of the public sector at **the federal level alone**, and will make assumptions of private investment and consumer spending based on certain percentages of American Gross Domestic Product (GDP).

To begin, we'll start with the federal budget, FY2018, using figures sourced from the Congressional Budget Office.[626]

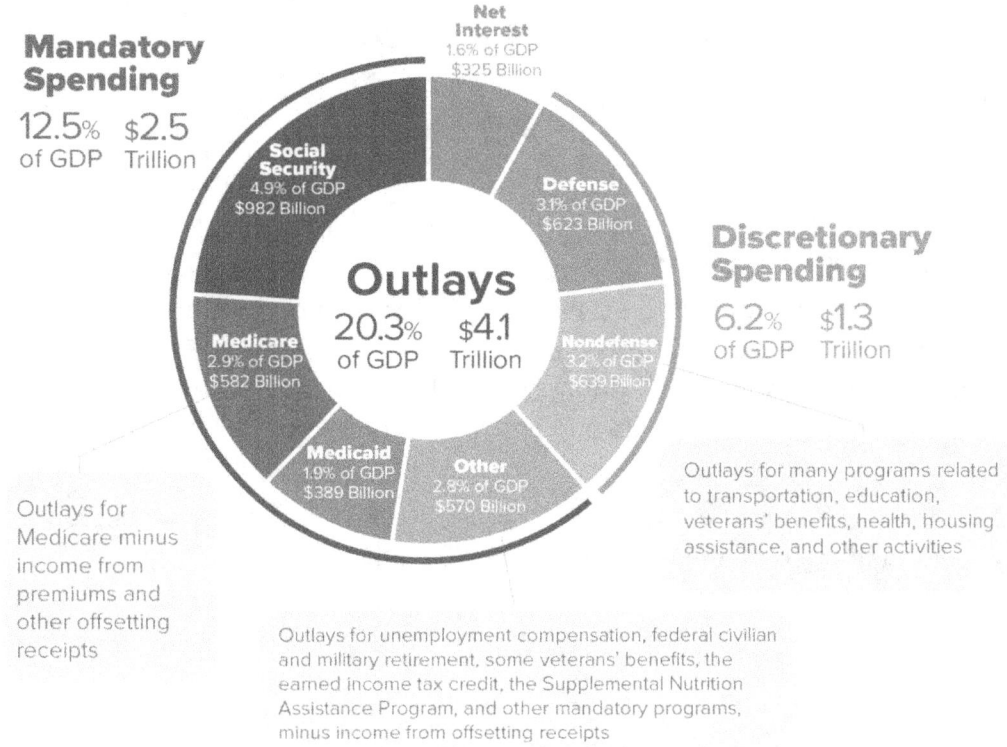

Of these outlays, Social Security, Medicare and Medicaid are (at least in theory) supposed to be funded through separate revenue-raising measures (payroll taxes (FICA)) but are often included in total federal expenditures. This model acknowledges the comparatively poor value American public healthcare costs represent compared to the rest of the developed world (we generally pay more per-person for public healthcare programs that are only available to the poor and elderly than other nations do for healthcare that covers *everyone* in their society).

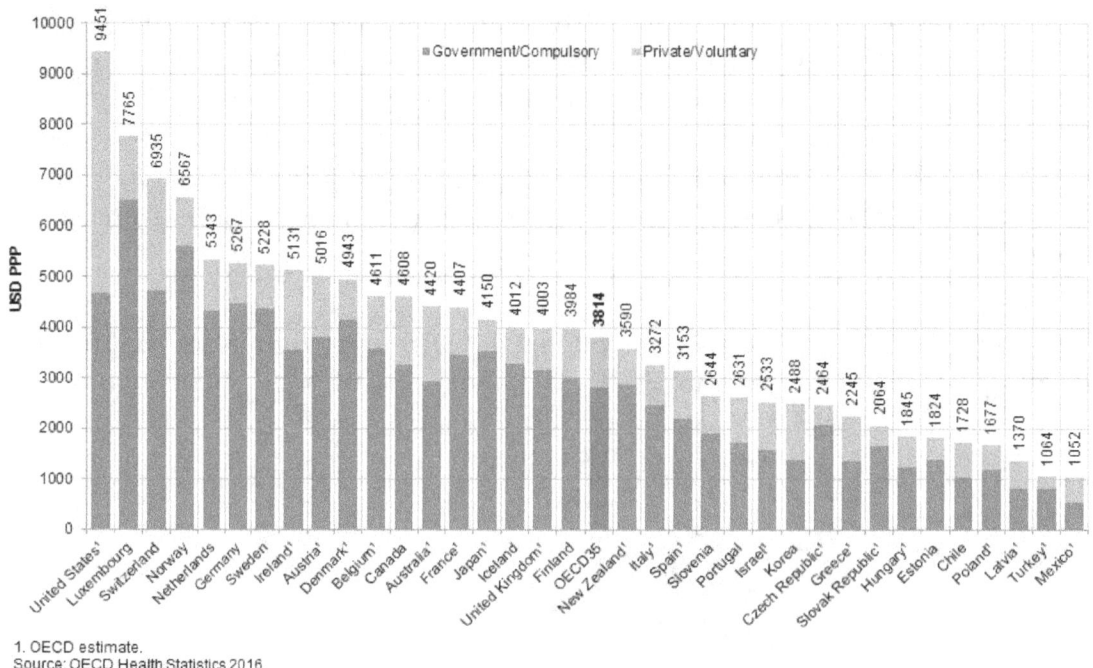

1. OECD estimate.
Source: OECD Health Statistics 2016.

However, this model's focus will not center on the over-payment of public health programs – nor the current state or solvency of Social Security. Instead, it will focus primarily on the expenditures and functions of the United States Federal Government as it exists today. In doing so, we'll also need to take a look at the revenue it earns over a fiscal year (again in this case for 2018).

For FY2018, the U.S. raised a total of $3.3 trillion – some $800 billion less than it spent (meaning there remained a "deficit" of $800 billion).

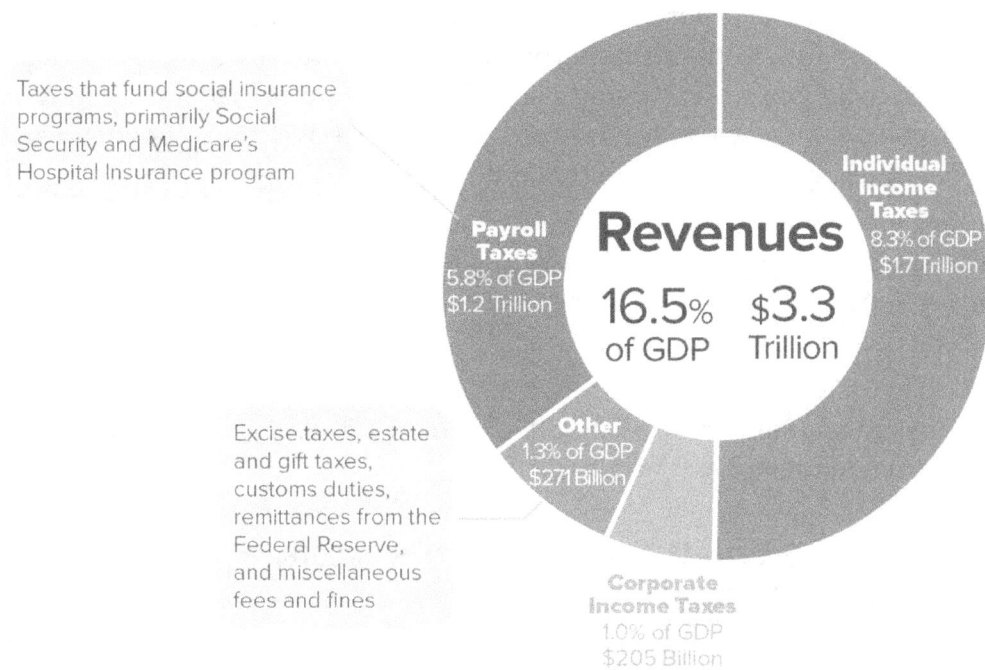

Of this revenue raised, $1.2 trillion was for the "trust-funded" programs of Social Security, Medicare and Medicaid (which we're not including in this analysis). As these programs are supposed to be funded through separate payroll taxes, we will break down revenue allocations/deficits by funding source and intended use:

Medicare, Medicare, Medicaid and Social Security:

$1.2 trillion raised.
$1.95 trillion spent.
Deficit: -$750 billion

Other mandatory, defense and discretionary spending:

$2.176 trillion raised.
$2.157 trillion spent.
Surplus: $19 billion.

For this analysis, we will assume that the deficit comes primarily from a lack of payroll taxes to fund FICA-related trust programs – not general revenue for the operating functions of the Federal Government. To fill this gap, this model suggests raising payroll taxes accordingly alongside honest and effective investigations into why American public healthcare problems are so exorbitantly overpriced compared to their counterparts abroad. (Hint: it's drug + insurance lobbies, along with oppressive medical school costs). For other federal accounting we will focus on three key areas: defense spending, discretionary spending, and additional revenue, and proceed in that order.

Defense spending. One of the most surprising aspects of the federal budget is how many expenditures outside of the "defense budget" are actually exclusively defense-related costs. For example:

- Veterans affairs, at $186.5 billion for FY2018, is paid out of the discretionary budget.[627]

- Military retiree pay, at $54.7 billion for FY2018, is paid out of the discretionary budget.[628]

- Several other defense-related expenditures ($10.6 billion for nuclear weapons[629], $44.1 billion for the Department of Homeland Security[630] (which includes Coast Guard and is indisputably a "defense-oriented" expenditure), clandestine intelligence operations ($81+ billion)[631], and foreign military assistance are all paid for out of the discretionary budget.

- Of the $325 billion in national debt interest payments for FY2018 (the total debt of which is $23 trillion), at least 26% is due from the $6.3 trillion the U.S. government has borrowed to fund wars in Iraq and Afghanistan.[632] The nonprofit *War Resisters League* estimates that war costs comprise as much as 80% of our national debt, although it acknowledges most sources estimate the figure to be approximately half.[633] We will assume a figure of 50% in this analysis, arriving at $162.5 billion.

This would come to a total of an *additional **$539.4 billion*** in defense-related costs that are billed as "non-defense." Therefore, our total "defense" spending is actually about $1.163 trillion, not the ~$623 billion it's annually billed as.

Conventional wisdom frequently touches on how much more our nation spends on "defense" compared to the rest of the world – exceeding the total of the next eight nations *combined*, even though we count six of the eight as allies.

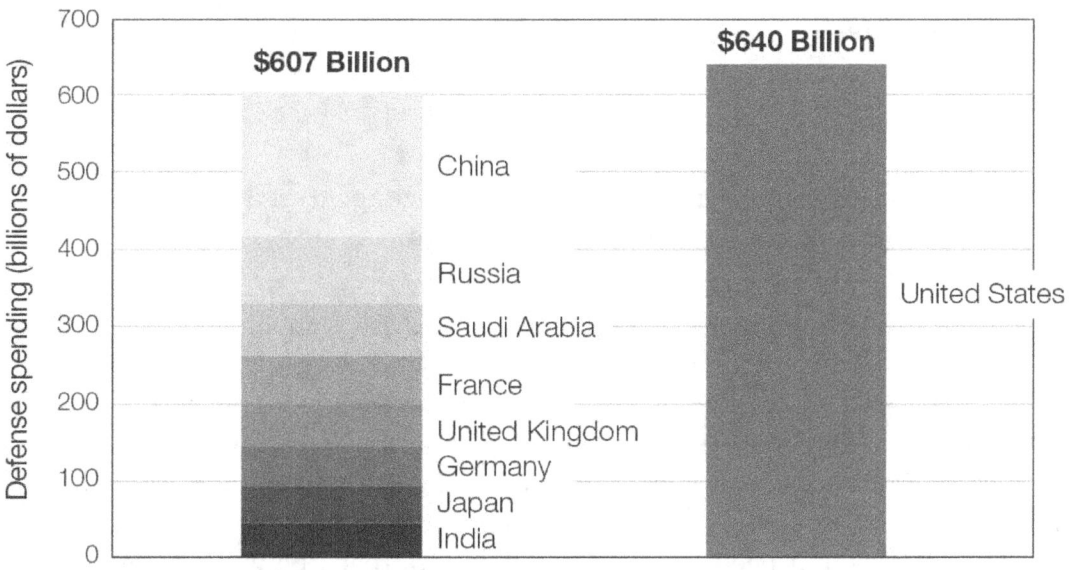

Source: Stockholm International Peach Research Institute, SIPRI Military Expenditure Database, April 2014. Compiled by PGPF. Note: Figures are in 2013 U.S. dollars converted from local currencies using market exchange rates.

Yet this figure only counts our *stated* defense spending, whereas our total spending is nearly **twice that.** This model suggests that investing in a framework that dramatically reduces the underlying need for a global military empire is a far wiser expenditure than wasting endless trillions on military technologies that eventually become obsolete as future conflicts evolve. Consequently, this model assumes that through a mixture of honest accounting, scaling down the mandate and global scope of our military – and ensuring **all** military costs are paid through the defense budget – we could be able to pare off **$350 billion annually** from our $1.163 trillion annual military expenditure (a 30% reduction that would still make us spend more than the next eight nations combined).

Subtotal savings from total military spending: **$350 billion.**

Discretionary spending. While much of the discretionary budget are actually military-related expenditures, valid concerns remain about the scale of redundancy within federal bureaucracy. This writing does not maintain an ideological position on the scope of the Federal Government, yet it does suggest addressing manifestations of *operational* and *organizational* lapses through redundant services. For example:

- There are at least 65 agencies empowered with enforcing various aspects of federal law.[634]

- There are more than 451 independent federal agencies tasked with various aspects of societal management with the power to independently enact regulations across all areas of our society.[635]

- There are seventeen independent intelligence agencies within the United States Intelligence Community.[636]

In mention, it's important to emphasize that this isn't an indictment of the functions of government, the use of regulation, or the need for intelligence. Rather, the concern is that these agencies serve functions that are a) independently operated, b) redundant in focus, and c) largely shared by counterparts at the state level. This, in aggregate, presents an amplified cost basis for their functions that make the discretionary functions of government cost much more than they would need to if a streamlined approach between federal and state services were facilitated.

Indeed, the aggregate sum of the discretionary spending of the Federal Government, at some $6.3 trillion per decade, is effectively the cost of implementing Universal Energy in entirety. It's difficult to see how that has delivered a comparable social value.

This model suggests that through bureaucratic restructuring to consolidate functions and reduce redundancy, end socially harmful and wasteful programs like the "War on Drugs" and place greater reliance on partnerships with agencies at the state level, that we could reduce the discretionary spending of the Federal Government by approximately 25%. This would raise a total of **$159.75 billion.**

Additional revenue. While this writing (and I as an author) remain politically nonpartisan, I should state for the record that I believe that our current state of taxation is neither fair nor of high value, considering what we actually get for our tax dollar. It's not a question of preferring high or low taxes, *it's a question of what value we obtain per tax dollar spent.*

However, if we are going to make an honest attempt to both reallocate our federal spending towards more socially beneficial focuses and pare down our national debt, the concept of increasing certain taxes is worth considering in abstract. Yet in doing so, this model will not focus on income taxes directly, and instead will seek to raise revenue through additional sales on tobacco, alcohol, marijuana (legalized) and luxury goods. Here's how this could look:

Vice excise taxes: the federal cigarette tax today is $1.01 per pack, and brings in $14 billion for the federal government.[637]

The federal tax on alcohol is $13.50 per proof-gallon of distilled spirits (roughly 21 cents per ounce of alcohol), $18 per barrel of beer (roughly 10 cents per ounce), and $1.07 per gallon of wine (roughly 8 cents per ounce).[638] These taxes collect approximately $10 billion per year for the federal government.

As marijuana is illegal on the federal level, the federal government does not collect a tax on its illicit sales, although the tax revenue raised in states it has been legalized are significant (roughly $1 billion for FY2018).[639]

Cognizant of these figures, this model would suggest considering the following excise tax increases:

- Increasing federal tobacco taxes by 300%, raising a total of $42 billion per year (assuming consistent demand).

- Alcohol, in all forms, would be taxed per proof gallon, which is the method proposed by the Congressional Budget Office. Recommending a flat tax of $16 per proof gallon, they concluded that such a tax would equal about 25 cents per ounce of alcohol (29.57ml).[640] This tax increase was estimated to net an additional $9 billion per year to total $19 billion on all alcohol sales taxes. As with tobacco, this model would suggest a 300% increase, arriving at a tax of $48 per proof gallon. Assuming consistent demand and sales volume, this would raise $57 billion annually.

- Marijuana would be made completely legal in this model and would be taxed and in the same capacity as alcohol and tobacco. Current sales models are difficult to predict because marijuana is presently classified as an illegal narcotic. Yet a 2005 report endorsed by more than 50 distinguished economists (including Milton Friedman) concluded that legalizing marijuana would save $13.7 billion in enforcement / prosecution / incarceration costs, while approximately $34.3 billion would be yielded by tax revenue of its nationwide sale.[641] This would come to a total of $48 billion.

Luxury excise tax: this model would also suggest the possibility of modest taxes on high-end luxury items under the reasoning that if one is wealthy enough to afford a private jet, exotic sports car or luxury yacht, a 10% tax increase would not be a prohibitively difficult expenditure. The U.S. luxury goods market in 2018 was $191.8 billion. A 10% tax of that sum would net $19.1 billion.

Potential total raised through vice + luxury taxes: $166.1 billion.

Revenue Totals

Based on these proposed measures, we arrive to the following subtotals:

- A 30% reduction in total military spending would net **$350 billion.**

- A 25% reduction in discretionary spending would net **$159.75 billion.**

- Increased excise taxes on alcohol and tobacco, along with new excise taxes on marijuana and luxury goods would net **$166.1 billion.**

These measures, in total, would net **$675 billion annually**, which would be sufficient to fund Universal Energy in its current form for 10 years, which thereafter could be used to pay off our national debt.

Considering that this sum is merely half of our total military spending over the past decade, 15% of our total federal spending over the past decade, and less than 5% of our annual Gross Domestic Product, it's a small price to pay for a world without resource scarcity, resource conflict, climate change and the endless maladies brought by scarcity and the zero-sum games – and corresponding conflicts – it fuels.

A4: Universal Energy Cost Estimate

Universal Energy is modular by design. Each of its technologies can have the capability to connect to one another in any configuration desirable, thus any realistic cost estimate will be determinant on the ultimate nature of configuration. For sake of argument in providing a baseline figure, this model will make several assumptions about the scale and nature of implementation, cognizant of the implementation strategy for public resources and proposed legal and social mechanisms we reviewed earlier in the Appendix. Due to the nature of the assumptions required to make such an estimate, this model will be more generalized in approach as opposed to seeking exact estimates that could shift based on future circumstances, but we'll try to be as reliable as possible based on the factual data at hand, and seek further to overestimate than come up short.

Further, as electricity generation in most contexts is denoted in kilowatt-hours, we will use the **kilowatt / kilowatt-hour unit** respectively for **power** and **energy** generating capacity in this estimate. Proceeding forward, this estimate is broken into two distinct areas: electricity generation and resource production, which we'll cover in that order.

ELECTRICITY GENERATION

The United States currently consumes 4.17 trillion kilowatt-hours of electricity annually as of 2018 (4,171 terawatt-hours).[642] The initial goal of Universal Energy is to provide 300% of our national electricity consumption, which would be roughly **12.5 trillion kilowatt-hours** (12,500 terawatt-hours).

If we were to leave our current capacity intact (and gradually phase out old power systems, starting with the oldest and dirtiest first), Universal Energy would initially need to generate **8.34 trillion kilowatt-hours** of electricity. This figure can and should scale over time, but it functions as a sufficient target for now.

Further, this estimate will also include the necessary infrastructure to produce sufficient fuel, water and food to provide for our respective needs, as well as desalinate sufficient water to store sufficient energy in the National Aqueduct to comprise its battery and power-generating functions. These estimates will be assessed as a separate consideration on top of electricity generation within the resource production section.

Lastly, while this estimate will make assumptions about the nature of Universal Energy's implementation, it's important to note that we will be *minimizing* assumptions of future cost reductions as investment in Universal Energy's technologies expands. This price estimate, consequently, will be based on the estimated price *today* in "overnight costs" – the cost of the construction if it were purchased and deployed as a standalone unit. It will not assess cost reductions over time through improvements in manufacturing, lowered energy costs, subsidies or *levelized* costs – costs incurred and offset by revenue earned over the lifetime of an energy-generating asset.

This latter consideration is especially important, as *levelized* costs assume cost-mitigating factors such as energy sales in a commercial market, future subsidies, and, especially in the case of renewables, the application costs of continuous operation (of which renewables, especially solar, have less of than power sources that require fuel and active maintenance of moving parts). A $100,000 solar array, for instance, might cost $100,000 to buy and deploy, but when considering the benefit of its generated energy that doesn't require fuel to be purchased, lack of maintenance, etc., the *levelized cost* of that solar array may be far less over its lifetime. It's not so much of an accounting "trick" as it is a view of long-term accounting, but its functional result is to put cost figures in perspective, and, ultimately make them appear less than they would be if we were only considering the retail sales price of the technology. This estimate will forgo levelized cost estimates and simply look to the "sticker price" (overnight cost) of a technology as it exists today, and it will do so based on three reasons:

1. The goal of Universal Energy is to make energy and resources as inexpensive as possible – its 2 cents per kilowatt-hour target is roughly 85% lower than the commercial price of energy today, so levelized costs as assessed today would not be viable with such a dramatic reduction in energy costs.

2. Levelized cost estimates use myriad factors and assumptions ranging from the average commercial price of energy, subsidies and tax incentives, cost of labor, cost of materials and estimated operational lifetime – all of which vary wildly by region. Levelized costs, therefore, include several moving parts that are difficult and complex to assess on a nationwide scale. Further, they don't incorporate the possibility of new technologies. (The levelized cost of solar, for instance, doesn't incorporate municipal integration. The levelized cost of nuclear, further, is based on light-water reactors, and doesn't incorporate the emergence of new technologies such as thorium).

3. Levelized costs, finally, are assessed to show how much a technology costs over time, which spreads the cost of energy over that time period. This estimate seeks to focus on what it would cost to implement Universal Energy *today* – "sticker shock" and all – because it is the investment in the technologies, themselves, and their accompanying infrastructure that solves resource scarcity and climate change. The focus isn't on a capital investment that seeks a capital return (although it will ultimately do so), the focus, rather, is on a capital investment that seeks a *social* return – we get a future that's not dominated by resource conflict, ecological collapse and all of the humanitarian and environmental crises they spawn. This factor, along with the others aforementioned, make levelized costs less appropriate for inclusion in this estimate.

With this clarified, we'll proceed from here to review Universal Energy's cost estimate in full. As the ultimate appropriateness of each of the energy-generating technologies will vary based on geographical region, proximity to coastlines, highways and cities, we will break this estimate down in terms of **units of 100 billion kilowatt-hours generated per year,** roughly 1/85th of Universal Energy's foundational target.

We will refer to this 100 billion kilowatt-hour figure as an **"Energy Unit,"** which broken down on a daily basis, comprises 273.97 million kilowatt-hours generated per-day over a 365-day year.

With this established, we'll start our analysis with renewables.

Integrated Renewables

The key strategy of renewables within Universal Energy is municipal integration, as it avoids the highly expensive requirement of large-scale land purchases, especially in urban environments (where land is most scarce and most expensive). Most renewable advocates ignore this cost factor when making estimates, which while normally disingenuous is less of a concern for our purposes here. That makes baseline renewable estimates valid for use in this context.

Solar Power: According to the National Renewable Energy Laboratory,[643] the benchmark cost of commercial solar implementation can be as low as $1.44 per watt for a fixed-tilt utility-scale system exceeding 2 megawatts in size (alternating current). Because we are seeking implementation in municipal infrastructure and not buying land, we will assume this figure for benchmark cost estimates. In doing so, we'll also be referring to our prior assumptions on made on page 52:

- An average of **five peak sun hours per day** in the U.S.

- One square foot of solar generates 18.7 watts under peak sun. That's 18.7 watt-hours per hour, 82.9 watt-hours per day, and 30 kilowatt-hours every year.

- One square meter of solar generates 201.28 watts under peak sun. That's 201.28 watt-hours per hour, 1,006.4 watt-hours per day, and 367.3 kilowatt-hours every year.

- At $1.44 per watt, one square foot of solar panels would cost $26.93. One square meter would cost $289.84.

A 100 billion kilowatt-hours per year Energy Unit translates to 273.97 million kilowatt-hours per day, or 273.97 billion watt-hours. Divided by 82.9 watt-hours per square foot of solar panel surface, and that comes to 3.3 billion square feet of solar panels (118.5 square miles / 306.91 square km). At a cost of $26.93 per square foot, that translates to $88.87 billion.

Total cost: **$88.87 billion** per 100 billion kilowatt-hour Energy Unit.

Wind Power: The nominal, non-levelized cost of wind power in the United States is estimated to be between $750-$950 per kilowatt of power-generating capacity, with a non-levelized cost of $0.07 per kilowatt-hour generated.[644] This figure, ostensibly, excludes the cost of land purchase, wiring and grid connection, which would present significant externalities on top of equipment purchases. While this estimate places greater emphasis on the promise of solar power within renewables (due to myriad factors ranging from limited locational deployment, unique impact on migratory wildlife, potential fire risks and damage during storms), its promise is nonetheless substantial in certain instances – all the more so if looked to as a supplemental energy source that's connected to the National Aqueduct or integrated within highway medians.

Assuming a figure reflecting the average cost of implementation (splitting the difference to arrive at $850 per kilowatt of power-generating capacity), we'll make the following assumptions when coming to a cost basis for wind power.

- **Capacity factor** is the *actual* output of a power source over a period of time as a proportion of the turbine's maximum capacity. For example: if a 1-megawatt turbine generates power at an average of 0.3 megawatts, its capacity factor is 30%. According to the Energy Information Administration, the average capacity factor for wind is approximately 34.6% for 2018.[645]

- To derive a 100 billion kilowatt-hour annual Energy Unit, we would need to generate 273.97 million kilowatt-hours per day.

Assuming the EIA's average capacity factor of 34.6% for wind turbines, we would need a daily energy-generating capacity of 790 million kilowatt-hours. Broken down over a 24-hour day, that's a power generating capacity of 32.91 million kilowatts per-hour.

At a cost of $850 per kilowatt generating capacity, that would cost **$27.97 billion** per 100 billion kilowatt-hour Energy Unit.

Liquid Fluoride Thorium Reactors

According to Robert Hargraves, author of *Thorium, Energy Cheaper than Coal* and a foremost expert on thorium energy, the cost of a 100-megawatt reactor is estimated to run $200 million.[646] This figure assesses end-unit manufacturing cost, *pre-learning ratio* (reductions in manufacturing costs every time the number of manufactured units doubles, due to process improvements, gleaned expertise, etc.). Naturally, this figure would differ in actual implementation as reactor cost would hinge on myriad factors: scalable generating capacity, non-included cost reductions in mass-manufacturing, regulatory requirements and ultimate funding sources, but it's an empirical baseline figure to begin our cost assessment.

Although LFTRs are more efficient than traditional nuclear reactors, we won't assess a higher operating capacity than the 92.5% of traditional nuclear power,[647] and include that figure for our estimations here. That would mean a 100-megawatt LFTR would generate 92,500 kilowatt-hours per hour (92.5 megawatt-hours). That translates to 2.22 million kilowatt-hours generated per day (810.3 million kilowatt-hours per year).

Sticking to our daily figure for sake of consistency, generating 273.97 million kilowatt hours per day would require 124 LFTRs. At a cost of $200 million per 100-megawatt reactor, that would come to a total of **$24.68 billion** per 100 billion kilowatt-hour Energy Unit.

The National Aqueduct

The National Aqueduct's electricity generation is comprised of three functions: internal turbines within pipelines, solar panels on top of pipeline arrays and hot water inside pipelines that itself has high potential for generating thermoelectric energy. As this system does not currently exist (outside of Lucid Energy's pipelines that, to date, do not have publicly released pricing models and do not come with integrated solar or thermoelectric functions), we'll refer to currently existing systems as starting points to derive cost estimates.

In doing so, we'll assume that the non-solarized aspects of the pipeline would cost similar to the largest oil pipelines today. According to the Oil and Gas Journal, oil pipelines cost an average of $6.5 million per mile to construct.[648]

This cost basis is broken down into four categories:

- Material - $894,139/mile. (13.62%)
- Labor - $2,781,619/mile. (42.36%)
- Miscellaneous - $2,547,600/mile.* (38.79%)
- ROW (Right of Way) and damages - $343,850/mile. (5.24%)

*'Miscellaneous' is defined as "Surveying, engineering, supervision, administration and overhead, regulatory filing fees, allowances for funds used during construction," which we'll presume includes land purchases alongside right-of-way (ROW) expenses.

With these costs in mind, we'll be making a few assumptions, mindful of the fact that National Aqueduct pipelines would be factory prefabricated, land wouldn't need to be purchased (as pipelines would be installed on publicly owned roads or under high voltage power lines) and regulatory approval would be streamlined. Cognizant of this, we will assume:

- That **materials** for the National Aqueduct will cost **four times higher** than for oil pipelines, as pipelines would include in-pipeline turbines + thermoelectric generators. That translates to an estimated $3.57 million/mile for material costs. This figure *does not* include the cost of solar panels.

- That **labor** for the National Aqueduct will cost **half** of oil pipelines as all aspects of the system would be factory prefabricated, coming to an estimated $1.39 million/mile.

- That **miscellaneous** costs would be **half** that of oil pipelines for the reasons listed above, coming to $1.2 million/mile.

- That **Right of Way/Damages** would **not be present** as well as the government wouldn't need to make right-of-way costs and factory prefabrication would dramatically reduce the number of damaged units compared to ad-hoc construction.

Combined, this provides an assumed cost estimate of **$6.16 million/mile** to construct National Aqueduct pipelines before solar panels are added (the cost of which was assessed above as $26.93 / square foot, or $289.84 per square meter).

With that established, let's determine how many miles of pipeline arrays we would conceptually require.

The U.S. consumes a total of 2,842 cubic meters of water per-person, per year, coming to 243.25 trillion gallons (920.8 billion cubic meters) across a society of 324 million people.[649] On a per-day basis, that comes to 667 billion gallons (2.53 billion cubic meters).

For initial deployment we will estimate that the National Aqueduct will store slightly less than one half of that daily volume of water (300 billion gallons – 1.135 billion cubic meters) at any given moment in time. 180 billion gallons (60%) would be stored in pipeline arrays, with the rest in storage tanks (681.36 billion cubic meters). The system would be constantly resupplied thereafter through coastal CHP Plants.

Based on these figures, we'll start our assessment first with cost, and then shift focus to calculating output.

Cost of Pipelines:

The volume of a 24" pipe is 23.5 gallons for every one foot of pipe, which translates to 124,080 gallons for every mile of pipeline[650] or 1.11 million gallons for an array of nine. (2,626 cubic meters per kilometer). If 180 billion gallons are stored in pipelines, that would require us to have 161,186 miles (259,404 km) of pipelines. (Assembled in arrays of six, that figure would drop to 26,864 miles (43,233 km)).

As each pipeline is estimated to run $6.16 million per mile, that span would cost **$996 billion**.

Cost of Storage:

Current estimates for commercial water tanks today come to around $1 per gallon[651] ($264.17 per cubic meter). However, National Aqueduct water storage tanks would differ from commercial storage tanks today in terms of insulation and electric UV sterilization, so we'll assess a 40% higher end-unit cost. This would come to roughly $1.40 per gallon ($369.84 per cubic meter).

As 60% of the 300 billion gallons within the National Aqueduct would be within pipeline arrays, the remaining water placed in storage would be 120 billion gallons (454.25 million cubic meters). At $1.40 per gallon, that comes to **$168 billion**.

Control:

As the National Aqueduct does not conceptually exist outside of this writing, effectively determining what it would cost to build the control component is prohibitively difficult. As such, we'll assume the cost of the control system and infrastructure would be **$30 billion.**

This would leave a non-solarized subtotal cost of **$1.194 trillion.**

With that established, we'll shift towards potential electricity generation.

Electricity due to internal water flow: according to Lucid Energy, a 24" pipe generates 18 kilowatts of power *per-turbine* with a flow rate of 24 million gallons per day[652] (90,849 cubic meters). Assuming a constant flow rate, over a 24-hour day, that comes to 423 kilowatt-hours generated per-turbine, per-day.

Lucid Energy's data suggests that maximum hydroelectric efficiency is turbine placement every 14 feet.[653] Over a pipeline span of 161,186 miles (259,404 km), that would involve use of 60.79 million turbines. At 423 kilowatt-hours generated per-turbine, per-day, with a 24 million gallon per day flow (90,849 cubic meters), this would come to **2.57 billion kilowatt-hours generated per day**, or **938.5 billion kilowatt-hours generated per year**. It's notable that the ultimate flow of the National Aqueduct would be significantly higher than 24 million gallons per day across the entire system, but we'll use this lower figure as a relative benchmark for electricity generation.

Electricity due to pipeline-mounted solar panels: We assessed earlier that solar panels generate 82.9 watt-hours per day, per square foot, at a cost of $26.93 per square foot (1,006.4 watt-hours per day, per square meter, at a cost of $289.84 per square meter). If pipelines were deployed in arrays of six (three on top of three), each 24" pipeline, assuming even spacing of about a foot and a half, would comprise 10 feet (3 meters).

10 feet, by a span distance of 26,864 miles, comes to a surface area of 1.418 billion square feet (131.73 million square meters). At 82.9 watt-hours per day, this would generate **117.6 million kilowatt-hours per day**, or **42.92 billion kilowatt-hours per year**, at an additional cost of cost of **$38.2 billion.**

Electricity due to hot water inside pipelines: To assess the potential energy in the hot water inside pipeline arrays and storage tanks, we'll base our calculations on the following assumptions: that the 300 billion gallons (1.135 billion cubic meters) stored in the National Aqueduct would be heated to 200 °F (94 °C), with a national average outside temperature of 55.7 °F (13.16 °C).[654]

According to Marlow Engineering, a leader in thermoelectric generating products for placement over hot pipelines, their 12" Powerstrap Generator outputs approximately 3 watts of power with a temperature differential of 94 °C to 13 °C.[655] Their 24" model does not have output figures available, but as their 12" model is roughly twice as powerful as their 6" model, we will assume their 24" model outputs roughly 6 watts of power at any given moment in time. As this system would operate 24 hours per day, we will assume each thermoelectric generator would output 144 watts per day, or 52.56 kilowatt-hours per year. Assuming further that we placed such thermoelectric generators in arrays of three (the maximum such units can operate in parallel),[656] each array would come to 432 watts per day, or 157.68 kilowatt-hours per year.

If these arrays of three were placed in between hydroelectric turbines (every 14 feet), we would employ the same number of thermoelectric arrays as hydroelectric turbines (60.79 million). This would translate to an output of **26.26 million kilowatt-hours per day**, or **9.59 billion kilowatt-hours per year**.

National Aqueduct Subtotals:

- Cost of system: $1.232 trillion (including solar).

- Total electricity output: 991 billion kilowatt-hours per year (2.715 billion kilowatt-hours per day).

Energy Unit Breakdown

Based on the analysis and assumptions made earlier, the electricity totals for Universal Energy are as follows. (One Energy Unit equals 100 billion kilowatt-hours generated annually).

- Integrated solar: $88.87 billion per Energy Unit

- Integrated wind: $27.87 billion per Energy Unit

- Liquid Fluoride Thorium Reactors: $24.68 billion per Energy Unit

- The National Aqueduct: $1.232 trillion, with annual energy generation output of 991 billion kilowatt-hours

Electricity Breakdown

In order to generate 8.34 trillion kilowatt-hours annually, we will make the following cost and deployment breakdown of Universal Energy's technologies. As mentioned previously, Universal Energy is modular by design, and can comprise most any configuration desirable. In this estimate, we will assume a total implementation of the National Aqueduct, which at 991 billion kilowatt-hours annually generated, leaves a remaining total of 7.35 trillion kilowatt-hours.

Of this remainder, half (3.68 trillion kilowatt-hours) would be provided by a backbone of LFTRs. The other half (3.68 trillion kilowatt-hours) would be comprised of solar (70% - 2.575 trillion kilowatt-hours) and wind (30% - 1.1 trillion kilowatt-hours).

Here's what that cost structure looks like:

Technology	Energy Units	Energy Unit Cost	Annual Output	Total Cost
National Aqueduct	N/A	N/A	991 billion kilowatt-hours	$1.232 trillion
LFTRs	36.8	$24.68 billion	3.68 trillion kilowatt-hours	$908.25 billion
Integrated Solar	25.75	$88.87 billion	2.575 trillion kilowatt-hours	$2.288.4 trillion
Integrated wind	11.03	$27.97 billion	1.103 trillion kilowatt-hours	$308.5 billion

Total Electricity Generation: **8.34 trillion kilowatt-hours**
Total Electricity Infrastructure Cost: **$4.74 trillion**

Resource Production

After covering the cost of electricity generation, we'll shift gears to the systems that synthesize water and fuel. In doing so, we won't be estimating their implementation in greater CHP Plants (which would be an ideal approach). This is because cost figures for CHP Plants are not yet present. Instead, we'll estimate the cost of building these systems on a standalone basis (with the exception of water desalination facilities without internal power plants) – even though this would translate to higher costs in this estimate.

As commercial resources, food and materials are not included in this estimate as the technical bar for implementation is far lower with Universal Energy's infrastructure in place. Additionally, their cost depends wholly on scale and sophistication, respectively, of the agricultural setup and material being produced.

Seawater Desalination

Most modern desalination facilities today are in the Middle East. Although they are capable of desalinating immense volumes of seawater, they generally are paired with internal power plants. This makes their construction significantly more expensive than desalination facilities would be within CHP Plants and makes it a bit tougher to determine standalone costs by themselves.

As the backbone of Universal Energy's desalination efforts is comprised of LFTRs, desalination facilities wouldn't need their own external power infrastructure in this model – nor would they need to consume as much additional energy. The non-radioactive heat exchangers of LFTRs should easily present sufficient cogenerative energy to desalinate seawater on a large scale, with low requirements for additional energy. Because of this, desalination plants will cost far less in the Universal Energy framework than they do today. However, we'll still need to make a few more assumptions to come to a realistic cost estimate. In doing so, we'll look to some of the larger desalination facilities operating today:

- The largest desalination facility in the world is currently the Ras Al Khair Plant in Saudi Arabia.[657] It has the capacity to produce 270.8 million gallons of water per day (1.025 million cubic meters) via both multistage flash and reverse osmosis. That translates to 98.8 billion gallons of water per year (375 million cubic meters).[658] It cost $7.2 billion to construct, and is also a 2,400 megawatt power plant.[659]

- The Jebel Ali facility in the United Arab Emirates outputs 140 million gallons of water per day via multistage flash distillation (530,000 cubic meters).[660] That translates to 51.1 billion gallons a year (193.4 million cubic meters). The facility cost $2.72 billion to construct, and is also a 1,400 megawatt power station.[661]

- The Fujairah power and desalination plant in the United Arab Emirates cost $1.2 billion to construct. It generates 656 megawatts of power and outputs 100 million gallons of water per day (378,500 cubic meters). Over a year, that comes to 36.5 billion gallons a year (138.17 million cubic meters).[662]

As noted above, an important component to using these facilities to create a cost estimate is the presence of power generation. The Fujairah facility only cost $1.2

billion to construct whereas Ras Al Khair cost $7.2 billion – but Ras Al Khair has a 2,400-megawatt generator that powers the facility and Fujairah's power plant only outputs 656 megawatts. The power generating potential of Ras Al Khair is nearly *four times higher*, but in terms of seawater desalination (270 million gallons daily versus 100 million), its output is only *2.7 times higher*. As Universal Energy's desalination facilities would come paired with LFTRs, our cost estimate must separate out the cost of traditional power generation.

To do so, we'll head over to the Energy Information Administration to get a general idea of the construction costs of a power plant.[663]

According to the EIA, a Natural Gas-fired Combined Cycle power plant (Adv Gas/Oil Comb Cycle CC) has an overnight cost of $1,080 per kilowatt for a 429 megawatt variant.[664] That means a 429 megawatt power plant would cost $463.2 million to construct, or roughly $1.08 million per megawatt.[665]

While construction costs likely vary in the Middle East, we'll nonetheless stick to this cost figure in the absence of more reliably specific data. Additionally, as the Ras Al Khair facility is **both** multistage flash and reverse osmosis (disproportionally increasing its cost), whereas Jebel Ali and Fujairah are strictly multistage flash, we'll only use Jebel Ali and Fujairah to estimate what a standalone desalination facility would cost if it didn't include a power plant.

Jebel Ali: $2.72 billion to construct with a 1,400-megawatt power station. Annual output: 51.1 billion gallons (193.4 million cubic meters).

At $1.08 million per megawatt, we'll estimate that $1.51 billion of the construction cost was for power generation. This would bring the estimated construction cost, sans-power, to $1.2 billion.

Desalination costs for one year of output: **$0.023 per gallon / $6.20 per cubic meter.**

Fujairah facility: $1.2 billion to construct with a 656-megawatt power station. Annual output: 36.5 billion gallons (138.17 million cubic meters)

At $1.08 million per megawatt, we'll estimate that $709 million of the construction cost was for power generation. This would bring the estimated construction cost, sans-power, to $493 million.

Desalination costs for one year of output: **$0.013 per gallon / $3.67 per cubic meter.**

Averaging these together, that comes to **$0.018 to desalinate a gallon of water** and **$4.94 for a cubic meter.**

We determined earlier that as the U.S. consumes 239.5 trillion gallons per year (920.8 billion cubic meters), which translates to 667 billion gallons of water per day. The National Aqueduct is intended to hold slightly less than half of that figure at any given moment in time (300 billion gallons), with 180 billion gallons (60%) in pipeline arrays, and the rest (40%) in storage tanks.

To ensure maximum effectiveness, we will assume an implementation capability sufficient to refill the National Aqueduct's capacity in full, twice over (600 billion gallons). At a price of 1.8 cents per gallon, constructing facilities with a capacity to desalinate 600 billion gallons of seawater would come to an estimated cost of **$10.8 billion.**

Hydrogen production

Analysts from the Department of Energy[666] estimate that hydrogen can be produced (factory gate price[667]) by way of water electrolysis for $3 per kilogram of contained hydrogen, at an energy price of $0.045 (4.5 cents) per kilowatt-hour.

As hydrogen's role in the Universal Energy framework is to produce fuel, we'll look at our domestic gasoline usage as a metric as opposed to overall petroleum consumption (which would still be helpful for lubricants and other synthetic materials). According to the Energy Information Administration, the U.S. consumed 142.86 billion gallons of gasoline in 2018.[668] Although this model envisions the majority of cars migrating to electric due to Universal Energy's material advancements, we'll still assess the cost of what it would take to have hydrogen replace gasoline in our society in terms of production.

As hydrogen production via electrolysis is measured in kilograms, we'll use specific energy to calculate our comparison.

Gasoline has a specific energy of 46.4 megajoules per kilogram.[669] One gallon of gasoline has a mass of roughly 2.8 kilograms. As such, 140.43 billion gallons of gasoline would have a mass of 393.2 billion kilograms. At 46.4 megajoules per kilogram, that comes to 8.47 billion megajoules.

Compressed hydrogen has a specific energy of 142 megajoules per kilogram.[670] To produce 8.47 billion megajoules of energy through hydrogen, we'd need 57.6 million kilograms of compressed hydrogen on an annual basis.

According to the Department of Energy, a hydrogen production facility today with an output of 50,000 kilograms of compressed hydrogen per day has a cost of $900 per kilowatt of system energy with a multiplier factor cost of 1.12 for installation, coming to $1,008 per kilowatt of system energy.[671] A 50,000 kilogram per day plant has a system energy of 113,125 kilowatts, which would make its estimated capital cost $114 million.

Dividing $114 million by 50,000 kilograms daily output, we'll assess that the capital costs of a hydrogen production plant are $2,280 per kilogram of daily production capability. As the United States would need 57.6 million kilograms of compressed hydrogen to replace gasoline in our society, at $2,280 per kilogram of daily production capacity, that comes to **$131.33 billion**.

Cost of Labor

Although labor cost was already included in the National Aqueduct's price estimate of $1.232 trillion, there are significant considerations that need to be given to the labor forces inherent to Universal Energy's implementation. These occur not only in terms of raw cost, but also the long-term management and upgrades as part of a long-term shift towards an advanced energy economy. In one vein of thinking, it would be of course possible to utilize the ranks of our uniformed service members and leverage their manpower and logistical expertise to build systems that *actually* present an effective defense against the underlying causes of conflict. Another vein of thinking would bear mention of the need to train larger segments of our workforce towards an advanced-manufacturing economy, which generates tangible wealth far more than service-based occupations. Another still might leverage the technically literate students/recent graduates of universities and trade schools as a primary option.

All of these approaches are valid, and each present unique options across the wide spectrum of vocational opportunities made possible by an investment in Universal Energy and the upgrades it installs on our economic foundations. As touched on within the *Collective Capitalism* section of this Appendix, an investment in such technologies on the scale proposed will create a tremendous demand for jobs across nearly every economic sector we have, which further does the same for all of the educational, support and sub-supporting positions that increase a collective quality of life for all of the above.

Rather than estimate the nuanced specifics of how much this labor would cost (as nearly all of it would hinge on assumptions), we'll instead take a more general approach and estimate labor costs as a percentage of the total framework. Recognizing that labor costs are already included within the "overnight costs" of each of the systems described herein, we'll estimate that the residual labor costs for all aspects of installation and management for the initial scale of implementation will come to **30%** of the total estimated $5 trillion "overnight" price tag for Universal Energy. This comes to a total estimated labor cost of **$1.5 trillion.**

GRAND TOTALS

Electricity generation and National Aqueduct	**$4.74 trillion for systems that annually generate 8.34 trillion kilowatt-hours.**
Cost of seawater desalination:	$10.8 billion
Cost of hydrogen production	$131.33 billion
Cost overrun buffer (5%)	$250 billion
Estimated costs of labor	$1.5 trillion

Grand Total: **$6.63 trillion**

Why this cost estimate is high

As this estimate hinged on several assumptions, care was taken to minimize any assumed reductions in future costs that would almost certainly be present once Universal Energy was implemented. The cost of any system at scale always costs less than the overnight cost, all the more so with systems implemented on a massive scale. There are several additional factors that would drive costs down even lower.

Learning Ratio: As we saw in Chapter Four (and elsewhere in this writing) learning ratio is the applied concept of 'learning by doing,' which means price reductions come through learned efficiencies and experience by building systems. The 'ratio' aspect of it is the reduction in price every time the number of produced units doubles. If it's a 10% ratio after the 100th produced unit, unit number 200 would cost 10% less than unit number 100. Unit number 400 would cost 10% less than unit number 200, and 20% less than unit number 100, and so on. If you recall back to the original invention of computers, flat screen televisions, smartphones, etc., the models we see today are vastly superior and less expensive than the initial releases they evolved from. Energy technologies are no different.

Further, learning ratio applies especially in the case of Universal Energy's because most of the framework's technologies are in their technical infancy and stand to enjoy substantial improvements through greater investment and research. Their overnight cost may be $5 trillion today, but over time – and especially with purchase orders on scale – that figure will drop as it has in every other industry. Yet as it's prohibitively difficult to accurately assess what these reductions might look like in actuality, they were not incorporated in the pricing estimate. However, in reality they would be significant.

CHP Plants: CHP Plants are the envisioned approach for large-scale implementation of Universal Energy's power, hydrogen fuel and fresh water resources. As they can use the waste/excess energy from one facility to power the functions of another in the same physical footprint, the energy costs to perform functions like water desalination and hydrogen production drop drastically. Just as importantly, the capital expenses of constructing power plants incorporates the cost of buying land. By building multiple systems within the same facility, the cost of land is proportionally shared – as are the costs of construction. This would make CHP Plants less expensive than the estimated costs to build each system standalone.

Energy cost reductions: Universal Energy's primary purpose is to generate an effectively unlimited amount of energy at a low enough cost to make possible the large-scale synthesis of critical resources and address climate change. Yet while this is intended to solve the core, pressing problems of our civilization, it also makes it a lot less expensive to do business and manufacture things. Energy costs are a huge component of a company's bottom line, especially in manufacturing – figures we assessed earlier to be hundreds of billions in aggregate.

If we're able to reach Universal Energy's target of 2 cents per kilowatt hour, that's hundreds of billions of dollars that businesses save when building products they take to market – systems behind Universal Energy being no exception. That's billions of dollars that longer need to be incorporated in the per-unit delivery cost of energy and resource production systems, which in turn presents billions of dollars in cost savings to their large-scale purchase and implementation.

Direct energy sales: even at drastically reduced rates of 2 cents per kilowatt-hour, the sale of 8.34 trillion kilowatt-hours returns a tidy sum – some $168.8 billion annually, 1.68 trillion per decade. Over time, this can and will offset the costs of Universal Energy's infrastructure. In conjunction with foreign sales through The Public Interest Company, this could reach a faster point of profitability as global adoption expands.

Reduced social afflictions: Universal Energy is designed to solve resource scarcity and climate change so that unlimited energy and resources in turn can solve the myriad social afflictions fueled by resource scarcity. These afflictions consume immense funds, time and concentration from our society: poverty, crime, economic depression, failing infrastructure, lost hope, lackluster employment and rampant drug addiction among them. All of these problems consume huge percentages of public budgets. As dramatically reduced energy and resource costs address these afflictions, the resources we presently devote to their mitigation can be spared in kind – saving even more money.

With the presence of these cost reductions in practice, it is possible if not likely that the present estimate for Universal Energy's implementation is skewed higher. In this case, any cost savings we can obtain along the way, should this model be implemented, would simply be "gravy" on top and allow us to increase any scale of implementation in kind.

A5: Citation Policy

As *The Next Giant Leap* covers a large scope of material that is technical in nature, maximizing readability is of paramount importance. As such, I've opted to adopt a citation style reflecting of that. In practice, the following citation conventions are applied as much as possible:

Articles, data sheets, journals, whitepapers, and documents are directly hyperlinked in their respective citation numbers, allowing easily readability. When possible, documents are cited with page numbers **directly embedded** in the URL.

For example: http://sitename.org/document_title/data.pdf#p53

In this link, p53 equals "page 53." As the "#" modifier in a URL points to a named link, the PDF file will load fine with that URL, enabling easy retrieval of the data in question. Further, virtual sources enable rapid "searching" via most web browsers of PDF viewers – providing easy retrieval of relevant data points. PDF + virtual copies of *The Next Giant Leap* will have direct links in the citations.

For whitepapers and data PDFs of a short or self-explanatory nature, or where the general facts are provided on the cover/summary/abstract, or documents designed to provide 'general conceptual information' as opposed to a specific factual citation, no page numbers are included in the URL structure. If you find a citation that you believe to be 1) broken (dead link), 2) unclear, 3) misinterpreted, please get in touch and let me know so it can be fixed.

A6: Source policy

Mindful of the focus this writing places on general readability and accessibility of concepts, *The Next Giant Leap* sources facts and data from a wide spectrum of sources under a transparent methodology.

While hard facts and technical parameters are nigh-exclusively sourced from objectively reputable sources (academic, public sector, industry whitepapers and data sheets), this writing also includes supporting facts from supplemental sources, including journalistic outlets, technical publications and, in very limited occurrences, opinion pieces.

These supplemental sources are included for readability – a primary reason people don't read academic journals is they're nearly impossible to read unless one is a professional academic; my audience is the collective and thus the citation policy of *The Next Giant Leap* is geared for the readability by the collective above all else.

Consequently, this writing will approach sources under the following reasoning:

Government sources: these sources include both domestic and international government agencies (Bureau of Labor and Statistics, Environmental Protection Agency, United Nations, World Health Organization, etc.). This writing considers these sources reliable and factual unless cause is presented to believe otherwise.

Scientific / technical media: these sources include media outlets dedicated to scientific / technical research (Scientific American, National Geographic, CNET, Wired, ARS Technica, etc.). This writing considers these sources reliable and factual unless cause is presented to believe otherwise.

Academic sources and journals: these sources include university publications (harvard.edu, etc.) and academic journals (*Nature*, etc.) and press releases by universities. This writing considers these sources reliable and factual unless cause is presented to believe otherwise.

Flagship journalism: these sources include mainstream news media outlets with established pedigrees. These include Associated Press/Reuters and any news outlets who retain membership in the AP. This can also include other media sources if the material is well-written and cited appropriately.

[Note: This author and writing refuses to entertain the false and disingenuous trend of declaring legitimate media "fake news" if they print facts inconvenient to an ideological or political narrative.]

Smaller-circulation high-brow journalism: these sources include media outlets with a solid degree of ethical integrity in journalism, but are smaller circulation. They include financial and industry publications, local newspapers, etc. Although they may have an ideological slant, their ethics in journalism remains high. This writing considers these sources reliable and factual, unless cause is presented to believe otherwise.

Ideological mouthpieces: these sources include broadcast platforms dedicated to a political ideology (Huffington Post, Fox News, CNN, National Review, The Nation.) Although some of these sources often have solid reporting and analysis, their ideological slant is severe enough to warrant their exclusion as sources in this writing, barring a few exceptions:

- In cases where the mouthpiece publishes material of uncanny excellence and factual support, an exception may be made, and the material may be included in one or two citations.

- In cases where the mouthpiece publishes material contrary to its ideological slant (Fox News supporting a traditionally liberal position or Huffington Post supporting a traditionally conservative position), the material may be included if it is itself well-reasoned and/or well-cited, but these circumstances are limited and generally only pertain to background information on social issues for ease of explanation.

- In cases where the mouthpiece is reporting in areas to which it has demonstrated significant expertise, the material may be included in a citation.

Barring these exceptions, ideological mouthpieces are generally excluded.

Wikipedia is cited primarily for "general background" information on concepts, locations, overviews of systems or historical events. Although its editorial team has proven quite adept at ensuring adherence to facts, it's open-edit policy makes it less desirable for direct citations of specific, novel facts. However, for general topics and scientific concepts, it is highly effective at providing background information for people unfamiliar with the subject material in question. For this reason, relevant Wikipedia articles may be included in certain citations to provide clearer context.

Think tanks are generally cited as reliable even if they have an ideological slant, as they have demonstrated a sufficient degree of ethics when reporting. This writing generally considers these sources reliable and factual, unless cause is presented to believe otherwise.

Industry publications and whitepapers are material presented by organizations with a vested commercial interest of the material they're reporting on. This information is generally considered reliable and factual if it is itself cited and provides calculations that can be independently verified.

Third-party blogs and statistics services: these sources include blogs like Nate Silver's Fivethirtyeight and statistics services like Statista and Brilliantmaps (for global population in urban environments). In the limited areas they are used, they are considered reliable and factual, barring any reason to believe otherwise.

A7: Retraction Policy

As *The Next Giant Leap* is the work of one individual, I have limited resources to make sure every aspect of the writing maintains 100% uniformity in formatting. Additionally, although all research was thorough and extensive, facts may change in the future, and it's possible certain elements of a fact were misinterpreted in the source they were included. In the event these are brought to light, changes will be made, but it is not feasible to issue public retractions for every change.

If you believe any information contained herein to be empirically incorrect, please get in touch with me and I will update the material accordingly.

A8: Copyright Policy

The purpose of *The Next Giant Leap* and Universal Energy is to make the world a better place. In conveyance of the ideas and technologies therein, multiple types of multimedia are used to help explain varied concepts, ideas, and systems. Whenever possible, care has been taken to commission original works to describe these nuanced details. Of works commissioned for this purpose, you are free to use them for any purpose – commercial or otherwise – provided that their source clearly indicates original basis for *The Next Giant Leap* at https://nextgiantleap.org. You need not ask permission for their re-use; however, if you would like high-definition images of any of the originally commissioned works used herein, please get in contact and I will work to provide them for you.

In regards to the copyright of this work, any part may be reproduced without permission so long as the reproduction is a) non-commercial, b) if commercial, that such reproduction is only a segment consisting of no more than 20% of the total commercial work and cited appropriately to https://nextgiantleap.org, **AND** c) the work in cases A **and** B are objectively in the public interest.

Use of non-commissioned multimedia:

In certain instances, this work uses original/copyrighted imagery to help explain concepts and ideas. This multimedia, without exception, is used in accordance with 17 U.S. Code § 107 – *Limitations on exclusive rights: fair use*, also known as "The Fair Use Doctrine," as they are, without exception, used only once, and further are used for purposes of public education in furtherance of the public interest as specified under 17 U.S.C § 107.

If you are a copyright holder of any of the copyrighted works used herein and wish to have them excluded from future use, simply get in touch with me and I will oblige and recommission original works under my own copyright based on original data. No intent is made to capitalize on any reserved copyright, if your work was included, it was done so exclusively because it was of requisite quality to help explain concepts in the public interest in accordance with the Fair Use Doctrine.

Special Thanks

Laura Yuan, first editor.

Emma Burns, technical artist.

Megan Tackett, confidant and critical analyst.

Scot Chrisman, cheerleader and a source of inspiration.

Amanda Jameson, second editor.

Michal Karcz, cover artist and visual powerhouse.

Kelsey Morrison, my lovely wife. Thank you deeply for your years of support.

Cited Facts and Sources

[1] Smithsonian Magazine. *"Greenleand lost 12.5 billion tons of ice in a single day."* 5 August, 2019. https://www.smithsonianmag.com/smart-news/greenland-lost-record-breaking-125-billion-tons-ice-single-day-180972808/

New York Times. *"Iceland Mourns Loss of a Glacier by Posting a Warning About Climate Change."* L. Holson. 19 August, 2019. https://www.nytimes.com/2019/08/19/world/europe/iceland-glacier-funeral.html

[2] BBC News: *"A looming mass extinction caused by humans."* G. Vince. 1 November 2012. http://www.bbc.com/future/story/20121101-a-looming-mass-extinction

[3] BBC News. *"Amazon Fires: Record number burning in Brazil rainforest – space agency."* 21 August, 2019. https://www.bbc.com/news/world-latin-america-49415973

[4] Axios magazine. *"By the numbers: unprecedented devastation of California's wildfires."* A. Freedman. 8 August, 2018. https://www.axios.com/california-wildfires-break-records-statistics-mendocino-complex-fire-ae073d56-b170-4086-b567-2cf7c3d23fd8.html

The Atlantic Magazine. *"California's wildfires are 500 percent larger due to climate change."* R. Meyer. 16 July, 2019. https://www.theatlantic.com/science/archive/2019/07/climate-change-500-percent-increase-california-wildfires/594016/

[5] BBC News. *"Australia fires: a visual guide to the bushfire crisis."* 6 January, 2020. https://www.bbc.com/news/world-australia-50951043. A point of note: several causes of these fires were direct human action, namely felony arson. This has been pointed to as reason to discredit climate change's role in Australian wildfires, which is both disingenuous and categorically false – arson, and man-caused fires have persisted for millennia, yet the reason such fires now encompass wide swaths of the Australian continent is because of the aggravating factors of climate change, namely drought and hotter conditions that enable wildfires to spread faster and burn drier material.

https://www.cbsnews.com/news/australia-fires-how-climate-change-has-intensified-the-deadly-bushfires/

[6] Council of Foreign Relations. *"A global report on the decline of democracy."* 17 April, 2018. https://www.foreignaffairs.com/press/2018-04-17/global-report-decline-democracy

[7] Washington Post. *"17 ways the unprecedented migrant crisis is reshaping our world."* A. Taylor, 20 June, 2015.
https://www.washingtonpost.com/news/worldviews/wp/2015/06/20/17-ways-the-unprecedented-migrant-crisis-is-reshaping-our-world/

[8] Brookings Institution. *"How to combat fake news and disinformation."* D. West. 18 December, 2017. https://www.brookings.edu/research/how-to-combat-fake-news-and-disinformation/

[9] "The Long Peace" is a term defining the post-WWII era of lasting peace after the defeat of Axis powers. It was first coined by John Lewis Gaddis in the *International Security* journal, Vol. 10, No. 4 (Spring, 1986), pp. 99-142. https://www.jstor.org/stable/2538951

[10] Union of Concerned Scientists. *"North Korea, reports and multimedia."* 18 December, 2017. https://www.ucsusa.org/resources/north-korea-basics

[11] At the time of this writing, the United States is either involved, or threatens to be involved, with state-level conflict against Iran, Venezuela, and North Korea, along with several other conflicts against non-state actors in furtherance of the "Global War on Terror."

Deutsche Welle. *"Nicolas Maduro tells military to 'be ready' for potential US military action."* 5 May, 2019. https://p.dw.com/p/3HwPL

BBC News. *"Iran attack: US Troops targeted with ballistic missiles."* 8 January, 2020. https://www.bbc.com/news/world-middle-east-51028954

[12] Radioactive dating tells us that modern humans evolved approximately 200,000 years ago, although some sources suggest as long as 300,000.
http://humanorigins.si.edu/human-characteristics/humans-change-world
Nature Journal. *"Oldest Homo sapiens fossil claim rewrites our species' history."* E. Callaway. 7 June, 2017. https://www.nature.com/news/oldest-homo-sapiens-fossil-claim-rewrites-our-species-history-1.22114

[13] Most sources estimate 1900 as having a population of just over 1.5 billion.
https://www.worldometers.info/world-population/world-population-by-year/

[14] U.N.: *"World population projected to reach 9.7 billion by 2050."*
http://www.un.org/en/development/desa/news/population/2015-report.htm

[15] Background Reading on Thomas Malthus' *Principle of Population* essay. https://en.wikipedia.org/wiki/An_Essay_on_the_Principle_of_Population

[16] Background Reading *The Limits To Growth* - 1972-era computer simulation of exponential economic and population growth with a finite resource supply. https://en.wikipedia.org/wiki/The_Limits_to_Growth

[17] U.N. Food and Agriculture Organization report. *"State of the world's forests, 2012."* http://www.fao.org/docrep/016/i3010e/i3010e.pdf#p23,p25,p28

[18] BBC: *"A looming mass extinction caused by humans."* G. Vince. 1 November 2012. http://www.bbc.com/future/story/20121101-a-looming-mass-extinction

[19] The Atlantic. "It Will Take Millions of Years for Mammals to Recover From Us". E. Young. 15 October, 2018. https://www.theatlantic.com/science/archive/2018/10/mammals-will-need-millions-years-recover-us/573031/

[20] U.N. report: *"General facts regarding world fisheries,"* 24 May 2010. http://www.un.org/depts/los/convention_agreements/reviewconf/FishStocks_EN_A.pdf

[21] National Geographic. "Seafood may be gone by 2048, Study Says." J. Roach. 2 November, 2006. https://news.nationalgeographic.com/news/2006/11/061102-seafood-threat.html

[22] World Wildlife Fund. *"Living Planet Report 2018: Aiming Higher"* p54.

[23] World Wildlife Fund. *"Living Planet Report 2018: Aiming Higher"* p4

[24] World Wildlife Fund. *"Living Planet Report 2018: Aiming Higher"* p4

[25] The Guardian. *"Humans just 0.01% of all life but have destroyed 83% of wild mammals – study."* D. Carrington. 21 May, 2018. https://www.theguardian.com/environment/2018/may/21/human-race-just-001-of-all-life-but-has-destroyed-over-80-of-wild-mammals-study

Proceedings of the National Academy of Sciences of the United States of America. *"The Biomass Distribution on Earth."* Y. Bar-On, R. Philips, R. Milo. 21 May, 2018. https://www.pnas.org/content/115/25/6506

[26] The Guardian. *"Humans just 0.01% of all life but have destroyed 83% of wild mammals – study."* D. Carrington. 21 May, 2018.
https://www.theguardian.com/environment/2018/may/21/human-race-just-001-of-all-life-but-has-destroyed-over-80-of-wild-mammals-study

Proceedings of the National Academy of Sciences of the United States of America. *"The Biomass Distribution on Earth."* Y. Bar-On, R. Philips, R. Milo. 21 May, 2018.
https://www.pnas.org/content/115/25/6506

[27] The Guardian. *"Humans just 0.01% of all life but have destroyed 83% of wild mammals – study."* D. Carrington. 21 May, 2018.
https://www.theguardian.com/environment/2018/may/21/human-race-just-001-of-all-life-but-has-destroyed-over-80-of-wild-mammals-study

Proceedings of the National Academy of Sciences of the United States of America. *"The Biomass Distribution on Earth."* Y. Bar-On, R. Philips, R. Milo. 21 May, 2018.
https://www.pnas.org/content/115/25/6506

[28] Intergovernmental Science-Policy Platform on Biodiversity and Ecosystem Services (IPBES). *"Nature's Dangerous Decline 'Unprecedented' Species Extinction Rates 'Accelerating'."* 6 May, 2019. P. 3.

[29] Intergovernmental Science-Policy Platform on Biodiversity and Ecosystem Services (IPBES). *"Nature's Dangerous Decline 'Unprecedented' Species Extinction Rates 'Accelerating'."* 6 May, 2019. P. 3.

[30] U.N. *"Water Cooperation facts and figures."* Archived: http://www.unwater.org/water-cooperation-2013/water-cooperation/facts-and-figures/en/

[31] World Health Organization. *"Lack of Sanitation for 2.4 Billion People is Undermining Health Improvements."* http://www.who.int/mediacentre/news/releases/2015/jmp-report/en/.

[32] World Health Organization. *"Lack of Sanitation for 2.4 Billion People is Undermining Health Improvements."* http://www.who.int/mediacentre/news/releases/2015/jmp-report/en/.

[33] S. Damkjaer, R. Taylor. *"The measurement of water scarcity: Defining a meaningful indicator."* Ambio vol. 46,5 (2017): 513-531. https://doi.org/10.1007/s13280-017-0912-z

United Nations. "Water Scarcity." https://www.unwater.org/water-facts/scarcity/

[34] United Nations. World Water Development Report 2015. https://www.un-ihe.org/sites/default/files/wwdr_2015.pdf#p2.

[35] Newsweek. *"The Race to Buy Up the World's Water."* J. Interlandi. 8 October, 2010. http://www.newsweek.com/race-buy-worlds-water-73893.

[36] Newsweek. *"The Race to Buy Up the World's Water."* J. Interlandi. 8 October, 2010. http://www.newsweek.com/race-buy-worlds-water-73893.

[37] B. Gaybullaev, C., Su-Chin. *"Changes in water volume of the Aral Sea after 1960"* 1 December, 2012. 10.1007/s13201-012-0048-z JO - Applied Water Science ER. https://doi.org/10.1016/j.catena.2020.104566

Background reading on the Aral Sea: https://en.wikipedia.org/wiki/Aral_Sea

[38] Lake Michigan has a surface area of 22,404 square miles (58,030 km^3), and Lake Erie has a volume of 116 cubic miles (480km^3). https://en.wikipedia.org/wiki/Lake_Michigan / https://en.wikipedia.org/wiki/Lake_Erie

[39] Background reading on the Aralkum Desert: https://en.wikipedia.org/wiki/Aralkum_Desert

[40] National Geographic. "Aral Sea's Eastern Basin is Dry for the First Time in 600 years." B. Clark-Howard. 2 October, 2014. https://www.nationalgeographic.com/news/2014/10/141001-aral-sea-shrinking-drought-water-environment/

[41] NASA. *"Study: Third of Big Groundwater Basins in Distress."* 16 June, 2015. https://www.jpl.nasa.gov/news/news.php?feature=4626

NASA. *"Getting at groundwater with gravity."* G. Hicks. Last updated 6 January, 2020. https://earthdata.nasa.gov/learn/sensing-our-planet/getting-at-groundwater-with-gravity

[42] NASA. *"Study: Third of Big Groundwater Basins in Distress."* 16 June, 2015. https://www.jpl.nasa.gov/news/news.php?feature=4626

NASA. *"Getting at groundwater with gravity."* G. Hicks. Last updated 6 January, 2020. https://earthdata.nasa.gov/learn/sensing-our-planet/getting-at-groundwater-with-gravity

[43] National Geographic. *"If You Think the Water Crisis Can't Get Worse, Wait Until the Aquifers are Drained."* D. Dimick. 21 August, 2014.

New York Times. *"Beneath California Crops, Groundwater Crisis Grows."* J. Gillis, M. Richtel. 5 April, 2015. https://www.nytimes.com/2015/04/06/science/beneath-california-crops-groundwater-crisis-grows.html?_r=0

[44] National Geographic. *"If You Think the Water Crisis Can't Get Worse, Wait Until the Aquifers are Drained."* D. Dimick. 21 August, 2014.

New York Times. *"Beneath California Crops, Groundwater Crisis Grows."* J. Gillis, M. Richtel. 5 April, 2015. https://www.nytimes.com/2015/04/06/science/beneath-california-crops-groundwater-crisis-grows.html?_r=0

Washington Post. *"California's Terrifying Climate Forecast: It Could Face Droughts Nearly Every Year."* D. Fears. 2 March, 2015. https://www.washingtonpost.com/news/energy-environment/wp/2015/03/02/californias-terrifying-forecast-in-the-future-it-could-face-droughts-nearly-every-year/?utm_term=.2c407783c5b8

[45] NCAR & UCAR News Center. "Climate change: Drought may threaten much of globe within decades." 19 October, 2010.
http://www.cgd.ucar.edu/cas/adai/news/Dai_Drought_UCAR.htm

[46] NASA. "Earth's Freshwater Future: Extremes of Flood and Drought." 13 June, 2019. https://climate.nasa.gov/news/2881/earths-freshwater-future-extremes-of-flood-and-drought/

[47] New York Times. *"Rural Water, Not City Smog, May be China's Pollution Nightmare."* C. Buckley, V. Piao. 11 April, 2016. https://www.nytimes.com/2016/04/12/world/asia/china-underground-water-pollution.html

South China Morning Post. *"80 per cent of groundwater in China's major river basins is unsafe for humans, study reveals."* https://www.scmp.com/news/china/policies-politics/article/1935314/80-cent-groundwater-chinas-major-river-basins-unsafe

D. Shemie, K. Vigerstol, M. Quan, et al. *"China Urban Water Blueprint."* The Nature Conservancy. 2016.
https://www.nature.org/content/dam/tnc/nature/en/documents/Urban_Water_Blueprint_Region_China.pdf

[48] New York Times. *"Rural Water, Not City Smog, May be China's Pollution Nightmare."* C. Buckley, V. Piao. 11 April, 2016. https://www.nytimes.com/2016/04/12/world/asia/china-underground-water-pollution.html

South China Morning Post. *"80 per cent of groundwater in China's major river basins is unsafe for humans, study reveals."* https://www.scmp.com/news/china/policies-politics/article/1935314/80-cent-groundwater-chinas-major-river-basins-unsafe

D. Shemie, K. Vigerstol, M. Quan, et al. *"China Urban Water Blueprint."* The Nature Conservancy. 2016. https://www.nature.org/content/dam/tnc/nature/en/documents/Urban_Water_Blueprint_Region_China.pdf

[49] NASA. *"NASA Satellites Unlock Secret to India's Vanishing Water."* 12 August, 2009. https://www.nasa.gov/topics/earth/features/india_water.html

New York Times. *"India's Water Crisis."* Editorial board. 3 May, 2016. https://www.nytimes.com/2016/05/04/opinion/indias-water-crisis.html

The Guardian. *"Armed guards at India's dams as drought grips country."* A. France-Presse. 2 May, 2016. https://www.theguardian.com/world/2016/may/02/armed-guards-at-indias-dams-as-drought-grips-country

[50] New York Times. *"India's Water Crisis."* Editorial board. 3 May, 2016. https://www.nytimes.com/2016/05/04/opinion/indias-water-crisis.html

The Guardian. *"Armed guards at India's dams as drought grips country."* A. France-Presse. 2 May, 2016. https://www.theguardian.com/world/2016/may/02/armed-guards-at-indias-dams-as-drought-grips-country

[51] The Times of India. *"80% of India's Surface Water May Be Polluted, Report by International Body Says."* S. Deyl. 28 June, 2015. https://timesofindia.indiatimes.com/home/environment/pollution/80-of-Indias-surface-water-may-be-polluted-report-by-international-body-says/articleshow/47848532.cms

[52] The Hindu. *"Untreated Sewage Flow is Killing Indian Rivers, Says New Study."* AFP. 6 March, 2013. http://www.thehindu.com/todays-paper/tp-in-school/untreated-sewage-flow-is-killing-indian-rivers-says-new-study/article4480149.ece

53 Forbes Magazine. *"Oil and gas giants spend millions lobbying to block climate change policies."* N. McCarthy. 25 March, 2019. https://www.forbes.com/sites/niallmccarthy/2019/03/25/oil-and-gas-giants-spend-millions-lobbying-to-block-climate-change-policies-infographic/

54 NASA. *"Scientific Consensus: Earth's Climate is Warming."* https://climate.nasa.gov/scientific-consensus/

55 NASA. *"Scientific Consensus: Earth's Climate is Warming."* https://climate.nasa.gov/scientific-consensus/

NASA. *"Do scientists agree on climate change?"* https://climate.nasa.gov/faq/17/do-scientists-agree-on-climate-change/

56 N. McCarthy. "Oil Firms Spend Millions on Climate Lobbying." 26 May, 2019. https://www.statista.com/chart/17467/annual-expenditure-on-climate-lobbying-by-oil-and-gas-companies/

57 Image Source: The Economist Magazine. Data sources: Corinne Le Quéré, et al (2018). Global Carbon Project (GCP); Carbon Dioxide Information Analysis Centre (CDIAC).

58 United Nations. World Meteorological Organization. *"Global Climate in 2015-2019: Climate change accelerates."* 22 September, 2019. https://public.wmo.int/en/media/press-release/global-climate-2015-2019-climate-change-accelerates

59 Image source: United States Environmental Protection Agency. *"Climate Change Indicators: U.S. and Global Temperature."* Data Source: NOAA (National Oceanic and Atmospheric Administration), 2016. https://www.epa.gov/climate-indicators/climate-change-indicators-us-and-global-temperature

60 NOAA (National Oceanic and Atmospheric Administration), 2016. National Centers for Environmental Information, 2016. Note from source: for more information, visit U.S. EPA's "Climate Change Indicators in the United States" at http://www.epa.gov/climate-indicators.

61 Image source: The European Space Agency Climate Change Initiative. https://www.nytimes.com/interactive/2017/09/22/climate/arctic-sea-ice-shrinking-trend-watch.html

62 Data source: National Snow and Ice Data Center and the Colorado Center for Astrodynamics Research. Image source: New York Times. *"In the Arctic, the old ice is disappearing."* J. White, K. Pierre-Louis. 14 May, 2018. https://www.nytimes.com/interactive/2018/05/14/climate/arctic-sea-ice.html

[63] OECD (Organization for Economic Co-operation and Development). *"The Economic Consequences of Climate Change."* 3 November, 2015. https://www.oecd.org/env/the-economic-consequences-of-climate-change-9789264235410-en.htm

[64] Versions 1.1+ of *The Next Giant Leap* do not feature detailed focus on the state of oil scarcity due to its post-COVID-19 price collapse. Oil scarcity is a future reality that humanity must contend with, but the purpose of this writing is to solve resource scarcity – not focus on the consequences of its existence. Consequently, the nature of future oil scarcity will be touched upon in future posts on nextgiantleap.org.

[65] Columbia University Earth Institute. *"Climate change to exacerbate rising food prices."* A. Mazhirov, 22 March, 2011.

[66] International Food Policy Research Institute. *"Food Security, Farming and Climate Change to 2050. Scenarios, Results, Policy Options."* December, 2010. http://ebrary.ifpri.org/utils/getfile/collection/p15738coll2/id/127066/filename/127277.pdf#23

[67] Oxfam Issue Briefing. *"Extreme Weather, Extreme Prices. The costs of feeding a warming world."* September 2012. https://blogs.ei.columbia.edu/2011/03/22/climate-change-to-exacerbate-rising-food-prices/

https://oxfamilibrary.openrepository.com/bitstream/handle/10546/241131/ib-extreme-weather-extreme-prices-05092012-en.pdf#p7

[68] Oxfam Issue Briefing. *"Extreme Weather, Extreme Prices. The costs of feeding a warming world."* September 2012. https://oxfamilibrary.openrepository.com/bitstream/handle/10546/241131/ib-extreme-weather-extreme-prices-05092012-en.pdf#p7

[69] U.N. High Commissioner for Refugees. *"Figures at a Glance."* http://www.unhcr.org/en-us/figures-at-a-glance.html.

[70] The impact of climate change and resource scarcity, beyond imminent or already-present incidents, reflects a timeline measured by years to decades. See: United Nations - Intergovernmental Panel on Climate Change reports https://www.ipcc.ch/reports/, especially *"Global Warming of 1.5 ºC"* https://www.ipcc.ch/sr15/

[71] "Why We Fight" https://nextgiantleap.org/mindset/why-we-fight

[72] "Why We Fight" https://nextgiantleap.org/mindset/why-we-fight

73 "Why We Fight" https://nextgiantleap.org/mindset/why-we-fight

74 United States Military – Joint Forces Command. *"The Joint Operating Environment – 2010."* 18 February, 2010. https://fas.org/man/eprint/joe2010.pdf

75 International Campaign to Abolish Nuclear Weapons: http://www.icanw.org/the-facts/nuclear-arsenals/.

76 Background reading on *Ohio*-class SSBN's. https://en.wikipedia.org/wiki/Ohio-class_submarine

77 Israel is widely believed to have second-strike capability through classified submarine developments. https://en.wikipedia.org/wiki/Dolphin-class_submarine

78 Picture of blast door in Minuteman-II missile silo, which has an intercontinental flight time of minutes. The caption is a play on Domino's Pizza, which reads "World-wide delivery in 30 minutes or less – or your next one is free." https://commons.wikimedia.org/wiki/File:World-wide_delivery_in_30_minutes_or_less.JPG

79 Reference.com. *"How Many Cities Are There In The World?"* https://www.reference.com/geography/many-cities-world-c25cce21cb142891.

80 International Campaign to Abolish Nuclear Weapons: http://www.icanw.org/the-facts/nuclear-arsenals/.

81 Meadows (2), Randers and Behrens, *The Limits to Growth,* Potomac Associates Press. http://collections.dartmouth.edu/teitexts/meadows/diplomatic/meadows_ltg-diplomatic.html

82 Background reading on aluminum's value. Slate Magazine, "Blogging the Periodic Table." http://www.slate.com/articles/health_and_science/elements/features/2010/blogging_the_periodic_table/aluminum_it_used_to_be_more_precious_than_gold.html

83 Background reading on the Hall–Héroult process to extract aluminum: https://en.wikipedia.org/wiki/Hall%E2%80%93H%C3%A9roult_process

84 Background reading on sugar taxation and its former status as a luxury commodity: https://en.wikipedia.org/wiki/History_of_sugar
https://en.wikipedia.org/wiki/Sugar_Act

[85] Background reading on Graphene. https://en.wikipedia.org/wiki/Graphene. (See Chapter Ten for a full breakdown in materials and recycling).

[86] U.S. Energy Information Administration. *"Most U.S. nuclear power plants were built between 1970 and 1990."* 27 April, 2017.
https://www.eia.gov/todayinenergy/detail.php?id=30972

[87] Electronic calculators emerged in the 1970's and weren't commercially prevalent until the second half of the decade. https://en.wikipedia.org/wiki/Calculator#1970s_to_mid-1980s

[88] Foreign Policy Magazine. *"Resource Wars: The New Landscape of Global Conflict."* M. Klare. May, 2001. https://www.foreignaffairs.com/reviews/capsule-review/2001-05-01/resource-wars-new-landscape-global-conflict

Global Policy Forum. *"The Dark Side of National Resources."*
https://www.globalpolicy.org/the-dark-side-of-natural-resources-st.html

[89] Energy Information Administration. *"What is U.S. electricity generation by energy source?"*
https://www.eia.gov/tools/faqs/faq.php?id=427&t=3

[90] Information on the number of Energy Utilities operating in the United States.
https://www.publicpower.org/system/files/documents/2018-Public-Power-Statistical-Report-Updated.pdf

[91] Federal Energy Regulatory Commission. *"Electric Power Markets."*
https://www.ferc.gov/market-assessments/mkt-electric/overview.asp

[92] Advisian Consulting. *"The Costs of Desalination."* D. Mishra. 15 February, 2018.
https://www.advisian.com/en-us/global-perspectives/the-cost-of-desalination

[93] Energy Information Administration data on electricity sales:
https://www.eia.gov/electricity/annual/html/epa_02_05.html

[94] Energy Information Administration. *"Average Price of Electricity to Ultimate Customers."*
https://www.eia.gov/electricity/annual/html/epa_02_04.html

[95] Background reading on Electrolysis: https://en.wikipedia.org/wiki/Electrolysis

[96] See Chapter Nine for details and citations (page 183)

[97] See Chapter Ten for details and citations (page 201)

[98] National Geographic. *"Renewable Energy Record Set in U.S."* S. Gibbens. 15 June, 2017. https://news.nationalgeographic.com/2017/06/solar-wind-renewable-energy-record/

[99] New York Times: *"Rooftop Solar Dims Under Pressure from Utility Lobbyists."* H. Tabuchi. 8 July, 2017. https://www.nytimes.com/2017/07/08/climate/rooftop-solar-panels-tax-credits-utility-companies-lobbying.html

[100] NPR: *"Texas Power Players Sit Out Political Opposition to Clean Power Plan."* 16 April, 2016. http://www.npr.org/2016/04/16/474462519/texas-power-players-sit-out-political-opposition-to-clean-power-plan

[101] International Electrotechnical Commission. *"Efficient Electrical Energy Transmission and Distribution."* (2007). https://basecamp.iec.ch/download/efficient-electrical-energy-transmission-and-distribution/#

[102] Splinter magazine. *"How much land is needed to power the U.S. with solar? Not much."* R. Wile. 5 January, 2015. https://splinternews.com/how-much-land-is-needed-to-power-the-u-s-with-solar-n-1793847493

[103] Electric Light & Power. *"Underground vs. Overhead: Power Line Installation-Cost Comparison and Mitigation."* 2 January, 2013. http://www.elp.com/articles/powergrid_international/print/volume-18/issue-2/features/underground-vs-overhead-power-line-installation-cost-comparison-.html

[104] Energy.gov. *"The Falling Price Of Utility-Scale Solar Photovoltaic (PV) Projects."* http://energy.gov/maps/falling-price-utility-scale-solar-photovoltaic-pv-projects

[105] U.S. Department of Energy. "Quadrennial Technology Review." Table 10.4: Range of materials requirements (fuel excluded) for various electricity generation technologies." September, 2015. https://nextgiantleap.org/sites/default/files/source_files/quadrennial-technology-review-2015.pdf

Forbes Magazine. "If Solar Panels Are So Clean, Why Do They Produce So Much Toxic Waste?" M. Shellenberger, contributor. 23 May, 2018. https://www.forbes.com/sites/michaelshellenberger/2018/05/23/if-solar-panels-are-so-clean-why-do-they-produce-so-much-toxic-waste/#175aff0e121c

[106] The Verge Magazine. *"More solar panels mean more waste and there's no easy solution."* A. Chen. 25 October, 2018. https://www.theverge.com/2018/10/25/18018820/solar-panel-waste-chemicals-energy-environment-recycling

[107] Background reading on solar shingle companies:

Tesla Solar Shingle: https://www.tesla.com/solarroof
RGS Solar Shingle: https://rgsenergy.com/
CertainTeed: https://www.certainteed.com/solar/
SunTegra: https://www.suntegrasolar.com/suntegra-shingles/

[108] Technical data sheet of the SunPower E20 solar panel: https://us.sunpower.com/sites/default/files/sunpower-e-series-commercial-solar-panels-e20-327-com-datasheet-505701-revh.pdf

[109] Energy Information Administration. *"How much electricity does an American home use?"* Last updated 26 October, 2018. https://www.eia.gov/tools/faqs/faq.php?id=97&t=3. (Note: this table has been updated to reflect an average use of 10,927 kWh annually, but the original 10,400 was included in formulas and will remain the benchmark until subsequent versions of this writing are released.)

[110] CleanTechnia. *"Solar Panels Do Work on Cloudy Days."* J. Richardson. 8 February, 2018. https://cleantechnica.com/2018/02/08/solar-panels-work-cloudy-days-just-less-effectively/

[111] Energy Information Administration. *"How Much Electricity Does an American Home Use?"* https://www.eia.gov/tools/faqs/faq.php?id=97&t=3

[112] Solar Window Power Model: https://solarwindow.com/powermodel/

[113] Solar Window Power Model: https://solarwindow.com/powermodel/

[114] CivicSolar. *"How Does Heat Affect Solar Panel Efficiencies?"* S. Fox. December, 2017. https://www.civicsolar.com/article/how-does-heat-affect-solar-panel-efficiencies

[115] Inhabitat Magazine. *"SOLAR ARK: World's Most Stunning Solar Building."* A. Kriscenski. 14 January, 2008.

[116] U.S. Department of Energy. *"Passive Solar Home Design."* https://www.energy.gov/energysaver/energy-efficient-home-design/passive-solar-home-design

[117] Lazard Insights. *"Levelized Cost of Energy and Levelized Cost of Storage 2018."* 8 November, 2018. https://www.lazard.com/perspective/levelized-cost-of-energy-and-levelized-cost-of-storage-2018/

[118] Energy Information Administration. "Use of Energy in the United States Explained." https://www.eia.gov/energyexplained/index.cfm?page=us_energy_commercial

[119] Background reading on the Sunshine Skyway Bridge: https://en.wikipedia.org/wiki/Sunshine_Skyway_Bridge

[120] Using Google Maps, I assessed the square footage of my local Lowe's parking lot to be roughly 150,000 square feet.

[121] CivicSolar. *"How Does Heat Affect Solar Panel Efficiencies?"* S. Fox. December, 2017. https://www.civicsolar.com/article/how-does-heat-affect-solar-panel-efficiencies

[122] Autoblog. *"Enlil urban turbine uses traffic to generate electricity."* 28 June, 2018. https://www.autoblog.com/2018/06/28/enl-l-urban-turbine-wind-from-traffic/

[123] National Academies of Sciences, Engineering, Medicine. National Academies Press. Assessing and Managing the Ecological Impacts of Paved Roads. Chapter 2: History and Status of the U.S. Road System. P. 41 https://www.nap.edu/read/11535/chapter/4

[124] National Academies of Sciences, Engineering, Medicine. National Academies Press. Assessing and Managing the Ecological Impacts of Paved Roads. Chapter 2: History and Status of the U.S. Road System. P. 41 https://www.nap.edu/read/11535/chapter/4

[125] Background reading on road lanes: https://en.wikipedia.org/wiki/Lane

[126] World Meteorological Organization. *"July matched, and maybe broke, the record for the hottest month since analysis began."* 1 August, 2019. https://public.wmo.int/en/media/news/july-matched-and-maybe-broke-record-hottest-month-analysis-began

[127] The New York Times. *"How Hot Was It in Australia? Hot Enough to Melt Asphalt."* M. Astor. 7 January, 2018. https://www.nytimes.com/2018/01/07/world/australia/heat-wave.html

[128] Background reading on "aquaplaning" in vehicles. https://en.wikipedia.org/wiki/Aquaplaning

[129] Eniscuola Energy and Environment. *"Road runoff and environmental pollution."* 22 March, 2017. http://www.eniscuola.net/en/2017/03/22/road-runoff-environmental-pollution/

[130] Forbes Magazine. *"The Reason Renewables Can't Power Modern Civilization Is Because They Were Never Meant To."* M. Shellenberger. 6 May, 2019.

https://www.forbes.com/sites/michaelshellenberger/2019/05/06/the-reason-renewables-cant-power-modern-civilization-is-because-they-were-never-meant-to/#4932ce85ea2b

[131] J. Am. Chem. Soc. 2012, 134, 4, 1895–1897. Published 6 January, 2012
https://doi.org/10.1021/ja209759s

Background reading on the Ambri corporation. https://ambri.com/

[132] Nature Energy 4, 495–503 (2019). *"Building aqueous K-ion batteries for energy storage."* Jiang, L., Lu, Y., Zhao, C. et al. doi:10.1038/s41560-019-0388-0.
https://www.nature.com/articles/s41560-019-0388-0#citeas

[133] Journal of Materials Chemistry. *"Emergence of rechargeable seawater batteries."* S. Senthilkumar, W. Go, J. Han, et al. 7 October, 2019.
https://pubs.rsc.org/en/content/articlelanding/2019/ta/c9ta08321a#!divAbstract

[134] Physics World. *"Water-based batteries enable a green energy future."* J. Lewis. 28 May, 2019.
https://physicsworld.com/a/water-based-batteries-enable-a-green-energy-future/

[135] Background reading on Molten-salt batteries. https://en.wikipedia.org/wiki/Molten-salt_battery

[136] Background reading on solar-thermal collectors:
https://en.wikipedia.org/wiki/Solar_thermal_collector

[137] Wikipedia Entry on Miami-Dade County. https://en.wikipedia.org/wiki/Miami-Dade_County,_Florida

[138] Miami-Dade County. Energy outlook assessment. Page 2.
https://www.miamidade.gov/greenprint/planning/library/milestone_one/energy.pdf#p2

[139] EIA information on power plants nationwide.
https://www.eia.gov/tools/faqs/faq.php?id=65&t=2

[140] Information on the number of Energy Utilities operating in the United States.
https://www.publicpower.org/system/files/documents/2018-Public-Power-Statistical-Report-Updated.pdf

[141] Energy.gov information on the number of power lines in the U.S.
http://energy.gov/articles/top-9-things-you-didnt-know-about-americas-power-grid

[142] Reuters. *"Power reliability will cost Americans more."* 13 September, 2011. http://www.reuters.com/article/us-utilities-sandiego-blackout/insight-power-reliability-will-cost-americans-more-idUSTRE78C4UG20110913

[143] Los Angeles Times. *"Blackout losses could top $100 million."* R Marosi. S Allen. 9 September, 2011. http://latimesblogs.latimes.com/lanow/2011/09/blackout-losses-could-top-100-million.html?trac=lat-pick

[144] Reuters. *"Power reliability will cost Americans more."* 13 September, 2011. http://www.reuters.com/article/us-utilities-sandiego-blackout/insight-power-reliability-will-cost-americans-more-idUSTRE78C4UG20110913

[145] Washington Post. *"141 Deaths Later, Heat Wave Appears Over."* O. Munoz. 28 July, 2006. http://usatoday30.usatoday.com/weather/news/2006-07-26-power-problems_x.htm

[146] Reuters. *"Power reliability will cost Americans more."* 13 September, 2011. http://www.reuters.com/article/us-utilities-sandiego-blackout/insight-power-reliability-will-cost-americans-more-idUSTRE78C4UG20110913

[147] Background reading on base load power: https://en.wikipedia.org/wiki/Base_load

https://energyeducation.ca/encyclopedia/Baseload_power

[148] Energy Information Administration. *"What is U.S. electricity generation by energy source?"* https://www.eia.gov/tools/faqs/faq.php?id=427&t=3

[149] Foundation for Water & Energy Education. *"How a Hydroelectric Project Can Affect a River."* https://fwee.org/environment/how-a-hydroelectric-project-can-affect-a-river/

Union of Concerned Scientists. *"Environmental Impacts of Natural Gas."* https://www.ucsusa.org/clean-energy/coal-and-other-fossil-fuels/environmental-impacts-of-natural-gas

[150] Energy Information Administration. *"Most coal-fired electric capacity was built before 1980."* 28 June, 2011. https://www.eia.gov/todayinenergy/detail.php?id=1990

[151] Quartz Magazine. *"Most coal-fired power plants in the US are nearing retirement age."* T. Woody. 12 March, 2013. https://qz.com/61423/coal-fired-power-plants-near-retirement/

[152] Forbes Magazine. *"The Thing About Thorium: Why The Better Nuclear Fuel May Not Get a Chance."* M. Katusa. 16 February, 2012.
https://www.forbes.com/sites/energysource/2012/02/16/the-thing-about-thorium-why-the-better-nuclear-fuel-may-not-get-a-chance/#79d522b51d80

M. Baker Schaffer, RAND Corporation. *"Abundant thorium as an alternative nuclear fuel Important waste disposal and weapon proliferation advantages"* 30 May, 2018.
https://web.mit.edu/mission/www/m2018/pdfs/japan/thorium.pdf

World Nuclear Association. *"Thorium."* February, 2017. https://www.world-nuclear.org/information-library/current-and-future-generation/thorium.aspx

[153] AIP Conference Proceedings 1659, 040001 (2015) *"Advantages of Liquid Fluoride Thorium Reactor in Comparison with Light Water Reactor."*

[154] See pages 107-115 for detailed breakdown on countries investing in thorium

[155] Background reading on cold fusion. https://en.wikipedia.org/wiki/Cold_fusion

[156] U.S. Department of Energy. "Quadrennial Technology Review." Table 10.4: Range of materials requirements (fuel excluded) for various electricity generation technologies." September, 2015. https://nextgiantleap.org/sites/default/files/source_files/quadrennial-technology-review-2015.pdf

Center for Environmental Progress. *"Are we headed for a solar waste crisis?"* J. Desai, M. Nelson. 21 June, 2017. http://environmentalprogress.org/big-news/2017/6/21/are-we-headed-for-a-solar-waste-crisis

[157] Background facts about the Topaz Solar Farm.
http://www.firstsolar.com/Resources/Projects/Topaz-Solar-Farm

[158] Exelon Corp. *"Limerick Generating Station."*
http://www.exeloncorp.com/locations/power-plants/limerick-generating-station

Background reading on LGS: https://en.wikipedia.org/wiki/Limerick_Generating_Station

[159] U.S. Nuclear Regulatory Commission. *"Palo Verde Nuclear Generating Station, Unit 1."*
https://www.nrc.gov/info-finder/reactors/palo1.html
Additional background reading:
https://en.wikipedia.org/wiki/Palo_Verde_Nuclear_Generating_Station

[160] Proceedings of the National Academy of Sciences. C. Clack, S. Qvist, J. Apt, et al. *"Evaluation of a proposal for reliable low-cost grid power with 100% wind, water, and solar."* PNAS June 27, 2017 114 (26) 6722-6727; first published June 19, 2017 https://doi.org/10.1073/pnas.1610381114

[161] C. Bahri, A. Majid, W. Al-Areqi. *"Advantages of Liquid Fluoride Thorium Reactor in Comparison with Light Water Reactor."* Nuclear Science Program, School of Applied Physics, Universiti Kebangsaan Malaysia. https://aip.scitation.org/doi/pdf/10.1063/1.4916861

[162] C. Bahri, A. Majid, W. Al-Areqi. *"Advantages of Liquid Fluoride Thorium Reactor in Comparison with Light Water Reactor."* Nuclear Science Program, School of Applied Physics, Universiti Kebangsaan Malaysia. https://aip.scitation.org/doi/pdf/10.1063/1.4916861

[163] MIT Technology Review. *"Meltdown-Proof Nuclear Reactors get a Safety Check in Europe."* R. Martin. 4 September, 2015. https://www.technologyreview.com/s/540991/meltdown-proof-nuclear-reactors-get-a-safety-check-in-europe/

[164] Machine Design. *"Thorium: A Safe Form of Clean Energy?"* K. Sorensen. 16 March, 2010. https://www.machinedesign.com/energy/thorium-safe-form-clean-energy

World Nuclear Association. *"Thorium."* February, 2017. https://www.world-nuclear.org/information-library/current-and-future-generation/thorium.aspx

[165] U.S. Department of Energy. Office of Scientific and Technical Information. *"Fast Spectrum Molten Salt Reactor Options."* 1 July, 2011. https://www.osti.gov/biblio/1018987

M. Baker Schaffer, RAND Corporation. *"Abundant thorium as an alternative nuclear fuel Important waste disposal and weapon proliferation advantages"* 30 May, 2018. https://web.mit.edu/mission/www/m2018/pdfs/japan/thorium.pdf

[166] C. Bahri, A. Majid, W. Al-Areqi. *"Advantages of Liquid Fluoride Thorium Reactor in Comparison with Light Water Reactor."* Nuclear Science Program, School of Applied Physics, Universiti Kebangsaan Malaysia. https://aip.scitation.org/doi/pdf/10.1063/1.4916861

[167] World Nuclear Association. *Information Library: Uranium Enrichment.* http://www.world-nuclear.org/information-library/nuclear-fuel-cycle/conversion-enrichment-and-fabrication/uranium-enrichment.aspx. Last updated date: June, 2018.

[168] A. Ahmad, E. McClamrock, A. Glaser. *"Neutronics calculations for denatured molten salt reactors: Assessing resource requirements and proliferation-risk attributes."* Annals of Nuclear Energy. Volume 75. January, 2015. Pages 261-267. https://doi.org/10.1016/j.anucene.2014.08.014

M. Schaffer. *"Abundant thorium as an alternative nuclear fuel: Important waste disposal and weapon proliferation advantages."* Journal of Energy Policy, Volume 60. September, 2013, pages 4-12. https://doi.org/10.1016/j.enpol.2013.04.062

The Economist Magazine. *"Asgard's fire. Thorium, an element named after the Norse god of thunder, may soon contribute to the world's electricity supply."* 12 April, 2014. https://www.economist.com/science-and-technology/2014/04/12/asgards-fire

[169] UxC SMR Research Center. *"SMR Design Profile."* 3 April, 2013. https://www.uxc.com/smr/uxc_SMRDetail.aspx?key=LFTR

[170] S. Lam. *"Economics of Thorium and Uranium Reactors."* 30 April, 2013. http://pages.hmc.edu/evans/LamThorium.pdf

[171] Background information on LFTRs, including historical progress: https://en.wikipedia.org/wiki/Thorium-based_nuclear_power#Background_and_brief_history

[172] LibreTexts. *"10.2: Fission and Fusion."* A. Soult. 2 November, 2019. https://chem.libretexts.org/Courses/University_of_Kentucky/UK%3A_CHE_103_-_Chemistry_for_Allied_Health_(Soult)/Chapters/Chapter_10%3A_Nuclear_and_Chemical_Reactions/10.2%3A_Fission_and_Fusion

[173] Background reading on fusion energy: https://en.wikipedia.org/wiki/Fusion_power#Deuterium,_tritium

[174] Forbes. *"The Thing About Thorium: Why The Better Nuclear Fuel May Not Get A Chance."* M. Katusa. 16 February, 2012. https://www.forbes.com/sites/energysource/2012/02/16/the-thing-about-thorium-why-the-better-nuclear-fuel-may-not-get-a-chance/

[175] American Physical Society. *"Liquid Fuel Nuclear Reactors."* R. Hargraves, R. Moir. January, 2011. https://www.aps.org/units/fps/newsletters/201101/hargraves.cfm

World Nuclear Association. *"Thorium."* February, 2017. https://www.world-nuclear.org/information-library/current-and-future-generation/thorium.aspx

[176] Background reading on Pressurized Water Reactors: https://en.wikipedia.org/wiki/Pressurized_water_reactor

[177] Background reading on Light Water Reactors: https://en.wikipedia.org/wiki/Light-water_reactor

[178] Background reading on Heavy Water Reactors:
https://en.wikipedia.org/wiki/Pressurized_heavy-water_reactor

[179] Background reading on Molten Salt Reactors (generic):
https://en.wikipedia.org/wiki/Molten_salt_reactor

[180] World Nuclear Association. *"The Nuclear Fuel Cycle."* Last updated date: March, 2017. http://www.world-nuclear.org/information-library/nuclear-fuel-cycle/introduction/nuclear-fuel-cycle-overview.aspx

[181] Hargreaves, Robert, PhD. *"THORIUM: Energy Cheaper than Coal."* July 25, 2012. pp 164

[182] Hargreaves, Robert, PhD. *"THORIUM: Energy Cheaper than Coal."* July 25, 2012. pp 164

[183] National Institute of Health. *"Lessons Learned from the Fukushima Nuclear Accident for Improving Safety of U.S. Nuclear Plants."* Nuclear and Radiation Studies Board; Division on Earth and Life Studies; National Research Council. 29 October, 2014. https://www.ncbi.nlm.nih.gov/books/NBK253931/

[184] World Nuclear Association. *"How is uranium made into nuclear fuel?"* No date or author provided. https://www.world-nuclear.org/nuclear-essentials/how-is-uranium-made-into-nuclear-fuel.aspx

[185] Background reading on Chernobyl disaster:
https://en.wikipedia.org/wiki/Chernobyl_disaster

[186] American institute of physics. *"Searching for lost WWII-era uranium cubes from Germany."* 1 May, 2019. https://phys.org/news/2019-05-lost-wwii-era-uranium-cubes-germany.html

[187] Discover Magazine. *"Why Aren't we using thorium in nuclear reactors?"* A. Hadhazy. 6 May, 2014. https://www.discovermagazine.com/the-sciences/why-arent-we-using-thorium-in-nuclear-reactors

110th Congress. (2008). S. 3680 (110th): Thorium energy independence and security act of 2008. Retrieved from website: http://www.govtrack.us/congress/bills/110/s3680/text

U.S. Department of Energy. *"Closing the Circle on the Splitting of the Atom."* January, 1996. https://www.energy.gov/sites/prod/files/2014/03/f8/Closing_the_Circle_Report.pdf

[188] World Nuclear Association. *"Plutonium."* https://www.world-nuclear.org/information-library/nuclear-fuel-cycle/fuel-recycling/plutonium.aspx

189 Globalsecurity.org *"Weapons of Mass Destruction – Gun Device."* https://www.globalsecurity.org/wmd/intro/gun-device.htm

190 Globalsecurity.org. *"Weapons of Mass Destruction – Implosion Device."* https://www.globalsecurity.org/wmd/intro/implosion-device.htm

191 *"Neptunium 237 and Americium: World Inventories and Proliferation Concerns."* D. Albright and K. Kramer. July 10, 2005. http://isis-online.org/uploads/isis-reports/documents/np_237_and_americium.pdf#p14

192 *"Neptunium 237 and Americium: World Inventories and Proliferation Concerns."* D. Albright and K. Kramer. July 10, 2005. http://isis-online.org/uploads/isis-reports/documents/np_237_and_americium.pdf#p14

193 Plutonium-239 is created via neutron capture of uranium-238, which is the most common form of uranium in the world (99%+), and is an aspect of the uranium fuel supply used in Pressurized Water Reactors. World Nuclear Association. *"Plutonium."* https://www.world-nuclear.org/information-library/nuclear-fuel-cycle/fuel-recycling/plutonium.aspx

194 Background reading on how hydrogen (thermonuclear) weapons work: https://en.wikipedia.org/wiki/Thermonuclear_weapon

195 Background reading on Teller-Ulam thermonuclear weapon design. https://en.wikipedia.org/wiki/Thermonuclear_weapon

196 University of Norte Dame lecture slides. *"The Hydrogen Bomb."* https://www3.nd.edu/~nsl/Lectures/phys20061/pdf/10.pdf

197 Common nuclear weapons yields: https://en.wikipedia.org/wiki/Nuclear_weapon_yield

198 International Campaign to Abolish Nuclear Weapons: http://www.icanw.org/the-facts/nuclear-arsenals/.

199 Minerals Education Coalition. Thorium. No date provided. https://mineralseducationcoalition.org/elements/thorium/

200 S. Kim, S. Hong, R. Park. *"Analysis of steam explosion under conditions of partially flooded cavity and submerged reactor vessel."* 5 July, 2018. https://www.hindawi.com/journals/stni/2018/3106039/

[201] Background reading on LFTR operation:
https://en.wikipedia.org/wiki/Liquid_fluoride_thorium_reactor

[202] Molten salts in LFTRs can take varied forms. This source provides background reading on different salt options and their benefits and drawbacks:
https://en.wikipedia.org/wiki/Molten_salt_reactor#Mixtures

[203] *"An Overview of Thorium Utilization in Nuclear Reactors and Fuel Cycles."* J. Maiorino, F. D'Auria,, R. Akbari-Jeyhouni. Respectively: Federal University of ABC, CECS, Av. dos Estados, 5001, Santo André-SP-Brazil, University of Pisa, GRNSPG-DESTEC, Largo L Lazzarini, Pisa, Italy, Amirkabir University of Technology, Department of Energy Engineering & Physics, Tehran, Iran.
https://www.researchgate.net/publication/325670177_An_Overview_of_Thorium_Utilization_in_Nuclear_Reactors_and_Fuel_Cycles

[204] International Atomic Energy Agency. *"Status Report – LFTR, Flibe Energy."* 28 July, 2016.
https://aris.iaea.org/PDF/LFTR.pdf

M. Baker Schaffer, RAND Corporation. *"Abundant thorium as an alternative nuclear fuel Important waste disposal and weapon proliferation advantages"* 30 May, 2018.
https://web.mit.edu/mission/www/m2018/pdfs/japan/thorium.pdf

[205] Watt-Logic. *"New project re-ignites European interest in thorium."* 10 September, 2017.
http://watt-logic.com/2017/09/10/thorium/

[206] American Physical Society. *"Liquid Fuel Nuclear Reactors."* R. Hargraves, R. Moir. January, 2011. https://www.aps.org/units/fps/newsletters/201101/hargraves.cfm

WIRED magazine. *"Uranium is So Last Century – Enter Thorium, the New Green Nuke."* R. Martin. 21 December, 2009. https://www.wired.com/2009/12/ff_new_nukes/

[207] Robertson, R. C. (June 1971). "Conceptual Design Study of a Single-Fluid Molten-Salt Breeder Reactor" (PDF). ORNL-4541. Oak Ridge National Laboratory.
https://energyfromthorium.com/pdf/ORNL-4541.pdf#p1

[208] Hargreaves, Robert, PhD. *"THORIUM: Energy Cheaper than Coal."* July 25, 2012. pp 177-257

[209] NASA Glenn Research Center / American Institute of Aeronautics and Astronautics. *"High Efficiency Nuclear Power Plants Using Liquid Fluoride Thorium Reactor Technology."* A. Juhaz, R. Rarick, R. Rangarajan. October, 2009. https://ntrs.nasa.gov/archive/nasa/casi.ntrs.nasa.gov/20090029904.pdf#p3

WIRED magazine. *"Uranium is So Last Century – Enter Thorium, the New Green Nuke."* R. Martin. 21 December, 2009. https://www.wired.com/2009/12/ff_new_nukes/

[210] International Atomic Energy Agency. *"Status Report – LFTR, Flibe Energy."* https://aris.iaea.org/PDF/LFTR.pdf

[211] American Physical Society. *"Liquid Fuel Nuclear Reactors."* R. Hargraves, R. Moir. January, 2011. https://www.aps.org/units/fps/newsletters/201101/hargraves.cfm

[212] U.S. Department of Energy. Office of Scientific and Technical Information. "Fast Spectrum Molten Salt Reactor Options." 1 July, 2011. https://www.osti.gov/biblio/1018987

[213] NASA Glenn Research Center / American Institute of Aeronautics and Astronautics. *"High Efficiency Nuclear Power Plants Using Liquid Fluoride Thorium Reactor Technology."* A. Juhaz, R. Rarick, R. Rangarajan. October, 2009. https://ntrs.nasa.gov/archive/nasa/casi.ntrs.nasa.gov/20090029904.pdf#p4

[214] World Nuclear Association. *Information Library: Uranium Enrichment.* http://www.world-nuclear.org/information-library/nuclear-fuel-cycle/conversion-enrichment-and-fabrication/uranium-enrichment.aspx. Last updated date: June, 2018.

Stanford University (coursework). *"Thorium Energy Viability."* J. Ting. 12 November, 2015. http://large.stanford.edu/courses/2015/ph240/ting1/

[215] Rare Earth Investing News. "Thorium: Rare Earth Liability or Asset?" M. Montgomery. 14 March, 2011. https://investingnews.com/daily/resource-investing/critical-metals-investing/rare-earth-investing/thorium-rare-earth-liability-or-asset/

[216] Hargreaves, Robert, PhD. *"THORIUM: Energy Cheaper than Coal."* July 25, 2012. pp 177-257

[217] Thorium MSR Foundation. *"The Flibe Energy LFTR49: the triple ace in nuclear GEN IV design."* T. Wolters. 24 July, 2016. https://articles.thmsr.nl/the-flibe-energy-lftr49-the-triple-ace-in-nuclear-gen-iv-design-ea9bffcd71dd

M. Schaffer. *"Abundant thorium as an alternative nuclear fuel: Important waste disposal and weapon proliferation advantages."* Journal of Energy Policy, Volume 60. September, 2013, pages 4-12. https://doi.org/10.1016/j.enpol.2013.04.062

[218] LERNER, L. (2012, JUNE 22). Nuclear fuel recycling could offer plentiful energy. Argonne National Laboratory News. Retrieved from http://www.anl.gov/articles/nuclear-fuel-recycling-could-offer-plentiful-energy

[219] U.S. Department of Energy. Office of Scientific and Technical Information. "Fast Spectrum Molten Salt Reactor Options." 1 July, 2011. https://www.osti.gov/biblio/1018987

[220] American Physical Society. *"Liquid Fuel Nuclear Reactors."* R. Hargraves, R. Moir. January, 2011. https://www.aps.org/units/fps/newsletters/201101/hargraves.cfm

[221] Caesum-137 has a half-life of 30.17 years. https://en.wikipedia.org/wiki/Caesium-137

[222] S.F. Ashley, B.A. Lindley, G.T. Parks, et al, *"Fuel Cycle modelling of open cycle thorium fueled nuclear energy systems."* Annals of Nuclear Energy. V. 69, pp.314-330. July, 2014.

[223] Background reading on LFTRs and environmental benefits:
APS Physics: Liquid Fuel Nuclear Reactors. R. Hargraves and R. Moir.
http://www.aps.org/units/fps/newsletters/201101/hargraves.cfm
https://en.wikipedia.org/wiki/Liquid_fluoride_thorium_reactor#advantages

[224] Oak Ridge National Laboratory. *"Preparation of high purity neptunium on multi-gram scale."* P. Pantz, W. Martin, G. Parker. ORNL-2642.
https://www.osti.gov/servlets/purl/4275225

R. E. Brooksbank and CD. Hilton, Recovery of Neptunium-237 from Fluorinator Ash in Metal Recovery Plant, ORNL-2515, (15 September, 1958).

[225] *"Neptunium 237 and Americium: World Inventories and Proliferation Concerns."* Table 1. D. Albright and K. Kramer. July 10, 2005. http://isis-online.org/uploads/isis-reports/documents/np_237_and_americium.pdf#p14

[226] *"Neptunium 237 and Americium: World Inventories and Proliferation Concerns."* D. Albright and K. Kramer. July 10, 2005. http://isis-online.org/uploads/isis-reports/documents/np_237_and_americium.pdf#p14

[227] Progress In Nuclear Energy. Volume 106. pp. 204-214. *"The U-232 production in thorium cycle."* A. Wojciechowski. July, 2018. https://doi.org/10.1016/j.pnucene.2018.03.011

[228] Langford, R. Everett (2004). Introduction to Weapons of Mass Destruction: Radiological, Chemical, and Biological. Hoboken, New Jersey: John Wiley & Sons. p. 85.

[229] Science and Global Security, Volume 9. *"U-232 and the proliferation resistance of u-233 in spent fuel."* J. Kang, F. Hippel. P.1. http://fissilematerials.org/library/sgs09kang.pdf

[230] Background reading on the REM unit of measuring radiation: https://en.wikipedia.org/wiki/Roentgen_equivalent_man

[231] Science and Global Security, Volume 9. *"U-232 and the proliferation resistance of u-233 in spent fuel."* J. Kang, F. Hippel. P.1. http://fissilematerials.org/library/sgs09kang.pdf

[232] Los Alamos National Laboratory. *"Benchmark Critical Experiments of Uranium-233 Spheres Surrounded by Uranium-235."* R. Brewer, 1995. P. 9. https://fas.org/sgp/othergov/doe/lanl/lib-www/la-pubs/00285681.pdf#p9

[233] Atomic Archive. *"Effects of radiation levels on the human body."* No date provided. http://www.atomicarchive.com/Effects/radeffectstable.shtml

[234] International Atomic Energy Agency. *"Safe Handling of Plutonium."* Safety series no 39, 1974. https://gnssn.iaea.org/Superseded%20Safety%20Standards/Safety_Series_039_1974.pdf

[235] D. Makowski. *"The impact of radiation on electronic devices with the specific consideration of neutron and gamma radiation monitoring."* Technical University of Lodz. P. 10. https://jra-srf.desy.de/e86/e575/e605/infoboxContent608/care-thesis-06-004.pdf#p10

[236] Wisconsin Project on Nuclear Arms Control. *"Nuclear Weapons Primer."* Figure 4-3, Weapon design and production: firing sets. https://www.wisconsinproject.org/nuclear-weapons/#implosion

[237] National Nuclear Laboratory, Government of the United Kingdom. *"The Thorium Fuel Cycle. An independent assessment by the UK National Nuclear Laboratory."* August, 2010. https://www.nnl.co.uk/wp-content/uploads/2019/01/nnl__1314092891_thorium_cycle_position_paper.pdf#p6

[238] "Neptunium 237 and Americium: World Inventories and Proliferation Concerns." D. Albright and K. Kramer. July 10, 2005. http://isis-online.org/uploads/isis-reports/documents/np_237_and_americium.pdf#p14

[239] Background reading on first-strike methodology. https://en.wikipedia.org/wiki/Pre-emptive_nuclear_strike

[240] Background reading on "Operation Outside the Box," an Israeli airstrike against a Syrian reactor suspected of nuclear weapons development. https://en.wikipedia.org/wiki/Operation_Outside_the_Box

[241] Pacific Northwest National Laboratory. "Seawater yields first grams of yellowcake." 13 June, 2018. S. Bauer. https://www.pnnl.gov/news/release.aspx?id=4514

[242] Science and Global Security, Volume 9. "U-232 and the proliferation resistance of u-233 in spent fuel." J. Kang, F. Hippel. P.1. http://fissilematerials.org/library/sgs09kang.pdf#p1

[243] Background reading on two lackluster tests with uranium-233 and neptunium-237:
http://nuclearweaponarchive.org/India/IndiaShakti.html
https://en.wikipedia.org/wiki/Operation_Teapot

[244] Background reading on two lackluster tests with uranium-233 and neptunium-237:
http://nuclearweaponarchive.org/India/IndiaShakti.html
https://en.wikipedia.org/wiki/Operation_Teapot

[245] Background reading on two lackluster tests with uranium-233 and neptunium-237:
http://nuclearweaponarchive.org/India/IndiaShakti.html
https://en.wikipedia.org/wiki/Operation_Teapot

[246] Diagram of Boiling Water Reactor. Georgia State University. http://hyperphysics.phy-astr.gsu.edu/hbase/NucEne/reactor.html

[247] J. Mena, P. Edmondson, L. Margetts, et al. "Characterisation of the spatial variability of material properties of Gilsocarbon and NBG-18 using random fields." Journal of Nuclear Materials 511. September, 2008. DOI: 10.1016/j.jnucmat.2018.09.008

Hargreaves, Robert, PhD. "THORIUM: Energy Cheaper than Coal." July 25, 2012. pp 177-257

Transatomic Power. Technical White Paper. V.1.0.1. 14 March, 2014. P. 9. https://nextgiantleap.org/sites/default/files/TAP_White_Paper.pdf#9

T. Fei, D. Ogata, K. Pham, et al. (16 May 2008). "*A modular pebble-bed advanced high temperature reactor.*" U.C. Berkeley Report UCBTH-08-001. 16 May, 2008. http://fhr.nuc.berkeley.edu/wp-content/uploads/2014/10/08-001_PB-AHTR_NE170_Design_Project_Rpt.pdf

M. Schaffer. *"Abundant thorium as an alternative nuclear fuel: Important waste disposal and weapon proliferation advantages."* Journal of Energy Policy, Volume 60. September, 2013, pages 4-12. https://doi.org/10.1016/j.enpol.2013.04.062

Wired Magazine. *"Uranium is so last century – enter thorium, the new green nuke."* R Martin. 21 December, 2009. https://www.wired.com/2009/12/ff-new-nukes/

[248] Hargreaves, Robert, PhD. *"THORIUM: Energy Cheaper than Coal."* July 25, 2012. pp 177-257

[249] Harvard Business Review. *"Profit from the learning curve."* W. Hirschmann. January, 1964. https://hbr.org/1964/01/profit-from-the-learning-curve

Investopedia. *"Learning Curve."* https://www.investopedia.com/terms/l/learning-curve.asp

[250] Background reading on Moore's Law. https://en.wikipedia.org/wiki/Moore%27s_law

[251] Hargreaves, Robert, PhD. *"THORIUM: Energy Cheaper than Coal."* July 25, 2012. pp 221

[252] Hargreaves, Robert, PhD. *"THORIUM: Energy Cheaper than Coal."* July 25, 2012. pp 220

[253] Oak Ridge National Laboratory. *"Molten Salt Reactor Experience Applicable to LS-VHTR Refueling."* C. Forsberg, ORNL, 18 April, 2006.

Background reading on experiments to power aircraft with atomic energy. https://en.wikipedia.org/wiki/Aircraft_Nuclear_Propulsion

[254] Oak Ridge National Laboratory. *"Molten Salt Reactor Experience Applicable to LS-VHTR Refueling."* C. Forsberg, ORNL, 18 April, 2006.

[255] Forbes. *"The Thing About Thorium: Why The Better Nuclear Fuel May Not Get A Chance."* M. Katusa. 16 February, 2012. https://www.forbes.com/sites/energysource/2012/02/16/the-thing-about-thorium-why-the-better-nuclear-fuel-may-not-get-a-chance/

[256] Oak Ridge National Laboratory. *"Fluoride-Salt-Cooled High-Temperature Reactors Overview."* https://www.ornl.gov/msr

Oak Ridge National Laboratory. *"Molten Salt Reactor Experience Applicable to LS-VHTR Refueling."* C. Forsberg, ORNL, 18 April, 2006. Also see: https://en.wikipedia.org/wiki/Molten-Salt_Reactor_Experiment for background reading and the National Laboratory Reports on MSRs here https://www.ornl.gov/content/national-laboratory-reports-fhrs

[257] Oak Ridge National Laboratory. *"Molten Salt Reactor Experience Applicable to LS-VHTR Refueling."* C. Forsberg, ORNL, 18 April, 2006.

[258] Oak Ridge National Laboratory. *"Update on SINAP TMSR Research."* https://public.ornl.gov/conferences/msr2016/docs/Presentations/MSR2016-day1-15-Hongjie-Xu-Update-on-SINAP-TMSR-Research.pdf#p3

[259] South China Morning Post. *"China hopes cold war nuclear energy tech will power warships, drones."* S. Chen. https://www.scmp.com/news/china/society/article/2122977/china-hopes-cold-war-nuclear-energy-tech-will-power-warships

China Daily. *"China among the countries looking to thorium as new nuclear fuel."* K. Wilson. 25 October, 2018.
http://www.chinadaily.com.cn/a/201810/25/WS5bd11cf3a310eff3032846d1.html

Next Big Future. *"China spending US$3.3 billion on molten salt nuclear reactors for faster aircraft carriers and in flying drones."* B. Wang. 6 December, 2017.
https://www.nextbigfuture.com/2017/12/china-spending-us3-3-billion-on-molten-salt-nuclear-reactors-for-faster-aircraft-carriers-and-in-flying-drones.html

[260] Next Big Future. *"China spending US$3.3 billion on molten salt nuclear reactors for faster aircraft carriers and in flying drones."* B. Wang. 6 December, 2017.
https://www.nextbigfuture.com/2017/12/china-spending-us3-3-billion-on-molten-salt-nuclear-reactors-for-faster-aircraft-carriers-and-in-flying-drones.html

[261] China Daily. *"China among the countries looking to thorium as new nuclear fuel."* K. Wilson. 25 October, 2018.
http://www.chinadaily.com.cn/a/201810/25/WS5bd11cf3a310eff3032846d1.html

South China Morning Post. *"China hopes cold war nuclear energy tech will power warships, drones."* S. Chen. https://www.scmp.com/news/china/society/article/2122977/china-hopes-cold-war-nuclear-energy-tech-will-power-warships https://www.nextbigfuture.com/2017/12/china-spending-us3-3-billion-on-molten-salt-nuclear-reactors-for-faster-aircraft-carriers-and-in-flying-drones.html

[262] China Daily. *"China among the countries looking to thorium as new nuclear fuel."* K. Wilson. 25 October, 2018.
http://www.chinadaily.com.cn/a/201810/25/WS5bd11cf3a310eff3032846d1.html

South China Morning Post. *"China hopes cold war nuclear energy tech will power warships, drones."* S. Chen. https://www.scmp.com/news/china/society/article/2122977/china-hopes-cold-war-nuclear-energy-tech-will-power-warships

https://www.nextbigfuture.com/2017/12/china-spending-us3-3-billion-on-molten-salt-nuclear-reactors-for-faster-aircraft-carriers-and-in-flying-drones.html

[263] Nuclear Research and Consultancy Group (Netherlands). *"NRG researches new nuclear reactor concept."* https://www.nrg.eu/about-nrg/news-press/detail/article/nrg-doet-onderzoek-voor-nieuw-type-kerncentrale-105.html

[264] The Thorium MSR Foundation. *"'Petten' has started world's first MSR-specific thorium fuel irradiation experiments in 45 years."* G. Zwartsenberg. 18 August, 2017.
https://articles.thmsr.nl/petten-has-started-world-s-first-thorium-msr-specific-irradiation-experiments-in-45-years-ff8351fce5d2

[265] The Thorium MSR Foundation. *"'Petten' has started world's first MSR-specific thorium fuel irradiation experiments in 45 years."* G. Zwartsenberg. 18 August, 2017.
https://articles.thmsr.nl/petten-has-started-world-s-first-thorium-msr-specific-irradiation-experiments-in-45-years-ff8351fce5d2

[266] BBC Future Now. *"Why India Wants to Turn Its Beaches Into Nuclear Fuel."* E. Gent. 18 October, 2018. http://www.bbc.com/future/story/20181016-why-india-wants-to-turn-its-beaches-into-nuclear-fuel

[267] Background reading on India's three-stage nuclear power program.
https://en.wikipedia.org/wiki/India%27s_three-stage_nuclear_power_programme

[268] Nuclear Asia. *"Criticality of prototype fast breeder reactor pushed back further."* 30 September, 2018. R. Shama. http://www.nuclearasia.com/news/criticality-prototype-fast-breeder-reactor-pushed-back/2453/

Times of India. *"Kalpakkam fast breeder reactor may achieve criticality in 2019."* PTI. 20 September, 2018. https://timesofindia.indiatimes.com/india/kalpakkam-fast-breeder-reactor-may-achieve-criticality-in-2019/articleshow/65888098.cms

[269] Times of India. *"Nuclear reactor at Kalpakkam: World's envy, India's pride."* https://timesofindia.indiatimes.com/india/nuclear-reactor-at-kalpakkam-worlds-envy-indias-pride/articleshow/59407602.cms

[270] Government of India. Department of Atomic Energy. *"Fast Breeder Programme: An Inevitable Option for Energy Security."* B. Raj. Director, Indira Gandhi Centre for Atomic Research. https://dae.nic.in/?q=node/208

[271] Background reading on India's three-stage nuclear power program. https://en.wikipedia.org/wiki/India%27s_three-stage_nuclear_power_programme

[272] Nuclear Engineering International Magazine. *"Russian Scientists Look to Thorium Reactors."* 29 January, 2018. https://www.neimagazine.com/news/newsrussian-scientists-look-to-thorium-reactors-6036775

Phys.org. *"Thorium reactors may dispose of enormous amounts of weapons-grade plutonium."* Tomsk Polytechnic University. 22 January, 2018. https://phys.org/news/2018-01-thorium-reactors-dispose-enormous-amounts.html

[273] New Atlas Magazine. *"Can Thorium Reactors Dispose of Weapons-Grade Plutonium?"* M. Irving. 22 January, 2018. https://newatlas.com/thorium-reactor-recycle-plutonium/53078/

[274] I. Shamanin, V. Grachev, Y. Chertkov, S. Bedenko, O. Mendoza, V. Knyshev. *"Neutronic properties of high-temperature gas-cooled reactors with thorium fuel."* Annals of Nuclear Energy, Volume 113, March 2018. https://www.sciencedirect.com/science/article/pii/S0306454917304358?via%3Dihub

[275] Thorium Energy World. *"Putin has Thorium Plans and Engages Russia's Vast Nuclear Establishment."* 31 August, 2016. http://www.thoriumenergyworld.com/news/putin-has-thorium-plans-and-engages-russias-vast-nuclear-establishment

[276] New York Times. *"Why 'Green' Germany Remains Addicted to Coal."* M. Eddy. 10 October, 2018. https://www.nytimes.com/2018/10/10/world/europe/germany-coal-climate.html

[277] New York Times. *"Why 'Green' Germany Remains Addicted to Coal."* M. Eddy. 10 October, 2018. https://www.nytimes.com/2018/10/10/world/europe/germany-coal-climate.html

[278] Background reading on the THTR-300 high-temperature reactor. https://en.wikipedia.org/wiki/THTR-300

[279] Nuclear Energy Institute. *"Statistics."* https://www.nei.org/resources/statistics

280 Tennessee Valley Authority. *"Watts Bar Nuclear Plant."* https://www.tva.gov/Energy/Our-Power-System/Nuclear/Watts-Bar-Nuclear-Plant

281 Washington Post. "It's the first new U.S. nuclear reactor in decades. And climate change has made that a very big deal." C. Mooney. 17 June, 2016. https://www.washingtonpost.com/news/energy-environment/wp/2016/06/17/the-u-s-is-powering-up-its-first-new-nuclear-reactor-in-decades/?utm_term=.64da71742e42

282 Westinghouse eVinci Microreactor. http://www.westinghousenuclear.com/New-Plants/eVinci-Micro-Reactor

283 Westinghouse eVinci Microreactor. http://www.westinghousenuclear.com/New-Plants/eVinci-Micro-Reactor

284 Pittsburgh Post-Gazette. *"Nuke on a truck: How Westinghouse is shrinking the nuclear power plant."* A. Litvak. 31 December, 2018. https://www.post-gazette.com/business/powersource/2018/12/31/westinghouse-microreactor-evinci-generator-nuscale-power-general-atomics/stories/201812300007

285 National Conference of State Legislatures. *"NuScale Power Overview."* M. McGough, CCO NuScale Power, Inc. October, 2013. http://www.ncsl.org/documents/environ/McGoughpresentation.pdf

286 NuScale. *"NuScale wins U.S. DOE funding for its SMR Technology."* https://www.nuscalepower.com/about-us/doe-partnership

287 The Oregonian. *"Corvallis nuclear power company NuScale passes regulatory milestone."* B. Hall. 3 May, 2018. https://www.oregonlive.com/business/index.ssf/2018/05/corvallis_nuclear_power_compan.html

288 Power Magazine. *"DOE Designates Part of UAMPS SMR Plant for Research, Self-Power."* S. Patel. 3 January, 2019. https://www.powermag.com/doe-designates-part-of-uamps-smr-plant-for-research-self-power/

289 General Atomics. *"Advanced Reactors."* http://www.ga.com/advanced-reactors

290 International Atomic Energy Agency. *"Advances in Small Modular Reactor Technology Developments."* https://aris.iaea.org/publications/smr-book_2016.pdf

291 International Atomic Energy Agency. *"Advances in Small Modular Reactor Technology Developments."* 2016. https://aris.iaea.org/publications/smr-book_2016.pdf

[292] Terrestrial Energy. Background on ISMR technology. https://www.terrestrialenergy.com/technology/versatile/

[293] Terrestrial Energy. ISMR overview. https://www.terrestrialenergy.com/

[294] Market Research Future. *"Micro Reactor Technology Market – Trends & Forecast, 2016-2022."* May, 2017. https://www.marketresearchfuture.com/reports/micro-reactor-technology-market-1128

[295] ThinkProgress. *"Taxpayers should not fund Bill Gates' nuclear albatross."* J. Romm. 4 February, 2019. https://thinkprogress.org/nuclear-power-is-so-uneconomical-even-bill-gates-cant-make-it-work-without-taxpayer-funding-faea0cdb60de/

[296] Popular Mechanics. *"The Alexandria Ocasio-Cortez 'Green New Deal' Wants to Get Rid of Nuclear Power. That's a Great Idea."* A. Thompson. 8 Feburary, 2019. https://www.popularmechanics.com/science/energy/a26255413/green-new-deal-nuclear-power/

[297] The Guardian. *"Don't believe the spin on thorium being a greener nuclear option."* E. Rees. 23 June, 2011. https://www.theguardian.com/environment/2011/jun/23/thorium-nuclear-uranium

[298] National Nuclear Laboratory, Government of the United Kingdom. *"The Thorium Fuel Cycle. An independent assessment by the UK National Nuclear Laboratory."* http://www.nnl.co.uk/media/1050/nnl__1314092891_thorium_cycle_position_paper.pdf

[299] National Nuclear Laboratory, Government of the United Kingdom. *"The Thorium Fuel Cycle. An independent assessment by the UK National Nuclear Laboratory."* http://www.nnl.co.uk/media/1050/nnl__1314092891_thorium_cycle_position_paper.pdf

[300] American Action Forum. *"The Costs And Benefits Of Nuclear Regulation."* S. Batkins, 8 September, 2018. https://www.americanactionforum.org/research/costs-benefits-nuclear-regulation/

[301] American Action Forum. *"Putting Nuclear Regulatory Costs in Context."* S. Batkins, P. Rossetti, D. Goldbeck. 12 July, 2017. https://www.americanactionforum.org/research/putting-nuclear-regulatory-costs-context/

[302] International Atomic Energy Agency. *"Construction time of PWR's."* P. Carajilescov. J. Moreira. 2011 International Nuclear Atlantic Conference - INAC 2011. https://inis.iaea.org/collection/NCLCollectionStore/_Public/42/105/42105221.pdf

[303] American Action Forum. *"Putting Nuclear Regulatory Costs in Context."* S. Batkins, P. Rossetti, D. Goldbeck. 12 July, 2017. https://www.americanactionforum.org/research/putting-nuclear-regulatory-costs-context/

[304] ARS Technica. *"Ballmer: iPhone has 'no chance' of gaining significant market share."* J. Hruska. 30 April, 2007. https://arstechnica.com/information-technology/2007/04/ballmer-says-iphone-has-no-chance-to-gain-significant-market-share/

[305] Computerworld. *CW@50: Data storage goes from $1m to 2 cents per gigabyte.* L. Mearian. 23 March, 2017. https://www.computerworld.com/article/3182207/cw50-data-storage-goes-from-1m-to-2-cents-per-gigabyte.html

[306] MIT Technology Review. *"Top U.S. Intelligence Official Calls Gene Editing a WMD Threat."* A. Regalado. 9 February, 2016. https://www.technologyreview.com/s/600774/top-us-intelligence-official-calls-gene-editing-a-wmd-threat/

Physics.org. *"Could CRISPR be used as a biological weapon?"* J. Revill. 31 August, 2017. https://phys.org/news/2017-08-crispr-biological-weapon.html

[307] Wired Magazine. *"Why It's So Hard To Wipe Out All of Syria's Chemical Weapons."* B. Barrett. 8 April, 2017. https://www.wired.com/2017/04/syria-sarin-chemical-weapons-chlorine/

Salt Lake City Desert News (sourcing Toronto Globe and Mail). *"Formula for Sarin is Simple"* https://www.deseretnews.com/article/411177/FORMULA-FOR-SARIN-IS-SIMPLE.html

[308] C. Bahri, A. Majid, W. Al-Areqi. *"Advantages of Liquid Fluoride Thorium Reactor in Comparison with Light Water Reactor."* Nuclear Science Program, School of Applied Physics, Universiti Kebangsaan Malaysia. https://aip.scitation.org/doi/pdf/10.1063/1.4916861

[309] LERNER, L. (2012, JUNE 22). Nuclear fuel recycling could offer plentiful energy. Argonne National Laboratory News. Retrieved from http://www.anl.gov/articles/nuclear-fuel-recycling-could-offer-plentiful-energy

[310] AIP Conference Proceedings 1659, 040001 (2015) *"Advantages of Liquid Fluoride Thorium Reactor in Comparison with Light Water Reactor."*

[311] AIP Conference Proceedings 1659, 040001 (2015) *"Advantages of Liquid Fluoride Thorium Reactor in Comparison with Light Water Reactor."*

[312] AIP Conference Proceedings 1659, 040001 (2015).

[313] Foundation for Economic Education. *"Solar Panels Produce Tons of Toxic Waste – Literally."* 18 November, 2019. https://fee.org/articles/solar-panels-produce-tons-of-toxic-waste-literally/

[314] U.S. Department of Energy. *"Quadrennial Technology Review."* Table 10.4: Range of materials requirements (fuel excluded) for various electricity generation technologies." September, 2015. https://nextgiantleap.org/sites/default/files/source_files/quadrennial-technology-review-2015.pdf

[315] U.S. Department of Energy. *"Quadrennial Technology Review."* Table 10.4: Range of materials requirements (fuel excluded) for various electricity generation technologies." September, 2015. https://nextgiantleap.org/sites/default/files/source_files/quadrennial-technology-review-2015.pdf

[317] Proceedings of the National Academy of Sciences. C. Clack, S. Qvist, J. Apt, et al. *"Evaluation of a proposal for reliable low-cost grid power with 100% wind, water, and solar."* PNAS June 27, 2017 114 (26) 6722-6727; first published June 19, 2017 https://doi.org/10.1073/pnas.1610381114

[318] New York Times. *"Why 'Green' Germany Remains Addicted to Coal."* M. Eddy. 10 October, 2018. https://www.nytimes.com/2018/10/10/world/europe/germany-coal-climate.html

[319] *"The benefits of nuclear flexibility in power system operations with renewable energy."* J. Jenkins, Z. Zhou, R. Ponciroli, R. Vilim, F. Ganda, F. Sisternes, A. Botterrud. Applied Energy 222. 15 July, 2018. https://www.sciencedirect.com/science/article/pii/S0306261918303180

"Real-world Challenges with a Rapid Transition to 100% Renewable Power Systems." C. Heuberger. N. MacDowell. Joule Volume 2, Issue 3. 21 March, 2018. https://www.sciencedirect.com/science/article/pii/S2542435118300485

[320] Nuclear Energy Institute. *"Only Nuclear Energy Can Save the Planet."* J. Goldstein, S. Qvist. 11 January, 2019. https://www.nei.org/news/2019/only-nuclear-energy-can-save-the-planet

[321] Strata policy research. *"U.S. Nuclear Power: Regulatory Barriers and Energy Potential."* August, 2017. https://www.strata.org/us-nuclear-power/

[322] Institute for Energy Research. *"Regulations Hurt Economics of Nuclear Power."* 19 January, 2018. https://www.instituteforenergyresearch.org/nuclear/regulations-hurt-economics-nuclear-power/

"Historical Construction Costs of global nuclear power reactors." J. Lovering, A. Yip, T. Nordhaus. 12 January, 2016.
https://www.sciencedirect.com/science/article/pii/S0301421516300106?via%3Dihub

[323]Wired Magazine. "How Boeing builds a 737 in just 9 days." J. Stewart. 27 September, 2016. https://www.wired.com/2016/09/boeing-builds-737-just-nine-days/
Timelapse of 737 construction: https://www.youtube.com/watch?v=liZ0WEEsuz4

The Post and Courier. "Pace of 787 Dreamliner production quickens and Boeing's North Charlson campus." D. Wren. 1 April, 2018.
https://www.postandcourier.com/business/pace-of-dreamliner-production-quickens-at-boeing-s-north-charleston/article_02b584ea-3384-11e8-a3d4-ff9ca36dc590.html

Boeing corp information on Everett production facility.
http://www.boeing.com/company/about-bca/everett-production-facility.page.

[324] R. Robertson.; R. Briggs, O. Smith. *"Two-Fluid Molten-Salt Breeder Reactor Design Study (Status as of January 1, 1968)."* (1970). ORNL-4528. Oak Ridge National Laboratory. doi:10.2172/4093364. https://www.osti.gov/biblio/4093364-two-fluid-molten-salt-breeder-reactor-design-study-status-january

R. Robertson. *"Conceptual Design Study of a Single-Fluid Molten-Salt Breeder Reactor"* ORNL-4541. Oak Ridge National Laboratory.
https://nextgiantleap.org/sites/default/files/source_files/ORNL-4541.pdf#p26

[325] D. LeBlanc. *"Molten salt reactors: A new beginning for an old idea."* Nuclear Engineering and Design. 240 (6): 1644. doi:10.1016/j.nucengdes.2009.12.033 / https://www.sciencedirect.com/science/article/pii/S0029549310000191

Transatomic Power. Technical White Paper. V.1.0.1. 14 March, 2014. P. 24.
https://nextgiantleap.org/sites/default/files/TAP_White_Paper.pdf#24

C. Sona, D. Gajbhiye, P. Hule, et al. *"High temperature corrosion studies in molten salt-FLiNaK."Journal of Corrosion Engineering, Science and Technology."* Vol 49, 2014 (issue 4). https://doi.org/10.1179/1743278213Y.0000000135

[326] C. Forsberg. *"Molten-Salt-Reactor Technology Gaps."* Oak Ridge National Laboratory. 10 February, 2006. https://nextgiantleap.org/sites/default/files/source_files/msrtg.pdf#p4

N. Brun, G. Hewitt, C. Markides. *"Transient freezing of molten salts in pipe-flow systems: applications to direct reactor auxiliary cooling systems (DRACS)."* Journal of Applied

Energy. Volume 186, part 1. 15 January, 2017. Pages 56-67. https://doi.org/10.1016/j.apenergy.2016.09.099 / https://www.sciencedirect.com/science/article/abs/pii/S0306261916314039

[327] Transatomic Power. Technical White Paper. V.1.0.1. 14 March, 2014. P. 9. https://nextgiantleap.org/sites/default/files/TAP_White_Paper.pdf#9

T. Fei, D. Ogata, K. Pham, et al. (16 May 2008). *"A modular pebble-bed advanced high temperature reactor."* U.C. Berkeley Report UCBTH-08-001. 16 May, 2008. http://fhr.nuc.berkeley.edu/wp-content/uploads/2014/10/08-001_PB-AHTR_NE170_Design_Project_Rpt.pdf

[328] Background reading on process milestones for LFTR development: https://en.wikipedia.org/wiki/Liquid_fluoride_thorium_reactor#Disadvantages

[329] M. Schaffer. *"Abundant thorium as an alternative nuclear fuel: Important waste disposal and weapon proliferation advantages."* Journal of Energy Policy, Volume 60. September, 2013, pages 4-12. https://doi.org/10.1016/j.enpol.2013.04.062

World Nuclear Association. *"Thorium."* February, 2017. https://www.world-nuclear.org/information-library/current-and-future-generation/thorium.aspx

[330] Nuclearpower.net *"Neutron Sources."* https://www.nuclear-power.net/nuclear-power/reactor-physics/atomic-nuclear-physics/fundamental-particles/neutron/neutron-sources/

[331] Background reading on AmBe neutron sources: https://en.wikipedia.org/wiki/Americium-241#Neutron_source

Background reading on startup neutron sources: https://en.wikipedia.org/wiki/Startup_neutron_source

[332] United States Nuclear Regulatory Commission. *"License-exempt consumer product uses of radioactive material."* https://www.nrc.gov/materials/miau/consumer-pdts.html#use

[333] Background reading on cryptographic hash comparisons: https://en.wikipedia.org/wiki/Comparison_of_cryptographic_hash_functions

[334] U.S. Geological Survey on global water data + statistics. https://water.usgs.gov/edu/earthhowmuch.html

[335] Background reading on electrolysis: https://energy.gov/eere/fuelcells/hydrogen-production-electrolysis

[336] Background reading on electrolysis: https://energy.gov/eere/fuelcells/hydrogen-production-electrolysis

[337] Background reading on MSFD systems: https://en.wikipedia.org/wiki/Multi-stage_flash_distillation

[338] Background reading on Multi-Stage Flash Distillation. https://en.wikipedia.org/wiki/Multi-stage_flash_distillation

[339] Background reading on countercurrent exchanges: https://en.wikipedia.org/wiki/Countercurrent_exchange

[340] Background reading on countercurrent exchanges: https://en.wikipedia.org/wiki/Countercurrent_exchange

[341] International Atomic Energy Agency. *"Introduction of Nuclear Desalination."* Technical Reports Series no 400. http://www-pub.iaea.org/MTCD/publications/PDF/TRS400_scr.pdf#p57

[342] International Desalination Association. *"Desalination by the Numbers."* No date provided, continually updated. http://idadesal.org/desalination-101/desalination-by-the-numbers/

[343] National Oceanic and Atmospheric Administration. "Is Sea Level Rising?" https://oceanservice.noaa.gov/facts/sealevel.html

[344] Alexandria Engineering Journal Volume 57, Issue 4, December 2018, Pages 2401-2413. *"Performance test of a sea water multi-stage flash distillation plant: Case study"* A. El-Ghonemy. https://www.sciencedirect.com/science/article/pii/S1110016817302697

[345] Röchling Industrial. *"Plastic for lightweight and corrosion resistant Subsea Equipment."* https://www.roechling.com/industrial/industries/oil-and-gas/subsea/

[346] Washington Post. *"Salt of the Sea, as Easy as Evaporation."* T. Haspel. 9 April, 2013. https://www.washingtonpost.com/lifestyle/food/salt-of-the-sea-as-easy-as-evaporation/2013/04/08/400f610e-9018-11e2-9cfd-36d6c9b5d7ad_story.html?utm_term=.631d3705738c

[347] U.S. Geological Survey. *"Mineral Commodity Summaries."* January, 2016. https://minerals.usgs.gov/minerals/pubs/commodity/salt/mcs-2016-salt.pdf

[348] Data from The Economist on the price of salt worldwide. http://www.economist.com/node/15276675

[349] Science Times. "Hydrogen is the most comment element: here's the reason why." R. Roy. 3 April, 2017. http://www.sciencetimes.com/articles/11524/20170403/hydrogen-is-the-most-common-element-heres-the-reason-why.htm

[350] Background reading on energy density of substances: https://en.wikipedia.org/wiki/Energy_density

[351] Virginia Tech University. *"Breakthrough in Hydrogen Fuel Production Could Revolutionize Alternative Energy Market."* 4 April, 2013. http://www.vtnews.vt.edu/articles/2013/04/040413-cals-hydrogen.html?utm_campaign=Argyle%2BSocial-2013-04&utm_content=shaybar&utm_medium=Argyle%2BSocial&utm_source=twitter&utm_term=2013-04-04-08-30-00

[352] Background reading on hydrogen production through fossil-fuel steam reformation. https://en.wikipedia.org/wiki/Hydrogen_production#Steam_reforming

[353] OPEC. *"Intervention by OPEC Secretary General to the 3rd Gas Summit of the GECF."* http://www.opec.org/opec_web/en/3180.htm

[354] Background reading on Hall-Heroult process to extract substances via electrolysis. https://en.wikipedia.org/wiki/Hall%E2%80%93H%C3%A9roult_process

[355] Energy Information Administration. *"Hydrogen explained."* https://www.eia.gov/energyexplained/hydrogen/production-of-hydrogen.php

[356] LibreTexts. *"Reaction of Main Group Elements with Hydrogen."* 9 November, 2019. https://chem.libretexts.org/Bookshelves/Inorganic_Chemistry/Modules_and_Websites_(Inorganic_Chemistry)/Descriptive_Chemistry/Main_Group_Reactions/Reactions_of_Main_Group_Elements_with_Hydrogen

[357] United States Department of Energy. Office of Energy Efficiency & Renewable Energy. *"Hydrogen Storage."* https://www.energy.gov/eere/fuelcells/hydrogen-storage

[358] Graphene is a single-atom-thick sheet of carbon-nanotube laid on a flat surface. It has unrivaled strength and conductivity to both heat and electricity. Background reading may be found here: https://en.wikipedia.org/wiki/Graphene

[359] Infographic on graphene: https://cleantechnica.com/files/2014/09/graphene.jpg

[360] "Batteries" technically require a chemical reaction to generate an electric charge. Graphene's storage potential is primarily capacitance. However, the phrase is interchangeable in the contemporary lexicon, so "battery" is used in this context with a degree of liberty.

[361] Phys.org *"Engineers Prove Graphene is the Strongest Material."* https://phys.org/news/2008-07-graphene-strongest-material.html

[362] W. Xiluan. G. Shi. "Flexible graphene devices related to energy conversion and storage." 7 January, 2015. https://www.researchgate.net/publication/270663402_Flexible_graphene_devices_related_to_energy_conversion_and_storage

[363] AIChE – The Global Home of Chemical Engineers. *"Where Do Chemical Engineers Fit into the Upstream Oil and Gas Industry?"* K. Horner. 7 December, 2010. https://www.aiche.org/chenected/2010/12/where-do-chemical-engineers-fit-upstream-oil-and-gas-industry

[364] OPEC. *"Intervention by OPEC Secretary General to the 3rd Gas Summit of the GECF."* http://www.opec.org/opec_web/en/3180.htm

[365] Background reading on synthetic hydrocarbons: https://en.wikipedia.org/wiki/Synthetic_fuel

[366] Background reading on fuel cell history: https://en.wikipedia.org/wiki/Fuel_cell#History

[367] Setra systems, Inc. https://www.setra.com/

[368] Background reading on initial directives on Combined Heat and Power - https://en.wikipedia.org/wiki/CHP_Directive
Background reading on cogeneration: https://en.wikipedia.org/wiki/Cogeneration

[369] Information on the number of Energy Utilities operating in the United States. https://www.publicpower.org/system/files/documents/2018-Public-Power-Statistical-Report-Updated.pdf

[370] Federal Energy Regulatory Commission. *"Electric Power Markets."* https://www.ferc.gov/market-assessments/mkt-electric/overview.asp

³⁷¹ Energy Information Administration. *"What is U.S. electricity generation by energy source?"* https://www.eia.gov/tools/faqs/faq.php?id=427&t=3

³⁷² General Electric. *"GE Global Power Plant Efficiency Analysis."* https://www.ge.com/reports/wp-content/themes/ge-reports/ge-power-plant/dist/pdf/GE%20Global%20Power%20Plant%20Efficiency%20Analysis.pdf
Hosted on-site: https://nextgiantleap.org/sites/default/files/source_files/ ge_efficiency.pdf

National Petroleum Council. *"Electric Generation Efficiency."* 18 July, 2007. http://www.npc.org/Study_Topic_Papers/4-DTG-ElectricEfficiency.pdf
Hosted on-site: https://nextgiantleap.org/sites/default/files/source_files/ 4-DTG-ElectricEfficiency.pdf

³⁷³ Background reading on cogeneration within contemporary power-generating systems: https://en.wikipedia.org/wiki/Cogeneration#Types_of_plants

³⁷⁴ Background reading on the Palo Verde Nuclear Generating Station. https://en.wikipedia.org/wiki/Palo_Verde_Nuclear_Generating_Station

³⁷⁵ National Renewable Energy Laboratory. *"REopt Optimizes Nuclear-Renewable Hybrid Energy Systems."* https://reopt.nrel.gov/projects/case-study-nuclear.html

³⁷⁶ Oak Ridge National Laboratory. *"Update on SINAP TMSR Research."* https://public.ornl.gov/conferences/msr2016/docs/Presentations/MSR2016-day1-15-Hongjie-Xu-Update-on-SINAP-TMSR-Research.pdf#p3

³⁷⁷ South China Morning Post. *"China hopes cold war nuclear energy tech will power warships, drones."* S. Chen. https://www.scmp.com/news/china/society/article/2122977/china-hopes-cold-war-nuclear-energy-tech-will-power-warships

China Daily. *"China among the countries looking to thorium as new nuclear fuel."* K. Wilson. 25 October, 2018.
http://www.chinadaily.com.cn/a/201810/25/WS5bd11cf3a310eff3032846d1.html

Next Big Future. *"China spending US$3.3 billion on molten salt nuclear reactors for faster aircraft carriers and in flying drones."* B. Wang. 6 December, 2017.
https://www.nextbigfuture.com/2017/12/china-spending-us3-3-billion-on-molten-salt-nuclear-reactors-for-faster-aircraft-carriers-and-in-flying-drones.html

378 Next Big Future. *"China spending US$3.3 billion on molten salt nuclear reactors for faster aircraft carriers and in flying drones."* B. Wang. 6 December, 2017. https://www.nextbigfuture.com/2017/12/china-spending-us3-3-billion-on-molten-salt-nuclear-reactors-for-faster-aircraft-carriers-and-in-flying-drones.html

379 Next Big Future. *"China spending US$3.3 billion on molten salt nuclear reactors for faster aircraft carriers and in flying drones."* B. Wang. 6 December, 2017. https://www.nextbigfuture.com/2017/12/china-spending-us3-3-billion-on-molten-salt-nuclear-reactors-for-faster-aircraft-carriers-and-in-flying-drones.html

380 Phys.org. *"Thorium reactors may dispose of enormous amounts of weapons-grade plutonium."* Tomsk Polytechnic University. 22 January, 2018. https://phys.org/news/2018-01-thorium-reactors-dispose-enormous-amounts.html

381 Masters, Gilbert (2004). Renewable and efficient electric power systems. New York: Wiley-IEEE Press. https://www.amazon.com/Renewable-Efficient-Electric-Power-Systems/dp/1118140621

382 Background reading on the CHP Directive: https://en.wikipedia.org/wiki/CHP_Directive

383 Government of the United Kingdom. *"Combined Heat and Power Quality Assurance Programme."* Last updated 17 October, 2019. https://www.gov.uk/guidance/combined-heat-power-quality-assurance-programme

384 Government of the United Kingdom. *"Combined Heat and Power Incentives."* Last updated 1 April, 2019. https://www.gov.uk/guidance/combined-heat-and-power-incentives

385 *"Advantages of liquid fluoride thorium reactor in comparison with light water reactor."* 29 April, 2015. AIP Conference Proceedings 1659, 040001 (2015); https://doi.org/10.1063/1.4916861

386 NASA. *"Scientific Consensus: Earth's Climate is Warming."* https://climate.nasa.gov/scientific-consensus/

387 Link to larger concept image of CHP Plant: http://nextgiantleap.org/sites/default/files/bookchapters/cogeneration/energy_plant.jpg

390 NASA. *"The Causes of Climate Change."* Last update 2 October, 2019. https://climate.nasa.gov/causes/

[391] NASA. Global Climate Change. Vital Signs of the Planet. *"Ice Sheets."* https://climate.nasa.gov/vital-signs/ice-sheets/

[392] NASA. *"The Effects of Climate Change."* https://climate.nasa.gov/effects/

[393] The Guardian. *"Americans 'under siege' from climate disinformation – former NASA chief scientist."* H. Devlin. 8 June, 2017. https://www.theguardian.com/science/2017/jun/08/americans-under-siege-from-climate-disinformation-former-nasa-chief-scientist

[394] CDP. *"CDP Carbon Majors Report 2017."* P. Griffin. July, 2017. https://www.cdp.net/en/articles/media/new-report-shows-just-100-companies-are-source-of-over-70-of-emissions

[395] Mining Congress Journal. *"Air pollution and the coal industry."* J. Allen Overton, Jr. 1966. P. 56. https://www.documentcloud.org/documents/6554117-Mining-Congress-Journal-August-1965-Air.html#document/p6/a536518

[396] NASA. *"Graphic: Carbon dioxide hits new high."* https://climate.nasa.gov/climate_resources/7/graphic-carbon-dioxide-hits-new-high/

[397] Forbes Magazine. *"Three Reasons Oil Will Continue To Run The World."* J. Clemente. 19 April, 2015. https://www.forbes.com/sites/judeclemente/2015/04/19/three-reasons-oil-will-continue-to-run-the-world/#530c4b7143f9

[398] Nature science journal. *"Sucking carbon dioxide from air is cheaper than scientists thought."* J. Tollefson. 7 June, 2018. https://www.nature.com/articles/d41586-018-05357-w

[399] Nature science journal. *"Commercial boost for firms that suck carbon from air."* D. Cressey. 14 October, 2015. https://www.nature.com/news/commercial-boost-for-firms-that-suck-carbon-from-air-1.18551

[400] D. Keith, G. Holmes, D. St. Angelo, K. Heidel. *"A Process for Capturing CO2 from the Atmosphere."* 7 June, 2018. https://doi.org/10.1016/j.joule.2018.05.006 https://www.cell.com/joule/fulltext/S2542-4351(18)30225-3

[401] Climate Engineering. *"Our technology."* https://carbonengineering.com/our-technology/

[402] Climeworks corporation. *"Our Products – Climeworks Plant."* https://www.climeworks.com/our-products/

[403] California Institute of Technology. *"Carbon Conversion."* 3 August, 2017. https://www.caltech.edu/about/news/carbon-conversion-79223

[404] DesignNews. *"CO2 Converted to Solid Carbon."* K. Clemens. 1 March, 2019. https://www.designnews.com/batteryenergy-storage/co2-converted-solid-carbon/206299923860346

[405] MIT Technology Review. *"Startups looking to suck CO2 from the air are suddenly luring big bucks."* J. Temple. 2 May, 2019. https://www.technologyreview.com/s/613447/startups-looking-to-suck-c02-from-the-air-are-suddenly-luring-big-bucks/

[406] Climate Engineering. *"Uses."* https://carbonengineering.com/uses/

[407] Phys.org. *"Pressure mounts on aviation industry over climate change."* S. Wolf, M. Abbugao. 9 June, 2019. https://phys.org/news/2019-06-pressure-mounts-aviation-industry-climate.html

[408] Nature science journal. *"Sucking carbon dioxide from air is cheaper than scientists thought."* J. Tollefson. 7 June, 2018. https://www.nature.com/articles/d41586-018-05357-w

[409] See Appendix Section A4 for a pricing breakdown of Universal Energy.

[410] Faucet Boss. *"What is the average flow rate of faucets?"* 27 November, 2018. https://www.faucetboss.com/faucet-flow-rate/

[411] Faucet Boss. *"What is the average flow rate of faucets?"* 27 November, 2018. https://www.faucetboss.com/faucet-flow-rate/

[412] USGS. *"What are zebra muscles and why should we care about them?"* https://www.usgs.gov/faqs/what-are-zebra-mussels-and-why-should-we-care-about-them?qt-news_science_products=0#qt-news_science_products

[413] Background reading on the Bath County Pumped Storage Station: https://en.wikipedia.org/wiki/Bath_County_Pumped_Storage_Station

[414] See Appendix, A4, on page 315 for a breakdown of power-generating capacity and costs thereof.

[415] National Academies of Sciences, Engineering, Medicine. National Academies Press. Assessing and Managing the Ecological Impacts of Paved Roads. Chapter 2: History and Status of the U.S. Road System. P. 41 https://www.nap.edu/read/11535/chapter/4

⁴¹⁶ Link to Lucid Energy corporation: http://lucidenergy.com/

⁴¹⁷ Slate magazine. *"Lake Mead Before and After the Epic Drought."* E. Holthaus. 25 July, 2014. http://www.slate.com/articles/technology/future_tense/2014/07/lake_mead_before_and_after_colorado_river_basin_losing_water_at_shocking.html.

⁴¹⁸ Engineering Toolbox. *"Specific Heat of Some Liquids and Fuels."* https://www.engineeringtoolbox.com/specific-heat-fluids-d_151.html

⁴¹⁹ Background reading on thermal energy storage: https://en.wikipedia.org/wiki/Thermal_energy_storage

⁴²⁰ Marlow Engineering. *"Technical Data Sheet for EHBMS."* https://cdn2.hubspot.net/hubfs/547732/Data_Sheets/EHBMS.pdf

⁴²¹ See Appendix, A4, on page 315 for a breakdown of power-generating capacity and costs thereof.

⁴²² Engineering toolbox is a fantastic site to help calculate various engineering-related formulas. In the case of water: https://www.engineeringtoolbox.com/energy-storage-water-d_1463.html

⁴²³ Dr. Dickson Despommier. *"The Vertical Farm."* 25 October, 2011. http://www.verticalfarm.com/

⁴²⁴ AG Funder News "The Economics of Local Vertical and Greenhouse Farming Are Getting Competitive." P. Tasgal, 3 April, 2019. https://agfundernews.com/the-economics-of-local-vertical-and-greenhouse-farming-are-getting-competitive.html

⁴²⁵ Tesla's automotive factory is 5.3 million square feet (~492,380 square meters). https://www.tesla.com/factory.

⁴²⁶ Government of the United Kingdom. Food Standards Agency. *"Arsenic in Rice."* https://www.food.gov.uk/safety-hygiene/arsenic-in-rice

National Institute of Health. *"Effect of growth promotants on the occurrence of endogenous and synthetic steroid hormones on feedlot soils and in runoff from beef cattle feeding operations."* Multiple authors. https://www.ncbi.nlm.nih.gov/pubmed/22242694

⁴²⁷ Vox Magazine. *"This company wants to build a giant indoor farm next to every major city in the world."* D. Roberts. 11 April, 2018. https://www.vox.com/energy-and-environment/2017/11/8/16611710/vertical-farms

[428] Vox Magazine. *"This company wants to build a giant indoor farm next to every major city in the world."* D. Roberts. 11 April, 2018. https://www.vox.com/energy-and-environment/2017/11/8/16611710/vertical-farms

[429] Agriculture, Technology and Investment News. *"Vertical farming startup plenty acquires bright aggrotech to scale."* E. Cosgrove. 13 June, 2017. https://agfundernews.com/breaking-vertical-farming-startup-plenty-acquires-bright-agrotech-scale.html

[430] Vox Magazine. *"This company wants to build a giant indoor farm next to every major city in the world."* D. Roberts. 11 April, 2018. https://www.vox.com/energy-and-environment/2017/11/8/16611710/vertical-farms

[431] Vox Magazine. *"This company wants to build a giant indoor farm next to every major city in the world."* D. Roberts. 11 April, 2018. https://www.vox.com/energy-and-environment/2017/11/8/16611710/vertical-farms

[432] Centers for Disease Control and Prevention. *"Multistate Outbreak of Listeriosis Linked to Whole Cantaloupes from Jensen Farms, Colorado."* 27 August, 2012.

[433] United States Department of Agriculture. *"Summary of Recall Cases in Calendar Year 2018."* https://www.fsis.usda.gov/wps/portal/fsis/topics/recalls-and-public-health-alerts/recall-summaries

[434] Food and Drug Administration. *"Consumer Info About Food From Genetically Engineered Plants."* 1 April, 2018. https://www.fda.gov/food/food-new-plant-varieties/consumer-info-about-food-genetically-engineered-plants

[435] Wired Magazine. *"Monsanto's newest GM crops may create more problems than they solve."* B. Keim. 2 February, 2015. https://www.wired.com/2015/02/new-gmo-crop-controversy/

[436] Michigan State University Extension. *"Urban farming practices developed in France in 1850 still are used in cities today."* R. Bell. 20 June, 2013. http://msue.anr.msu.edu/news/urban_farming_practices_developed_in_france_in_1850_still_are_used_in_citie

[437] Links to vertical farms, respectively in London, Chicago, Milan and Newark:
http://growup.org.uk/
http://chicagotonight.wttw.com/2014/06/23/vertical-farming-s-rise-chicago
https://inhabitat.com/bosco-verticale-in-milan-will-be-the-worlds-first-vertical-forest/
http://aerofarms.com/

438 AG Funder News "The Economics of Local Vertical and Greenhouse Farming Are Getting Competitive." P. Tasgal, 3 April, 2019. https://agfundernews.com/the-economics-of-local-vertical-and-greenhouse-farming-are-getting-competitive.html

439 Link to Bosco Verticale towers in Milan: https://inhabitat.com/bosco-verticale-in-milan-will-be-the-worlds-first-vertical-forest/

440 Dwell Magazine. *"New York City Passes Bill Requiring Green Roofs on New Buildings."* J. Tuohy. 23 April, 2019. https://www.dwell.com/article/new-york-city-requires-green-roofs-on-new-buildings-ede4deb8

441 Thomas Publishing Company. *"How Cities are Driving Growth and the Green Roofing Market."* 12 February, 2019. https://www.thomasnet.com/insights/how-cities-are-driving-growth-in-the-green-roofing-market/

442 Scientific American. *"Why are asthma rates soaring?"* V. Greenwood. 1 April, 2011. https://www.scientificamerican.com/article/why-are-asthma-rates-soaring/

443 Background multimedia on China's "Forest City." BBC. *"China's Forest city."* 5 July, 2017. http://www.bbc.com/news/av/world-asia-40498186/china-s-forest-city

444 Background multimedia on China's "Forest City." BBC. *"China's Forest city."* 5 July, 2017. http://www.bbc.com/news/av/world-asia-40498186/china-s-forest-city

445 Biofuels Digest. *"10 Top Strategic Investors in Biofuels & Materials."* 6 June, 2011. http://www.biofuelsdigest.com/bdigest/2011/06/06/10-top-strategic-investors-in-biofuels-materials/

446 Background reading on bioplastics: https://en.wikipedia.org/wiki/Bioplastic

447 Link to Algenol corporation: http://algenol.com/

448 Washington Post. *"A Promising Oil Alternative: Algae Energy."* 6 January, 2008. http://www.washingtonpost.com/wp-dyn/content/article/2008/01/03/AR2008010303907.html

449 Background reading on chlorella: https://en.wikipedia.org/wiki/Chlorella

450 Background reading on chlorella: https://en.wikipedia.org/wiki/Chlorella

[451] Belasco, Warren (July 1997). "Algae Burgers for a Hungry World? The Rise and Fall of Chlorella Cuisine". Technology and Culture. 38 (3): 608–34. https://www.jstor.org/stable/3106856

[452] Background reading on GMO controversies: https://en.wikipedia.org/wiki/Genetically_modified_food_controversies

[453] Background reading on Glyphosate: https://en.wikipedia.org/wiki/Glyphosate

[454] Background reading on glyphosate. https://en.wikipedia.org/wiki/Glyphosate

[455] Smithsonian Magazine. *"Scientists Turn Algae into Crude Oil in Less Than an Hour."* T. Nguyen. 31 December, 2013. https://www.smithsonianmag.com/innovation/scientists-turn-algae-into-crude-oil-in-less-than-an-hour-180948282/?no-ist

[456] B. Mooney. *"The second green revolution? Production of plant-based biodegradable plastics."* Biochem J (2009) 418 (2): 219–232. https://doi.org/10.1042/BJ20081769

Background reading on percent yield: https://en.wikipedia.org/wiki/Yield_(engineering)

[457] L. Christenson, R. Sims, *"Production and harvesting of microalgae for wastewater treatment, biofuels, and bioproducts."* Journal of Biotechnology Advances. Vol 29, Issue 6. December 2011. Pp 686-702. https://doi.org/10.1016/j.biotechadv.2011.05.015

Background reading on polymerization: https://en.wikipedia.org/wiki/Polymerization

[458] BBC. *"E. Coli Bacteria 'Can Produce Diesel Biofuel'."* R. Morelle. 22 April, 2013. http://www.bbc.com/news/science-environment-22253746

[459] MIT Technology Review. *"Making Gasoline from Bacteria."* N. Savage. 1 August, 2017. https://www.technologyreview.com/s/408334/making-gasoline-from-bacteria/

[460] V. Lorenzo. "Cleaning up behind us The potential of genetically modified bacteria to break down toxic pollutants in the environment." EMBO Rep. 2001 May 15; 2(5): 357–359. doi: 10.1093/embo-reports/kve100. 15 May, 2011. https://www.ncbi.nlm.nih.gov/pmc/articles/PMC1083894/

[461] Background reading on ocean garbage: https://en.wikipedia.org/wiki/Marine_debris

[462] J. Hopewell, R. Dvorak, E. Kosior. *"Plastics recycling: challenges and opportunities."* Philos Trans R Soc Lond B Biol Sci. 2009 Jul 27; 364(1526): 2115–2126. doi: 10.1098/rstb.2008.0311. https://www.ncbi.nlm.nih.gov/pmc/articles/PMC2873020/

Background reading on percent yield in chemistry. https://en.wikipedia.org/wiki/Yield_(chemistry). P

Plastics Technology. *"What's Your Production Efficiency?"* R. Hirschfeld. 7 January, 1999. https://www.ptonline.com/articles/what's-your-production-efficiency

[463] Plastics Technology. *"What's Your Production Efficiency?"* R. Hirschfeld. 7 January, 1999. https://www.ptonline.com/articles/what's-your-production-efficiency

[464] History of computing hardware (1960s – present): https://en.wikipedia.org/wiki/History_of_computing_hardware_(1960s%E2%80%93present)

[465] Chemical Engineering Magazine. *"Artificial Intelligence: A New Reality for Chemical Engineers."* M. Bailey. 1 February, 2019. https://www.chemengonline.com/artificial-intelligence-new-reality-chemical-engineers/

[466] Molecule.one – chemical synthesis company using machine learning.

[467] D-Wave systems. Background information on quantum computing. https://www.dwavesys.com/quantum-computing

[468] D-Wave systems. Background information on quantum computing. https://www.dwavesys.com/quantum-computing

[469] TechCrunch. *"The reality of quantum computing could be just three years away."* J. Shieber. 7 September, 2018. https://techcrunch.com/2018/09/07/the-reality-of-quantum-computing-could-be-just-three-years-away/

[470] Instructions per second: https://en.wikipedia.org/wiki/Instructions_per_second#Timeline_of_instructions_per_second

[471] Futurism. *"AI is learning quantum mechanics to design new molecules."* D. Robitzski. 22 November, 2019. https://futurism.com/the-byte/ai-quantum-mechanics-design-molecules

[472] Scientific American. *"Biofuel from Bacteria."* D. Biello. 1 April, 2010. https://www.scientificamerican.com/article/biofuel-from-bacteria/

Solar Energy Research Institute. *"Formation of Hydrocarbons by Bacteria and Algae."* T. Tornabene. December, 1980. https://www.nrel.gov/docs/legosti/old/999.pdf#p13

[473] N. Browning, R. Ramakrishnan, et al. *"Genetic Optimization of Training Sets for Improved Machine Learning Models of Molecular Properties."* J. Phys. Chem. Lett. 2018, 9, 22, 6480-6488 October 29, 2018 https://doi.org/10.1021/acs.jpclett.8b02956

[474] Scientific American. *"Making Plastic as Strong as Steel."* L. Greenemeier. 11 October, 2007. https://www.scientificamerican.com/article/making-plastic-as-strong/

[475] Link to Line-X product lines: http://www.linex.com/line-x-for-manufacturers

[476] Video demonstrations of Line-X spray: https://www.youtube.com/results?search_query=linex+demonstration

[477] Background reading on FR-4. https://en.wikipedia.org/wiki/FR-4

[478] Monarch Metal Fabrication. *"Types of Metal Strength."* C. Smith. 24 August, 2016. https://www.monarchmetal.com/blog/types-of-metal-strength/

[479] Plastics international datasheet on FR-4. https://www.plasticsintl.com/datasheets/Phenolic_G10_FR4.pdf ***BROKEN LINK

[480] Axion Structural Innovations. Background information on synthetic wood. http://www.axionsi.com/

[481] Journal of Materials Science and Nanomaterials. *"High-Temperature Structure Materials Beyond Nickel Base Supperalloy."* G. Ouyang. 15 October, 2017. https://www.omicsonline.org/open-access/high-temperature-structure-materials-beyond-nickel-base-superalloy.pdf

[482] Popular Mechanics. *"Scientists Invent a New Steel as Strong as Titanium."* W. Herkewitz. 4 February, 2015. https://www.popularmechanics.com/technology/news/a13919/new-steel-alloy-titanium/

[483] New Scientist. *"New alloys could lead to next generation of nuke plant metals."* J. Emspak. 18 March, 2016. https://www.newscientist.com/article/2081605-new-alloys-could-lead-to-next-generation-of-nuke-plant-metals/

[484] Columbia University News. *"Columbia Engineers Prove Graphene is the Strongest Material."* 21 July, 2008. http://www.columbia.edu/cu/news/08/07/graphene.html

485 Graphenea corporation. *"Properties of Graphene."* J. Fuente. https://www.graphenea.com/pages/graphene-properties#.XZ7FJudKhTY

486 Android Community. *"Samsung Producing Graphene, the Material for Flexible Displays."* N. Swanner. 4 April, 2014. https://androidcommunity.com/samsung-producing-graphene-the-material-for-flexible-displays-20140404/

487 ComputerWorld. *"Graphene sticky notes may offer 32GB capacity you can write on."* L. Mearian. 18 December, 2013. https://www.computerworld.com/article/2486937/graphene-sticky-notes-may-offer-32gb-capacity-you-can-write-on.html

488 MIT Technology Review. *"Graphene Antennas Would Enable Terabit Wireless Downloads."* D. Talbot. 5 March, 2013. https://www.technologyreview.com/s/511726/graphene-antennas-would-enable-terabit-wireless-downloads/

489 MIT Technology Review. *"Graphene Antennas Would Enable Terabit Wireless Downloads."* D. Talbot. 5 March, 2013. https://www.technologyreview.com/s/511726/graphene-antennas-would-enable-terabit-wireless-downloads/

490 MedGadget. *"Graphene: The Next Medical Revolution."* R. Peleg. 20 May, 2015. http://www.medgadget.com/2015/05/graphene-next-medical-revolution.html

491 Q. Ke. J. Wang. *"Graphene-based materials for supercapacitor electrodes – a review."* Journal of Materiomics, Volume 2, Issue 1. March, 2016. https://doi.org/10.1016/j.jmat.2016.01.001

492 Middle East Technical University, Turkey. Nanonotes. *"Lithium-Ion batteries vs. graphene batteries."* O. Kutun. 23 July, 2019. https://blog.metu.edu.tr/e207651/2019/07/23/lithium-ion-batteries-vs-graphene-batteries/

493 HyperTextbook. *"How thick is a sheet of printer paper"?* https://hypertextbook.com/facts/2001/JuliaSherlis.shtml

494 HyperTextbook. *"How thick is a sheet of printer paper"?* https://hypertextbook.com/facts/2001/JuliaSherlis.shtml

495 Stack Sports. *"How many acres in a football field?"* https://www.stack.com/a/how-many-acres-is-a-football-field

[496] Nanotechnology Journal. Volume 28, Number 44. *"Graphene supercapacitor with both high power and energy density."* H. Yang, S. Kannappan, A. Pandian, J. Hyung Jang, Y. Sung Lee and W. Lu. 4 October, 2017. http://iopscience.iop.org/article/10.1088/1361-6528/aa8948 Link two: https://www.ncbi.nlm.nih.gov/pubmed/28854156

[497] Union of Concern Scientists. "Electric Vehicles, batteries, cobalt, and rare earth metals." J. Goldman. 25 October, 2017. https://blog.ucsusa.org/josh-goldman/electric-vehicles-batteries-cobalt-and-rare-earth-metals

[498] Scientific American. *"More Recycling Won't Solve Plastic Pollution."* M. Wilkins. 6 July, 2018. https://blogs.scientificamerican.com/observations/more-recycling-wont-solve-plastic-pollution/

[499] WIRED magazine. *"High-powered Plasma Turns Garbage Into Gas."* D. Wolman. 20 January, 2012. https://www.wired.com/2012/01/ff_trashblaster/

[500] Background reading on gasification slag. https://en.wikipedia.org/wiki/Slag#Modern_uses

[501] Waldheim Consulting. *"Industrial Biomass Gasification Activities in Sweden 1997-2009."* ANNEX 1 to IEA Biomass Agreement Task 33 Country Report Sweden 2012

[502] Government of Sweden. *"The Swedish Recycling Revolution."* No date or author provided. https://sweden.se/nature/the-swedish-recycling-revolution/

[503] Table data source: Limpopo Eco-Industrial Park. http://limpopoecoindustrialpark.com/

[504] National Geographic. "Ocean Gyre." https://www.nationalgeographic.org/encyclopedia/ocean-gyre/

[505] Reuters. *"World's fish consumption unsustainable, U.N. warns."* T. Win. 9 July, 2018. https://www.reuters.com/article/us-global-fisheries-hunger/worlds-fish-consumption-unsustainable-u-n-warns-idUSKBN1JZ0YA

[506] Background reading on the Ocean Cleanup Project. https://www.theoceancleanup.com/

[507] Background reading on Circular Economies: https://en.wikipedia.org/wiki/Circular_economy

[508] Geissdoerfer, Martin; Savaget, Paulo; Bocken, Nancy M. P.; Hultink, Erik Jan (2017-02-01). *"The Circular Economy – A new sustainability paradigm?"* Journal of Cleaner Production. 143: 757–768. doi:10.1016/j.jclepro.2016.12.048.

509 Ellen MacArthur Foundation. *"Towards the Circular Economy: an economic and business rationale for an accelerated transition."* Vol 1. Release date 2013. https://www.ellenmacarthurfoundation.org/assets/downloads/publications/Ellen-MacArthur-Foundation-Towards-the-Circular-Economy-vol.1.pdf

510 Background reading on 3D printing. https://en.wikipedia.org/wiki/3D_printing

511 Singularity Hub. *"New progress in the biggest challenge with 3D printed organs."* E Gent. 7 May, 2019. https://singularityhub.com/2019/05/07/new-progress-in-the-biggest-challenge-with-3d-printed-organs/

512 University of Washington. "Transforming titanium with 3D printing." C. Yates. 28 October, 2019. https://www.engr.washington.edu/news/article/2019-10-28/transforming-titanium-with-3D-printing
Advanced Science News. "A new copper-titanium alloy enables 3D printing." M. Grolms. 3 February, 2020. https://www.advancedsciencenews.com/a-new-copper-titanium-alloy-enables-3d-printing/
MarkForged Corporation. "Complete 3D Metal Printing Solution." https://markforged.com/metal-x/

513 MarkForged Corporation. "Complete 3D Metal Printing Solution." https://markforged.com/metal-x/. See also:
University of Washington. "Transforming titanium with 3D printing." C. Yates. 28 October, 2019. https://www.engr.washington.edu/news/article/2019-10-28/transforming-titanium-with-3D-printing
Advanced Science News. "A new copper-titanium alloy enables 3D printing." M. Grolms. 3 February, 2020. https://www.advancedsciencenews.com/a-new-copper-titanium-alloy-enables-3d-printing/

514 Science Direct. *"Selective Laser Melting."* Myriad articles of varying dates. https://www.sciencedirect.com/topics/materials-science/selective-laser-melting

515 Background reading on prefabrication: https://en.wikipedia.org/wiki/Prefabrication

516 IBISWorld. *"Prefabricated Home Manufacturing In the US industry trends (2015-2020).* January, 2020. https://www.ibisworld.com/united-states/market-research-reports/prefabricated-home-manufacturing-industry/

517 Fixr magazine. *"How Much Does it Cost to Build a Single-Family Home?"* https://www.fixr.com/costs/build-single-family-house

[518] Wired Magazine. "How Boeing builds a 737 in just 9 days." J. Stewart. 27 September, 2016. https://www.wired.com/2016/09/boeing-builds-737-just-nine-days/
Timelapse of 737 construction: https://www.youtube.com/watch?v=liZ0WEEsuz4

[519] The Post and Courier. "Pace of 787 Dreamliner production quickens and Boeing's North Charlson campus." D. Wren. 1 April, 2018.
https://www.postandcourier.com/business/pace-of-dreamliner-production-quickens-at-boeing-s-north-charleston/article_02b584ea-3384-11e8-a3d4-ff9ca36dc590.html

Boeing corp information on Everett production facility.
http://www.boeing.com/company/about-bca/everett-production-facility.page.

[520] Seattle Times. "Flawed analysis, failed oversight: How Boeing, FAA certified the suspect 737 MAX flight control system." D. Gates. 17 March, 2019.
https://www.seattletimes.com/business/boeing-aerospace/failed-certification-faa-missed-safety-issues-in-the-737-max-system-implicated-in-the-lion-air-crash/

Forbes Magazine. "Ex-Boeing 737 MAX Engineer says team was pressured to cut costs as grounding continues." I. Togoh. 29 July, 2019.
https://www.forbes.com/sites/isabeltogoh/2019/07/29/ex-boeing-engineer-says-workers-were-pressured-to-keep-costs-down-amid-737-max-grounding/#59d12a1d2e9a
New York Times. "Boeing 737 MAX factory was plagued with problems, whistle-blower says." D Gates. 29 January, 2020. https://www.nytimes.com/2019/12/09/business/boeing-737-max-whistleblower.html

[521] WIRED magazine. *"Meet The Man Who Built a 30-Story Building in 15 days."* L. Hilgers. 25 September, 2012. https://www.wired.com/2012/09/broad-sustainable-building-instant-skyscraper/

[522] According to Budweiser and several sources, it takes Anheuser-Busch approximately 30 days to brew a bottle of beer. https://www.cnbc.com/2008/07/02/Brewing-the-King-of-Beers.html / https://thefederalist.com/2015/02/18/making-the-case-for-budweiser/

[523] PSE Consulting Engineers. *"What happens to used shipping containers?"*
https://www.structure1.com/projects/shipping-container-homes/what-happens-to-used-shipping-containers/

[524] Containerhomeplans.org. *"The Cheapest 5 Shipping Container Homes Ever Built."* 15 July, 2015. https://www.containerhomeplans.org/2015/07/the-cheapest-5-shipping-container-homes-ever-built/

525 Curbed magazine. *"This Ravishing Mod Pad is Actually a $40k Container Home."* J. Xie. 30 July, 2015. http://www.curbed.com/2015/7/30/9935524/best-shipping-container-homes

526 Singularity Hub. *"This 3D printed house goes up in a day for under $10,000."* V. Ramirez. 18 March, 2018. https://singularityhub.com/2018/03/18/this-3d-printed-house-goes-up-in-a-day-for-under-10000

527 Singularity Hub. *"This 3D printed house goes up in a day for under $10,000."* V. Ramirez. 18 March, 2018. https://singularityhub.com/2018/03/18/this-3d-printed-house-goes-up-in-a-day-for-under-10000

528 Singularity Hub. *"This 3D printed house goes up in a day for under $10,000."* V. Ramirez. 18 March, 2018. https://singularityhub.com/2018/03/18/this-3d-printed-house-goes-up-in-a-day-for-under-10000

529 Singularity Hub. *"See how this house was 3D printed in Just 24 hours."* V. Ramirez. 5 March, 2017. https://singularityhub.com/2017/03/05/watch-this-house-get-3d-printed-in-24-hours/

530 Inhabitat magazine. *"Dubai debuts world's first fully 3D-printed building."* C. DiStasio. 24 May, 2016. https://inhabitat.com/dubai-debuts-worlds-first-fully-3d-printed-building/

531 Singularity Hub. *"This 3D printed house goes up in a day for under $10,000."* V. Ramirez. 18 March, 2018. https://singularityhub.com/2018/03/18/this-3d-printed-house-goes-up-in-a-day-for-under-10000

532 PBS. *"Breaking poverty: Crime, poverty often linked."* J. Mitchell, The Philadelphia Tribune. 18 September, 2018. https://whyy.org/articles/breaking-poverty-crime-poverty-often-linked/

533 Congress for the New Urbanism. *"The Little House That Could."* R. Steuteville. 3 February, 2016. https://www.cnu.org/publicsquare/little-house-could

534 Disasters Emergency Committee. *"Haiti Earthquake Facts and Figures."* No date or author provided. https://www.dec.org.uk/articles/haiti-earthquake-facts-and-figures

535 NBC News Investigations. *"What Does Haiti Have to Show for $13 Billion in Earthquake Aid?"* T. Connor, H. Rappleye and E. Angulo. 12 January, 2015. https://www.nbcnews.com/news/investigations/what-does-haiti-have-show-13-billion-earthquake-aid-n281661

536 Y Charts. *"U.S. Existing Single-Family Home Average Sales Price."* https://ycharts.com/indicators/us_existing_singlefamily_home_average_sales_price

[537] CNN Money. *"It's Getting More Expensive to be a Renter."* K. Vasel. 21 May, 2015. http://money.cnn.com/2015/05/21/real_estate/rent-prices-rising/

[538] Endhomelessness.org. "Homelessness In America" Facts, figures and statistics. https://endhomelessness.org/homelessness-in-america/

[539] Huffington Post. *"U.S. Spending Historic Amount Fighting Homelessness, and it's Working: Report."* R. Couch. 3 April, 2015. https://www.huffingtonpost.com/2015/04/03/homelessness-report-2015_n_6987576.html

[540] U.S. Department of Defense. *"DOD Releases Fiscal Year 2020 Budget Proposal."* 19 March, 2020. https://www.defense.gov/Newsroom/Releases/Release/Article/1782623/dod-releases-fiscal-year-2020-budget-proposal/

[541] National Coalition for the Homeless. *"Substance Abuse and Homelessness."* July 2009. https://www.nationalhomeless.org/factsheets/addiction.pdf

[542] Nature Journal. *"Oldest Homo sapiens fossil claim rewrites our species' history."* E. Callaway. 7 June, 2017. https://www.nature.com/news/oldest-homo-sapiens-fossil-claim-rewrites-our-species-history-1.22114

[543] American Society of Civil Engineers. *"America's Infrastructure Grades Remain Near Failing."* 9 March, 2017. https://www.asce.org/templates/press-release-detail.aspx?id=24013

[544] Craftsman Books. *"2017 National Building Cost Manual"* 41st edition. Edited by B. Moselle. https://www.craftsman-book.com/media/static/previews/2017_NBC_book_preview.pdf

[545] CleanTechnica. *"Scandanavia is Home to Heavy-Duty Electric Construction Equipment & Truck Development."* S. Hanley. 30 January, 2018. https://cleantechnica.com/2018/01/30/scandinavia-home-heavy-duty-electric-construction-equipment-truck-development/

[546] Craftsman Books. *"2017 National Building Cost Manual"* 41st edition. Edited by B. Moselle. https://www.craftsman-book.com/media/static/previews/2017_NBC_book_preview.pdf

[547] Government of Maryland. *"William Preston Lane Jr. Memorial (Bay) Bridge."* http://www.mdta.maryland.gov/toll_facilities/wpl.html

[548] Figure sourced according to Google Maps

[549] Time estimation from Google Maps.

[550] Colorado Department of Transportation. *"Eisenhower Tunnel Traffic Counts."* https://www.codot.gov/travel/eisenhower-tunnel/eisenhower-tunnel-traffic-counts.html

[551] Colorado Department of Transportation. Eisenhower tunnel total traffic counts. https://www.codot.gov/travel/eisenhower-tunnel/eisenhower-tunnel-traffic-counts.html

[552] Femern Inc. https://femern.com/en/News-and-press/2018/December/German-approval-of-the-Fehmarnbelt-tunnel-ready-to-be-signed

[553] Information on Fehmarn Belt Fixed Link:
Railway Gazette. *"Fehmarn Belt Tunnel Contract Awards Authorized."* 4 March, 2016. http://www.railwaygazette.com/news/infrastructure/single-view/view/fehmarn-belt-tunnel-contract-awards-authorised.html

Information on Sheikh Rashid bin Saeed Crossing:

Industry Tap. *"World's Longest and Tallest Single-Arch Bridge Coming to Dubai in 2015."* D. Schilling. 7 November, 2013.

[554] Background reading on Channel rail tunnel, Seikan tunnel and Gotthard Base tunnel:
https://en.wikipedia.org/wiki/Channel_Tunnel
https://en.wikipedia.org/wiki/Seikan_Tunnel
https://en.wikipedia.org/wiki/Gotthard_Base_Tunnel

[555] Background reading on the English Channel:
https://en.wikipedia.org/wiki/English_Channel#Nature

[556] Background reading on submerged floating tunnels:
https://en.wikipedia.org/wiki/Submerged_floating_tunnel

[557] WIRED magazine. *"Yes, a 'Submerged Floating Bridge' is a Reasonable Way to Cross a Fjord.'* A. Marshall. 14 July, 2016. https://www.wired.com/2016/07/submerged-floating-bridge-isnt-worst-idea-norways-ever/#slide-2

[558] Background reading on internet backbones:
https://en.wikipedia.org/wiki/Internet_backbone

[559] Freepress. *"Save the Internet."* Background information on Net Neutrality. https://www.savetheinternet.com/net-neutrality

Vice Magazine. *"Republican Anti-Net Neutrality Crusade Advances in Congress."* S. Gustin. 11 February, 2016. https://motherboard.vice.com/en_us/article/jpgm83/republicans-in-congress-push-back-against-net-neutrality

[560] Fiber Optic Solutions. *"How fast fiber optic cable speed is."* 26 April, 2018. http://www.fiber-optic-solutions.com/fast-fiber-optic-cable-speed.html

[561] Inhabitat Magazine. *"One9: Nine-Story Prefab Apartment Tower was Installed in Just Five Days."* L. Wang. 25 July, 2014. https://inhabitat.com/one9-nine-story-prefab-apartment-tower-was-installed-in-just-five-days/?variation=d

[562] Kansas City Star. "MAC Properties Opens New Imported Modular Apartments in Midtown KC". D. Stafford. 21 June, 2017. http://centric.build/kcstar-2017june21/

[563] National Association of Home Builders. *"How Long Does it Take to Build a Single-Family Home?"* N. Zhao. 17 August, 2015. http://eyeonhousing.org/2015/08/how-long-does-it-take-to-build-a-single-family-home/

[564] CityMetric. *"Three Million People Move to Cities Every Week. So How Can Cities Plan for Migrants?"* M. Collyer. 3 December, 2015. http://www.citymetric.com/skylines/three-million-people-move-cities-every-week-so-how-can-cities-plan-migrants-1546

[565] New York Post. *"New York's Modular Building Revolution is Here."* S. Lubell. 13 September, 2018. https://nypost.com/2018/09/13/new-yorks-modular-building-revolution-is-here/

[566] Wired Magazine "The World's Tallest Modular Building May Teach Cities to Build Cheaper Housing." E. Stinton Design. 21 November, 2016. https://www.wired.com/2016/11/cities-can-learn-worlds-tallest-modular-building/

[567] Background reading on megacities: https://en.wikipedia.org/wiki/Megacity

[568] Background reading on quality of life indexes. https://en.wikipedia.org/wiki/Where-to-be-born_Index

[569] Background reading on wealth inequality in the United States: https://en.wikipedia.org/wiki/Wealth_inequality_in_the_United_States

570 U.N. *"World's Population Increasingly Urban With More Than Half Living in Urban Areas."* 10 July, 2014. http://www.un.org/en/development/desa/news/population/world-urbanization-prospects-2014.html

571 U.N. *"World's Population Increasingly Urban With More Than Half Living in Urban Areas."* 10 July, 2014. http://www.un.org/en/development/desa/news/population/world-urbanization-prospects-2014.html

572 Background reading on self-driving car initiatives: https://en.wikipedia.org/wiki/Self-driving_car

573 Fast Company Magazine. *"Humans were to blame in Google self-driving car crash, police say."* 5 April, 2018. https://www.fastcompany.com/40568609/humans-were-to-blame-in-google-self-driving-car-crash-police-say

574 Forbes Magazine. *"Uber will resume testing self-driving cars."* D. Silver. 3 November, 2018. https://www.forbes.com/sites/davidsilver/2018/11/03/uber-will-resume-self-driving-car-testing-in-pennsylvania/#12b330b63d7e

575 Federal Highway Administration. *Office of Highway Policy Information – Highway Statistics 2016.* https://www.fhwa.dot.gov/policyinformation/statistics/2016/vm1.cfm

576 Forbes Magazine, citing Carinsurance.com. *"How Many Times Will You Crash Your Car?"* D. Toups. 27 July, 2011. https://www.forbes.com/sites/moneybuilder/2011/07/27/how-many-times-will-you-crash-your-car/

577 CDC. *"Impaired Driving: Get the Facts."* No author or publication date provided. https://www.cdc.gov/motorvehiclesafety/impaired_driving/impaired-drv_factsheet.html

578 Background reading on maglev transportation: https://en.wikipedia.org/wiki/Maglev

579 Washington Post. *"Why the United States will never have high-speed rail."* M. McArdle. 12 February, 2019. https://www.washingtonpost.com/opinions/2019/02/13/why-united-states-will-never-have-high-speed-rail/

580 SpaceX. Whitepaper on Hyperloop Alpha. http://www.spacex.com/sites/spacex/files/hyperloop_alpha-20130812.pdf

581 Space-X. Hyperloop Alpha. https://www.spacex.com/sites/spacex/files/hyperloop_alpha.pdf

582 WIRED magazine. *"Hyperloop's First Real Test is a Whooshing Success."* A. Davies. 12 July, 2017. https://www.wired.com/story/hyperloop-one-test-success/

583 Two companies currently working on Hyperloop prototypes:
Virgin Hyperloop One: https://hyperloop-one.com/
Hyperloop Transportation Technologies: https://www.hyperloop.global/

584 Cirrus Aircraft. Innovation and Smart Safety.
https://cirrusaircraft.com/innovation/smart-safety/

585 Dan Johnson Aviation. *"Crumple Zones Coming to Light Aircraft."* 12 August, 2014.
https://www.bydanjohnson.com/crumple-zones-coming-to-light-aircraft/

586 CNET. *"RC Car Transforms Into a Quadcopter."* M. Starr. 28 May, 2013.
https://www.cnet.com/news/rc-car-transforms-into-a-quadcopter/

587 New York Times. *"Think Amazon's Drone Delivery Idea is a Gimmick? Think Again."* F. Manjoo. 10 August, 2016. https://www.nytimes.com/2016/08/11/technology/think-amazons-drone-delivery-idea-is-a-gimmick-think-again.html

588 Flight Deck Fiend. *"Can A Plane Land Automatically?"*
https://www.flightdeckfriend.com/can-a-plane-land-automatically

589 Futurism Magazine. *"Wireless Charing Tech Lets Drones Stay Aloft Indefinitely."* K. Houser. 7 January, 2019. https://futurism.com/drone-charging-mid-flight

590 The Verge. *"REL's Skylon Spaceplane Aims to Take on SpaceX With a Reusable Rocket Design."* J. Emspak. 8 March, 2016.

NASA. *"Audacious & Outrageous: Space Elevators."* 7 September, 2000.
https://science.nasa.gov/science-news/science-at-nasa/2000/ast07sep_1

591 Background reading on the Kardashev Scale.
https://en.wikipedia.org/wiki/Kardashev_scale

592 Background reading on the extensions to the original Kardashev Scale.
https://en.wikipedia.org/wiki/Kardashev_scale#Extensions_to_the_original_scale

Barrow, John (1998). Impossibility: Limits of Science and the Science of Limits. Oxford: Oxford University Press. p. 133. ISBN 978-0198518907.

[593] Background reading on the extensions to the original Kardashev Scale. https://en.wikipedia.org/wiki/Kardashev_scale#Extensions_to_the_original_scale

Barrow, John (1998). Impossibility: Limits of Science and the Science of Limits. Oxford: Oxford University Press. p. 133. ISBN 978-0198518907.

[594] Background reading on Maslow's Hierarchy of Needs. https://www.simplypsychology.org/maslow.html

[595] Space.com. *"How many stars are in the Milky Way?"* E. Howell. 30 March, 2018. https://www.space.com/25959-how-many-stars-are-in-the-milky-way.html

[596] NASA. *"Hubble Reveals Observable Universe Contains 10 Times More Galaxies Than Previously Thought."* 13 October, 2016. https://www.nasa.gov/feature/goddard/2016/hubble-reveals-observable-universe-contains-10-times-more-galaxies-than-previously-thought

[597] Space.com. *"How Many Stars Are In the Universe?"* E. Howell. 18 May, 2017. https://www.space.com/26078-how-many-stars-are-there.html

[598] Breakdown of the U.S. Federal Budget. https://en.wikipedia.org/wiki/United_States_federal_budget

[599] Statista. *"U.S. military spending from 2000 to 2018."* https://www.statista.com/statistics/272473/us-military-spending-from-2000-to-2012/ (don't let link fool you, it goes to 2018).

[600] Brown University, Watson Institute of International and Public Affairs. *"United States Budgetary Costs and Obligations of Post-9/11 Wars through FY:2020."* N. Crawford. 13 November, 2019. https://watson.brown.edu/costsofwar/files/cow/imce/papers/2019/US%20Budgetary%20Costs%20of%20Wars%20November%202019.pdf

[601] Statista. *"Total interest expense on debt held by the public of the United States from 2011 to 2018"*. https://www.statista.com/statistics/246439/interest-expense-on-us-public-debt/

[602] Bloomberg. *"F-35 Program Costs Jump to $406.5 Billion in Latest Estimate."* A. Capaccio. 10 July, 2017. https://www.bloomberg.com/news/articles/2017-07-10/f-35-program-costs-jump-to-406-billion-in-new-pentagon-estimate

[603] Reuters / Associated Press. *"U.S. Army fudged its accounts by trillions of dollars, auditor finds."* 19 August, 2016. S. Paltrow. https://www.reuters.com/article/us-usa-audit-army/u-s-army-fudged-its-accounts-by-trillions-of-dollars-auditor-finds-idUSKCN10U1IG

[604] PBS News Hour. *"This is how Internet speed and price in the U.S. compares to the rest of the world"* H. Yi. 26 April, 2015. https://www.pbs.org/newshour/world/internet-u-s-compare-globally-hint-slower-expensive

[605] Forbes Magazine. *"Holding U.S. Treasury's?" Beware: Uncle Sam can't account for $21 trillion."* L. Kotlikoff. 9 January, 2019. https://www.forbes.com/sites/kotlikoff/2019/01/09/holding-u-s-treasuries-beware-uncle-sam-cant-account-for-21-trillion/

[606] United States Census. *"Household Income: 2017."* G. Guzman. September, 2018. https://www.census.gov/content/dam/Census/library/publications/2018/acs/acsbr17-01.pdf

[607] Tax Foundation. *"The U.S. Tax Burden on Labor, 2019."* R. Bellafiore. https://taxfoundation.org/us-tax-burden-on-labor-2019/. (Note: we use median figures for income because the average is disproportionally skewed based on the ultra-wealthy, yet use average tax basis because it's applied evenly based on income from even IRS brackets, and not all states have the same taxes nor tax rates, making average required for a national analysis).

[608] In the United States, taxes tend to be aggregated in averages as tax rates vary wildly by state and occupation. A self-employed person in coastal California would pay substantially income higher taxes (base rate + 15.3% FICA + up to 12.3% CA income tax) than a W2 employee living in rural Washington State (base rate + 6.2% FICA + 0% income tax). However, in terms of raw income, the average is disproportionally weighed higher due to outlying wage earners ("the 1%"). However, average tax rates do not necessarily reflect this as taxes are both paid on a progressive scale, which does not subject investment income (capital gains) to the same taxes as wage income. Consequently, for ease of explanation, it's easiest and most accurate to use median income + average tax rates to reflect the basis of the standard American household.

[609] USA Today. *"Cost of feeding a family of four: $146 to $289 a week."* N. Hellmich. 1 May, 2013. https://www.usatoday.com/story/news/nation/2013/05/01/grocery-costs-for-family/2104165/

[610] AAA Newsroom. *"Your Driving Costs."* E. Edmonds. 12 September, 2018. https://newsroom.aaa.com/auto/your-driving-costs/

[611] U.S. Department of Transportation. Federal Highway Administration. *"Average Annual Miles per Driver by Age Group."* 29 March, 2018. https://www.fhwa.dot.gov/ohim/onh00/bar8.htm

[612] Energy Information Administration. *"How much electricity does an American home use?"* Last updated: 2 October, 2019. https://www.eia.gov/tools/faqs/faq.php?id=97&t=3

[613] Energy Information Administration. *"Average Price of Electricity to Ultimate Customers."* https://www.eia.gov/electricity/annual/html/epa_02_04.html

[614] Rocket Mortgage. *"Are You Average? Here's What The Typical U.S. Household Spends On Utility Bills Each Year?"* https://www.rockethq.com/learn/personal-finances/average-cost-of-utilities

[615] Rocket Mortgage. *"Are You Average? Here's What The Typical U.S. Household Spends On Utility Bills Each Year?"* https://www.rockethq.com/learn/personal-finances/average-cost-of-utilities

[616] Energy Information Administration. *"Commercial Buildings Energy Consumption Survey (CBECS)."* J. Michaels. https://www.eia.gov/consumption/commercial/data/2012/

[617] Energy Information Administration. *"Commercial Buildings Energy Consumption Survey (CBECS). Table E5. Electricity consumption (kWh) by end use, 2012"* Released May, 2016. https://www.eia.gov/consumption/commercial/data/2012/c&e/pdf/e5.pdf

[618] Energy Information Administration. *"Average Price of Electricity to Ultimate Customers."* https://www.eia.gov/electricity/annual/html/epa_02_04.html

[619] Energy Information Administration. *"Commercial Buildings Energy Consumption Survey (CBECS). Table E8. Natural gas consumption and conditional energy intensities (cubic feet) by end use, 2012"* Released May, 2016. https://www.eia.gov/consumption/commercial/data/2012/c&e/pdf/e8.pdf

[620] Bureau of Transportation Statistics. *"Transportation Statistics Annual Report, 2018. Table 1-1."* https://www.bts.dot.gov/sites/bts.dot.gov/files/docs/browse-statistical-products-and-data/transportation-statistics-annual-reports/Preliminary-TSAR-Full-2018-a.pdf

[621] Statista. *"U.S. motor gasoline and distillate fuel oil consumption by the transportation sector from 1992 to 2018."* https://www.statista.com/statistics/189410/us-gasoline-and-diesel-consumption-for-highway-vehicles-since-1992/

[622] Energy Information Administration. *"Gasoline and Diesel Fuel Update."* Figures accurate as of November 5, 2019. https://www.eia.gov/petroleum/gasdiesel/

[623] Forbes Magazine. *"Why the Tax Cuts and Jobs Act (TCJA) Led to Buybacks Rather Than Investment."* A. Knott. 21 February, 2019.
https://www.forbes.com/sites/annemarieknott/2019/02/21/why-the-tax-cuts-and-jobs-act-tcja-led-to-buybacks-rather-than-investment/

[624] Minnesota Post. *"History lessons: understanding the decline in manufacturing."* L. Johnson. 22 February, 2012. https://www.minnpost.com/macro-micro-minnesota/2012/02/history-lessons-understanding-decline-manufacturing/

[625] Reuters. *"Haiti reconstruction cost may near $14 billion: IADB."* P. Fletcher. 16 February, 2010. https://www.reuters.com/article/us-quake-haiti-cost-idUSTRE61F43Z20100216

[626] Congressional Budget Office. *"The Federal Budget in 2018."*
https://www.cbo.gov/system/files/2019-06/55342-2018-budget.pdf

[627] U.S. Department of Veterans Affairs. *"President Trump Seeks $12B Increase in FY2019 VA Budget to Support Nation's Veterans."* 12 February, 2018.
https://www.va.gov/opa/pressrel/pressrelease.cfm?id=4007

[628] Military Retirement Fund Audited Financial Report, Fiscal Year 2018.
https://comptroller.defense.gov/Portals/45/Documents/afr/fy2018/DoD_Components/2018_AFR_MRF.pdf

[629] Congressional Research Service. *"Energy and Water Development Appropriations: Nuclear Weapons Activities."* 29 July, 2019. https://fas.org/sgp/crs/nuke/R44442.pdf

[630] Department of Homeland Security. *"Administration's Fiscal Year 2018 Budget Request Advances DHS Operations."* 23 May, 2017.
https://www.dhs.gov/news/2017/05/23/administrations-fiscal-year-2018-budget-request-advances-dhs-operations

[631] Federation of American Scientists. *"Intelligence Budget Data."* No update date provided.
https://fas.org/irp/budget/

[632] Brown University, Watson Institute of International and Public Affairs. *"United States Budgetary Costs and Obligations of Post-9/11 Wars through FY:2020."* N. Crawford. 13 November, 2019.
https://watson.brown.edu/costsofwar/files/cow/imce/papers/2019/US%20Budgetary%20Costs%20of%20Wars%20November%202019.pdf

Brown University, Watson Institute for International and Public Affairs. *"Costs of War."*
https://watson.brown.edu/costsofwar/costs/economic

[633] War Resisters League. *"U.S. Federal Budget 2020 Fiscal War – Where Your Income Tax Money Really Goes."* https://www.warresisters.org/sites/default/files/fy2020pie_chart-hi_resb.pdf

[634] Discover Policing. *"Types of Law Enforcement Agencies."* https://www.discoverpolicing.org/explore-the-field/types-of-law-enforcement-agencies/

[635] United States Government Federal Register. https://www.federalregister.gov/agencies

[636] Background reading: United States Intelligence Community. https://en.wikipedia.org/wiki/United_States_Intelligence_Community

[637] Tax Foundation. *"How High Are Cigarette Taxes in Your State?"* J. Cammenga. 10 April, 2019. https://taxfoundation.org/2019-state-cigarette-tax-rankings/

[638] Tax Policy Center. *"Key Elements of the U.S. Tax System."* https://www.taxpolicycenter.org/briefing-book/what-are-major-federal-excise-taxes-and-how-much-money-do-they-raise

[639] Forbes Magazine. *"Which States Made the Most Tax Revenue From Marijuana in 2018?"* N. McCarthy. 26 March, 2019. https://www.forbes.com/sites/niallmccarthy/2019/03/26/which-states-made-the-most-tax-revenue-from-marijuana-in-2018-infographic/#733ad63e7085

[640] Congressional Budget Office. *"Increase All Taxes on Alcoholic Beverages to $16 per Proof Gallon."* https://www.cbo.gov/budget-options/2013/44854

[641] J. Minron, Department of Economics, Harvard University. *"The Budgetary Impacts of Drug Prohibition."* February, 2010. https://scholar.harvard.edu/files/miron/files/budget_2010_final_0.pdf

[642] Energy Information Administration. *"Sales of Electricity to Ultimate Customers – total by end-use sector, 2005-2015."* https://www.eia.gov/electricity/annual/html/epa_02_05.html

[643] National Renewable Energy Laboratory. *"U.S. Solar Photovoltaic System Cost Benchmark: QA, 2018."* R. Fu, D. Feldman, R. Margolis. https://www.nrel.gov/docs/fy19osti/72399.pdf

[644] Lawrence Berkeley National Laboratory. *"Report Confirms Wind Technology Advancements Continue to Drive Down Wind Energy Prices."* J. Chao. 23 August, 2018. https://newscenter.lbl.gov/2018/08/23/report-confirms-wind-technology-advancements-continue-to-drive-down-wind-energy-prices/

645 Energy Information Administration. Table 6.07.B. Capacity Factors for Utility Scale Generators Primarily Using Non-Fossil Fuels. August, 2019. https://www.eia.gov/electricity/monthly/epm_table_grapher.php?t=epmt_6_07_b

646 Hargreaves, Robert, PhD. *"THORIUM: Energy Cheaper than Coal."* July 25, 2012. pp 220

647 Energy Information Administration. Table 6.07.B. Capacity Factors for Utility Scale Generators Primarily Using Non-Fossil Fuels. August, 2019. https://www.eia.gov/electricity/monthly/epm_table_grapher.php?t=epmt_6_07_b

648 Oil & Gas Journal. *"Crude Oil Pipeline Growth, Revenues Surge; Construction Costs Mount."* http://www.ogj.com/articles/print/volume-112/issue-9/special-report-pipeline-economics/crude-oil-pipeline-growth-revenues-surge-construction-costs-mount.html

649 Scientific American. *"How Much Water Do Nations Consume?"* M. Fischetti. 21 May, 2012. https://www.scientificamerican.com/article/graphic-science-how-much-water-nations-consume/

650 Torrent Engineering and Equipment. *"Pipeline Volume Capacities."* http://www.torrentee.com/pdf/Pipe_Volume_Capacity_Table_Jun-02.pdf

651 Government of Michigan. Water tank prices and data per gallon. https://www.michigan.gov/documents/Vol2-35UIP11Tanks_121080_7.pdf

652 Lucid Energy Corp. *"How it works."* http://lucidenergy.com/how-it-works/

653 Lucid Energy Corp. *"How it works."* http://lucidenergy.com/how-it-works/

654 U.S. Climate Data. https://www.usclimatedata.com/

655 Marlow Engineering. *"Technical Data Sheet for EHBMS."* https://cdn2.hubspot.net/hubfs/547732/Data_Sheets/EHBMS.pdf

656 Marlow Engineering. *"Technical Data Sheet for EHBMS."* https://cdn2.hubspot.net/hubfs/547732/Data_Sheets/EHBMS.pdf

657 Background reading on Ras Al-Khair Power and Desalination Plant. https://en.wikipedia.org/wiki/Ras_Al-Khair_Power_and_Desalination_Plant

658 Background reading on Ras Al-Khair Power and Desalination Plant. https://en.wikipedia.org/wiki/Ras_Al-Khair_Power_and_Desalination_Plant

[659] Oxford Business Group. *"Saudi Arabia Expands its Desalination Capacity."* No author or date provided. https://www.oxfordbusinessgroup.com/analysis/world-leader-efforts-under-way-expand-desalination-capacity

[660] The National magazine. *"UAE's Largest Power and Desalination Plant Opens at Jebel Ali."* C. Simpson. 9 April, 2013. https://www.thenational.ae/uae/uae-s-largest-power-and-desalination-plant-opens-at-jebel-ali-1.455481

[661] WaterWorld. *"Dubai Opens UAE's Largest Desalination Plant."* T. Freyberg. 9 April, 2013. http://www.waterworld.com/articles/2013/04/dubai-opens-uaes-largest-desalination-plant.html

[662] Power Engineering International. *"Desalination: A First in Fujairah."* A. Hoel. 1 August, 2004. http://www.powerengineeringint.com/articles/print/volume-12/issue-8/features/desalination-a-first-in-fujairah.html

[663] Energy Information Administration. *"Cost and Performance Characteristics of New Generating Technologies, Annual Energy Outlook 2017."* January, 2017. https://www.eia.gov/outlooks/aeo/assumptions/pdf/table_8.2.pdf

[664] General Electric. *"Tour a Combined Cycle Power Plant."* https://powergen.gepower.com/resources/knowledge-base/combined-cycle-power-plant-how-it-works.html

[665] Background reading on overnight cost: https://en.wikipedia.org/wiki/Overnight_cost

[666] Department of Energy. *"Current (2009) State-of-the-art Hydrogen Production Cost Estimate Using Water Electrolysis."* 9 September, 2009. https://www.hydrogen.energy.gov/pdfs/46676.pdf

[667] Background reading on factory gate price: http://www.investorwords.com/9656/factory_gate_price.html

[668] Energy Information Administration. Frequently Asked Questions. https://www.eia.gov/tools/faqs/faq.php?id=23&t=10

[669] Background reading on energy density of gasoline: https://en.wikipedia.org/wiki/Energy_density

[670] Background reading on energy density of hydrogen: https://en.wikipedia.org/wiki/Energy_density

[671] Department of Energy. Scientific Analysis. *"Techno-economic Analysis of PEM electrolysis for hydrogen production."* W. Colella. B. James. J. Moton. G. Saur. T. Ramsden. 27 February, 2014.
https://energy.gov/sites/prod/files/2014/08/f18/fcto_2014_electrolytic_h2_wkshp_colella1.pdf#p10

www.ingramcontent.com/pod-product-compliance
Lightning Source LLC
Chambersburg PA
CBHW080551240526
45466CB00031B/2953